T0203100

FUNDAMENTAL NUMBER THEORY

WITH APPLICATIONS

SECOND EDITION

DISCRETE
MATHEMATICS
AND
ITS APPLICATIONS

Series Editor
Kenneth H. Rosen, Ph.D.

Continued Titles

DISCRETE MATHEMATICS AND ITS APPLICATIONS

Series Editor KENNETH H. ROSEN

FUNDAMENTAL NUMBER THEORY
WITH APPLICATIONS

SECOND EDITION

RICHARD A. MOLLIN

University of Calgary

Alberta, Canada

CRC Press

Taylor & Francis Group

Boca Raton London New York

CRC Press is an imprint of the
Taylor & Francis Group, an **informa** business

A CHAPMAN & HALL BOOK

CRC Press
Taylor & Francis Group
6000 Broken Sound Parkway NW, Suite 300
Boca Raton, FL 33487-2742

First issued in paperback 2019

© 2008 by Taylor & Francis Group, LLC
CRC Press is an imprint of Taylor & Francis Group, an Informa business

No claim to original U.S. Government works

ISBN-13: 978-1-4200-6659-3 (hbk)
ISBN-13: 978-0-367-38776-1 (pbk)

Library of Congress Cataloging-in-Publication Data

Mollin, Richard A., 1947-
   Fundamental number theory with applications / Richard A. Mollin. -- 2nd ed.
      p. cm. -- (Discrete mathematics and its applications ; 47)
   Includes bibliographical references and index.
   ISBN 978-1-4200-6659-3 (hardback : alk. paper)
   1. Number theory. I. Title. II. Series.

QA241.M598 2008
512.7--dc22   2007050650

Visit the Taylor & Francis Web site at
http://www.taylorandfrancis.com

and the CRC Press Web site at
http://www.crcpress.com

Fundamental Number Theory

with Applications

SECOND EDITION

Richard A. Mollin

Dedicated to the memory of Irving Kaplansky.

Contents

Preface

The second edition of the original introductory undergraduate text for a one-semester first course in number theory is redesigned to be more accessible and far reaching in its coverage from a truly "fundamental" perspective. This means that virtually all "advanced" material has been removed in favour of more topics at the elementary level, not included in the first edition. For instance, we have removed the algebraic number theory, elliptic curves, the (ideal-theoretic) continued fraction factoring algorithm, applications to quadratic orders, including ideals, the advanced material on quadratic polynomials, and applications to quadratics. There will be a second volume to be published that will have advanced material for a *second course* in number theory.

The background on arithmetic of the integers has been moved from the main text to Appendix A, and the discussion of complexity to Appendix B. More elementary material has been added, including partition theory and generating functions, combinatorial number theory, an expanded and more involved discussion of random number generation, more applications to cryptology, primality testing, and factoring. As well, there is an expanded coverage of Diophantine equations from a more elementary point of view, including a section for Legendre's Theorem on the equation $ax^2 + by^2 + cz^2 = 0$, and an expanded view of Bachet's equation $y^2 = x^3 + k$. Moreover, the coverage of sums of two, three, and four squares has been revised completely to concentrate on criteria for representation, and more on the total number of primitive representations, deleting the extensive coverage of the total number of imprimitive representations from the first edition. For sums of two squares, applications from continued fraction theory, not covered in the first edition, is discussed in detail. Sums of cubes is given a separate section, also not covered in the first edition. That rounds out Chapter Six on additivity.

The numbering system has been changed from the three-level approach (such as Theorem 1.2.3) to an easier, more standard two-level approach (such as Theorem 1.2). The use of footnotes has been curtailed in this edition. For instance, the mini-biographies are placed in highlighted boxes as sidebars to reduce distraction and impinging on text of footnote usage. Footnotes are employed only when no other mechanisms will work. Also, the Bibliography contains the page(s) where each entry is cited, another new inclusion, which helps the reader see the relevance of each such reference to the specific material in the text.

Other than the addition of Appendices A–B, as noted above, we retain the appendices from the first edition on primes and least primitive roots, indices, and the ABC conjecture, but have deleted the more specialized appendices on tables of special primes, Cunningham factorizations, pseudoprimes, Carmichael numbers, and values of some arithmetic functions. Also, although we also deleted the appendix from the first edition on the prime number theorem, we have included a section (§1.9) on distribution of primes that is more extensive, informative, and perhaps one of the few aspects of the main text that retains a flavour of being "advanced," yet accessible via the method of presentation. Furthermore, we have added Appendix F on *Primes is in P*, to delineate the recently discovered unconditional deterministic polynomial-time algorithm for primality testing that is indeed "advanced". However, this is worth the inclusion, at the end of the text, for its impressive implications made available for the more adventurous reader, perhaps interested in going on to a second course in number theory.

The list of symbols is a single page of the most significant ones in use. The index has over thirteen hundred entries presented in such a fashion that there is maximum cross-referencing to ensure that the reader will find data with ease.

There are nearly 400 exercises in this edition, and there are nearly seventy mini-biographies. Also, the more challenging exercises are marked with the ☆ symbol. As with the first edition, solutions of the odd-numbered exercises are included at the end of the text, and a solutions manual for the even-numbered exercises is available to instructors who adopt the text for a course. As usual, the website below is designed for the reader to access any updates and the e-mail address below is available for any comments.

◆ **Acknowledgments** The author is grateful for the proofreading done by the following people, each of whom lent their own valuable time: John Burke (U.S.A.), Jacek Fabrykowski (U.S.A.), Bart Goddard (U.S.A.), and Thomas Zaplachinski (Canada), a former student, now cryptographer.

December 7, 2007

website: http://www.math.ucalgary.ca/~ramollin/

e-mail: ramollin@math.ucalgary.ca

Chapter 1

Arithmetic of the Integers

> *Philosophy is written in the great books which ever lies before our eyes — I mean the universe... This book is written in mathematical language and its characters are triangles, circles and other geometrical figures, without whose help...one wanders in vain through a dark labyrinth.*
>
> **Galileo Galilei (1564–1642)**, Italian astronomer and physicist

In this introductory chapter, we discover the arithmetic underlying the integers and the tools to manipulate them. The reader should be familiar with the basic notation, symbols, set theory, and background in Appendix A.

1.1 Induction

An essential tool in number theory, which allows us in this section to prove the base representation theorem, is the following.

◆ Principle of Mathematical Induction — PMI

Suppose that $S \subseteq \mathbb{N}$ and both (a) and (b) below hold.

(a) $1 \in S$, and

(b) If $n > 1$ and $n - 1 \in S$, then $n \in S$.

Then $S = \mathbb{N}$.

In other words, the Principle of Mathematical Induction says that any subset of the natural numbers that contains 1, and can be shown to contain $n > 1$ whenever it contains $n - 1$ must be $\mathbb{N}$. Part (a) is called the *induction step*, and the assumption that $n - 1 \in S$ is called the *induction hypothesis*. First, one establishes the induction step, then assumes the induction hypothesis and

1

proves the conclusion, that $n \in S$. Then we simply say that *by induction, $n \in S$* for all $n \in \mathbb{N}$. This principle is illustrated in the following two results.

Theorem 1.1 | **A Summation Formula** |

For any $n \in \mathbb{N}$,

$$\sum_{j=1}^{n} j = \frac{n(n+1)}{2}.$$

Proof. If $n = 1$, then $\sum_{j=1}^{n} j = 1 = n(n+1)/2$, and the induction step is secured. Assume that

$$\sum_{j=1}^{n-1} j = (n-1)n/2,$$

the induction hypothesis. Now consider

$$\sum_{j=1}^{n} j = n + \sum_{j=1}^{n-1} j = n + (n-1)n/2,$$

by the induction hypothesis. Hence,

$$\sum_{j=1}^{n} j = [2n + (n-1)n]/2 = (n^2 + n)/2 = n(n+1)/2,$$

as required. Hence, by induction, this must hold for all $n \in \mathbb{N}$.   □

Theorem 1.2 | **A Geometric Formula** |

If $a, r \in \mathbb{R}$, $r \neq 1$, $n \in \mathbb{N}$, then $\sum_{j=0}^{n} ar^j = \frac{a(r^{n+1}-1)}{r-1}$.

Proof. If $n = 1$, then

$$\sum_{j=0}^{n} ar^j = a + ar = a(1+r) = a(1+r)(r-1)/(r-1) = a(r^2-1)/(r-1) =$$

$$a(r^{n+1}-1)/(r-1),$$

which is the induction step. By the induction hypothesis, we get

$$\sum_{j=0}^{n+1} ar^j = ar^{n+1} + \sum_{j=0}^{n} ar^j = ar^{n+1} + a(r^{n+1}-1)/(r-1) = a(r^{n+2}-1)/(r-1),$$

as required.   □

The sum in Theorem 1.2 is called a *geometric sum* where a is the *initial term* and r is called the *ratio*.

Now we look at a classical problem involving rabbits as a vehicle for introducing a celebrated sequence that lends itself very well as an application of induction.

◆ The Rabbit Problem

Suppose that a male rabbit and a female rabbit have just been born. Assume that any given rabbit reaches sexual maturity after one month and that the gestation period for a rabbit is one month. Furthermore, once a female rabbit reaches sexual maturity, it will give birth every month to exactly one male and one female. Assuming that no rabbits die, how many male/female pairs are there after n months?

We will use the symbol F_n to denote the number of pairs of rabbits at month n, while M_n denotes the number of pairs of mature rabbits at month n, and I_n the number of immature rabbits at month n. Then

$$F_n = M_n + I_n.$$

Therefore, we have $F_1 = F_2 = 1$, and for any $n \geq 3$, $M_n = F_{n-1}$, and $I_n = M_{n-1}$, since every newborn pair at time n is the product of a mature pair at time $n-1$. Thus,

$$F_n = F_{n-1} + M_{n-1}.$$

Moreover, $M_{n-1} = F_{n-2}$. Thus, we have

$$F_n = F_{n-1} + F_{n-2}, \tag{1.1}$$

for any $n \geq 3$, which generates the *Fibonacci Sequence* — *see Biography 1.1*. (A research journal devoted entirely to the study of such numbers is the *Fibonacci Quarterly*.)

Biography 1.1 Fibonacci (ca.1180–1250) *was known as* Leonardo of Pisa, *the son of an Italian merchant named Bonaccio. He had an Arab scholar as his tutor while his father served as consul in North Africa. Thus, he was well educated in the mathematics known to the Arabs. Fibonacci's first and certainly his best-known book is* Liber Abaci *or* Book of the Abacus *first published in 1202, which was one of the means by which the Hindu-Arabic number system was transmitted into Europe. However, only the second edition, published in 1228, has survived. In this work, Fibonacci included work on geometry, the theory of proportion, and techniques for determining the roots of equations. Also included in his book was the rabbit problem described above.*
Perhaps his most prominent work, Liber Quadratorum *or* Book of Square Numbers, *published in 1225, contains some sophisticated contributions to number theory. Fibonacci dedicated this book to his patron, Holy Roman Emperor* Friedrich II *of Germany.*

We now prove a result, as an application of induction, attributed to Binet (see Biography 1.2 on the next page), that links the Fibonacci sequence with the famous *golden ratio*:

$$\mathfrak{g} = \frac{1 + \sqrt{5}}{2}, \tag{1.2}$$

Theorem 1.3 $\boxed{\textbf{Binet's Formula}}$

F_n *is the n-th Fibonacci number for any* $n \in \mathbb{N}$, *and*

$$\mathfrak{g}' = \frac{1 - \sqrt{5}}{2}$$

is the conjugate of the golden ratio, then

$$F_n = \frac{1}{\sqrt{5}}[\mathfrak{g}^n - \mathfrak{g}'^n] = \frac{\mathfrak{g}^n - \mathfrak{g}'^n}{\mathfrak{g} - \mathfrak{g}'},$$

Proof. We use induction. If $n = 1$, then

$$\frac{1}{\sqrt{5}}\left[\mathfrak{g}^n - \mathfrak{g}'^n\right] = \frac{1}{\sqrt{5}}\left[\frac{1 + \sqrt{5}}{2} - \frac{1 - \sqrt{5}}{2}\right] = \frac{\sqrt{5}}{\sqrt{5}} = 1 = F_n.$$

Assume that $F_n = \frac{1}{\sqrt{5}}[\mathfrak{g}^n - \mathfrak{g}'^n]$, which is the induction hypothesis, from which we have

$$F_{n+1} = F_n + F_{n-1} = \frac{1}{\sqrt{5}}\left[\mathfrak{g}^n - \mathfrak{g}'^n\right] + \frac{1}{\sqrt{5}}\left[\mathfrak{g}^{n-1} - \mathfrak{g}'^{n-1}\right],$$

and by factoring out appropriate powers, this is equal to

$$\frac{1}{\sqrt{5}}\left[\mathfrak{g}^{n-1}(1 + \mathfrak{g}) - \mathfrak{g}'^{n-1}(1 + \mathfrak{g}')\right].$$

By Exercise 1.1, $1 + \mathfrak{g} = \mathfrak{g}^2$. It may be similarly verified that $1 + \mathfrak{g}' = \mathfrak{g}'^2$. Hence,

$$F_{n+1} = \frac{1}{\sqrt{5}}[\mathfrak{g}^{n+1} - \mathfrak{g}'^{n+1}] = \frac{\mathfrak{g}^{n+1} - \mathfrak{g}'^{n+1}}{\mathfrak{g} - \mathfrak{g}'},$$

since $\mathfrak{g} - \mathfrak{g}' = \sqrt{5}$. $\square$

The following result is a fascinating relationship between the golden ratio and Fibonacci numbers that is a consequence of the above.

Corollary 1.1 $\boxed{\textbf{Asymptotic Behaviour of Fibonacci Numbers}}$

If F_n *denotes the n-th Fibonacci number and* $\mathfrak{g}$ *denotes the golden ratio, then*

$$\lim_{n \to \infty} \frac{F_{n+1}}{F_n} = \mathfrak{g}.$$

Proof. By Theorem 1.3,

$$\lim_{n\to\infty}\frac{F_{n+1}}{F_n}=\lim_{n\to\infty}\frac{\mathfrak{g}^{n+1}-\mathfrak{g}'^{n+1}}{\mathfrak{g}^n-\mathfrak{g}'^n}=\mathfrak{g}\lim_{n\to\infty}\frac{\mathfrak{g}^n}{\mathfrak{g}^n-\mathfrak{g}'^n}-\mathfrak{g}'\lim_{n\to\infty}\frac{\mathfrak{g}'^n}{\mathfrak{g}^n-\mathfrak{g}'^n}=$$

$$\mathfrak{g}\lim_{n\to\infty}\frac{1}{\mathfrak{g}^n/\mathfrak{g}^n-\mathfrak{g}'^n/\mathfrak{g}^n}-\mathfrak{g}'\lim_{n\to\infty}\frac{1}{\mathfrak{g}^n/\mathfrak{g}'^n-\mathfrak{g}'^n/\mathfrak{g}'^n}=\mathfrak{g}\lim_{n\to\infty}1-0=\mathfrak{g},$$

since $\mathfrak{g}/\mathfrak{g}'>1$. □

Biography 1.2 Jacques Philippe Marie Binet (1786–1856) *was born on February 2 of 1786 in Rennes, Bretagne, France. After completing his education in 1806, he became a teacher at École Polytechnique in 1807. By 1816, after some other appointments, he became an inspector of studies at École Polytechnique, and by 1823 had been appointed to the astronomy chair at the Collège de France, which he held for more than 3 decades. For political reasons he was dismissed as inspector of studies on November 13, 1830. He is probably best known for his work on matrix theory, especially the rule for multiplying matrices, which was used later by Cayley, for instance, in extending the theory. He contributed to number theory as well, especially in the early 1840s. The paper that contains the formula with his name was published in 1843 — see [6]. However, as often happens with mathematical discoveries, it had already been discovered earlier. Indeed, de Moivre [13] had discovered it over a century earlier and in greater generality. Binet published in areas other than mathematics such as astronomy and physics, with a list of over 50 publications in total to his credit. He died on May 12, 1856 in Paris, France.*

The exercises at the end of this section contain numerous problems related to Fibonacci numbers and their generalizations. This includes links to the golden ratio and other values for the reader to get a better appreciation of these numerical sequences and their properties.

We now look at another interesting problem as an application of induction, a puzzle developed by François Édouard Anatole Lucas (see Biography 1.18 on page 63).

◆ Tower of Hanoi Problem

Assume that there are three vertical posts and $n \geq 1$ rings, all of different sizes, concentrically placed on one of the posts from largest on the bottom to smallest on the top. In other words, no larger ring is placed upon a smaller one. The object of the game is to move all rings from the given post to another post, subject to the following rules:

[1] Only one ring may be moved at a time.
[2] A ring may never be placed over a smaller ring.

We now use induction to show that the number of moves to transfer n rings from one post to another is $2^n - 1$.

Let $N(n)$ be the minimum number of moves required to do the above. First, we show that

$$N(n+1) = 2N(n) + 1.$$

To move the $(n+1)$-st (largest) ring to the destination post after $n \in \mathbb{N}$ rings have been moved there, we first move the rings to the unoccupied post, which requires $N(n)$ moves. Then we move the $(n+1)$-st ring to the destination post (one move). Finally, we move the original n rings back to the destination post, requiring another $N(n)$ moves for a total of $2N(n) + 1$ moves. Now, we use induction on n.

If $n = 1$, then $N(1) = 1$, and if $n = 2$, then $N(2) = 3 = 2^2 - 1$. Assume that the result holds for k such that $1 \leq k \leq n$. Hence,

$$N(n+1) = 2N(n) + 1 =$$

$$2(2^n - 1) + 1 = 2^{n+1} - 1.$$

Here is another tantalizing question that we can use the above to solve.

Ancient folklore tells us that monks in a temple tower were given 64 rings at the beginning of time. They were told to play the above game, and that the world would end when they were finished. Assume that the monks worked in shifts twenty-four hours per day, moving one ring per second without any errors. How long does the world last?

The answer is approximately 5,849,420,458 centuries!

Now we provide some further applications to induction by introducing the sequences related to Lucas, and their relationship with the Fibonacci sequence.

◆ The Lucas Sequence

The Lucas sequence for any $n \in \mathbb{N}$ is given by

$$L_n = \mathfrak{g}^n + \mathfrak{g}'^n, \tag{1.3}$$

where $\mathfrak{g}$ is the golden ratio introduced in (1.2) on page 4.

Now we show how the Lucas and Fibonacci sequences are related. Although this following result does not use induction directly, it does employ Theorem 1.3, which does use induction.

Theorem 1.4 | **Lucas and Fibonacci Relationship**

For any $n \in \mathbb{N}$,

$$L_{n+1} = F_{n+2} + F_n.$$

Proof. By the definition of the Fibonacci numbers, $F_{n+2} + F_n = F_{n+1} + 2F_n$, and by Theorem 1.3,

$$F_{n+1} + 2F_n = \frac{\mathfrak{g}^{n+1} - \mathfrak{g}'^{n+1}}{\mathfrak{g} - \mathfrak{g}'} + 2\frac{\mathfrak{g}^n - \mathfrak{g}'^n}{\mathfrak{g} - \mathfrak{g}'} = \frac{\mathfrak{g}^{n+1} + 2\mathfrak{g}^n - \mathfrak{g}'^{n+1} - 2\mathfrak{g}'^n}{\mathfrak{g} - \mathfrak{g}'} =$$

$$\mathfrak{g}^{n+1} + \mathfrak{g}'^{n+1} + \frac{\mathfrak{g}^{n+1} - \mathfrak{g}'^{n+1}}{\mathfrak{g} - \mathfrak{g}'} + 2\frac{\mathfrak{g}^n - \mathfrak{g}'^n}{\mathfrak{g} - \mathfrak{g}'} - (\mathfrak{g}^{n+1} + \mathfrak{g}'^{n+1}) =$$

$$L_{n+1} + \frac{\mathfrak{g}^{n+1} + 2\mathfrak{g}^n - \mathfrak{g}'^{n+1} - 2\mathfrak{g}'^n - (\mathfrak{g} - \mathfrak{g}')(\mathfrak{g}^{n+1} + \mathfrak{g}'^{n+1})}{\mathfrak{g} - \mathfrak{g}'},$$

so we need only show that the numerator of the second summand is zero. We have, after rearranging terms and using the fact that $\mathfrak{g}\mathfrak{g}' = -1$,

$$\mathfrak{g}^{n+1} + 2\mathfrak{g}^n - \mathfrak{g}'^{n+1} - 2\mathfrak{g}'^n - (\mathfrak{g} - \mathfrak{g}')(\mathfrak{g}^{n+1} + \mathfrak{g}'^{n+1}) =$$

$$\mathfrak{g}^n(1 + \mathfrak{g} - \mathfrak{g}^2) - \mathfrak{g}'^n(1 + \mathfrak{g}' - \mathfrak{g}'^2) = 0,$$

since the last quantities in the brackets are zero by Exercise 1.1. □

The following, which is a simple consequence of Theorem 1.3, is sometimes given as the definition of the Lucas sequence. Our approach is more focused on the golden ratio. Indeed Equation (1.3) is often called the Binet formula for Lucas numbers.

Corollary 1.2 *For any $n \in \mathbb{N}$, $L_{n+2} = L_{n+1} + L_n$.*

Proof. By Theorem 1.4, for any $n \in \mathbb{N}$,

$$L_{n+2} = F_{n+3} + F_{n+1} = (F_{n+2} + F_{n+1}) + (F_n + F_{n-1}) = L_{n+1} + L_n,$$

as required. □

What the definition of Lucas numbers and Corollary 1.2 tell us is that $L_1 = 1$, $L_2 = 3$, and $L_n = L_{n-1} + L_{n-2}$ for any $n \geq 3$.

The reader may find it an interesting exercise to take the result of Corollary 1.2 as the definition of Lucas numbers, and prove that the Binet formula given in Equation (1.3) follows from it.

In the exercises at the end of this section is developed a generalization of the Fibonacci and Lucas sequences, called two-term recurrence sequences that will give the reader a deeper insight into the properties resulting from the recursive rule that each term is the sum of the two preceding terms (in particular, see Exercise 1.11 on page 12).

Now we proceed to establish a major goal of this section as follows. We know how to represent numbers in base 10 such as $2037 = 2\cdot10^3 + 0\cdot10^2 + 3\cdot10^1 + 7\cdot10^0$. Also, modern computers use base 2 arithmetic, and we may represent any $n \in \mathbb{N}$ in the form

$$n = \sum_{j=0}^{m} a_j 2^j,$$

where the $a_j \in \{0, 1\}$ are called *bits*, being the contraction of **bi**nary digi**ts**. For instance, in the base 10 example above, 2037 may be represented as

$$1\cdot2^{10} + 1\cdot2^9 + 1\cdot2^8 + 1\cdot2^7 + 1\cdot2^6 + 1\cdot2^5 + 1\cdot2^4 + 0\cdot2^3 + 1\cdot2^2 + 0\cdot2^1 + 1\cdot2^0.$$

To simplify such representations, we use a notation with a base subscript that is self explanatory. For instance, from the above discussion

$$(2037)_{10} = (11111110101)_2,$$

where the powers of the base ascend from zero on the right to the maximum nonzero power on the left. However, there is nothing special about the base elements 2 and 10 beyond our familiarity with them. The ancient Babylonians, for instance, were familiar with base 60 (sexagesimal) representations and the ancient Mayans used base 20. What is key to all of this is that each such representation is unique irrespective of the base under consideration.

Theorem 1.5 | **The Base Representation Theorem**

Let $b > 1$ be any integer. Then, for any $n \in \mathbb{N}$, there exists a nonnegative integer m such that

$$n = \sum_{j=0}^{m} a_j b^j, \text{ where } 0 \le a_j < b \text{ for each } j = 0, 1, \ldots, m \text{ and } a_m \ne 0.$$

Furthermore, this representation is unique, called the representation of n to base b, a_j are the base-b digits of n, and the above representation is denoted by $n = (a_m a_{m-1} \ldots a_0)_b$.

Proof. Let $r_b(n)$ denote the number of representations of n to base b. Then the result we must prove is that $r_b(n) = 1$.

Claim 1.1 $r_b(1) \ge r_b(n) \ge r_b(b^n)$

By Exercise 1.4, $b^n > n$, so there exists a $t \in \mathbb{N}$ such that $b^n = t + n$. Suppose that, for some $k \ge 0$, we have $n + t = \sum_{j=0}^{k} a_j b^j$ with $0 \le a_j < b$, where we may assume without loss of generality that $a_k \ne 0$. If all $a_j = 0$ for $j = 0, 1, 2, \ldots, k - 1$, then $b^n = n + t = a_k < b$, a contradiction. Hence, there is a smallest nonnegative integer $\ell < k$ such that $a_\ell \ne 0$. We have,

$$n + t - 1 = \sum_{j=\ell}^{k} a_j b^j - 1 = \sum_{j=\ell}^{k-1} a_j b^j + (a_k - 1)b^k + b^k - 1 =$$

$$\sum_{j=\ell}^{k-1} a_j b^j + (a_k - 1)b^k + \sum_{j=0}^{\ell-1} (b - 1)b^j,$$

where the last equality follows from Theorem 1.2.

We have shown that for any given representation of $n + t$ to base b, we may produce a representation of $n + t - 1$ (distinct from any other given such representation produced by another representation of $n + t$).

By induction, we infer that

$$r_b(b^n) = r_b(n+t) \leq r_b(n+t-1) \leq \cdots \leq r_b(n) \leq r_b(n-1) \leq \cdots \leq r_b(1),$$

which yields Claim 1.1.

Therefore, since b^n is such a representation of itself,

$$1 \leq r_b(b^n) \leq r_b(n) \leq r_b(1) = 1,$$

whence $r_b(n) = 1$.   □

The ability to represent integers uniquely to any base is a highly useful tool and its applicability will be apparent throughout.

We close with definitions and results that are extremely valuable in number theory.

Definition 1.1 | Factorial Notation! |

If $n \in \mathbb{N}$, then $n!$ (read "enn factorial") is the product of the first n natural numbers. In other words,

$$n! = \prod_{i=1}^{n} i.$$

We agree, by convention, that $0! = 1$, namely, multiplication of no factors yields the identity.

Definition 1.2 | Binomial Coefficients |

If $k, n \in \mathbb{Z}$ with $0 \leq k \leq n$, then the symbol $\binom{n}{k}$ (read "n choose k") is given by

$$\binom{n}{k} = \frac{n!}{k!(n-k)!},$$

the binomial coefficient.

The reader may use induction to prove that $\binom{n}{k} \in \mathbb{N}$ as an exercise. The important fundamental result involving binomial coefficients that we will need in the text is the following.

Theorem 1.6 | The Binomial Theorem |

Let $x, y \in \mathbb{R}$, and $n \in \mathbb{N}$. Then

$$(x+y)^n = \sum_{i=0}^{n} \binom{n}{i} x^{n-i} y^i.$$

Proof. We use the Principle of Mathematical Induction on n to prove this. If $n = 1$, then

$$(x + y)^n = (x + y)^1 = \sum_{i=0}^{1} \binom{1}{i} x^{1-i} y^i = \binom{1}{0} x^1 y^0 + \binom{1}{1} x^0 y^1,$$

which secures the induction step.

Assume the induction hypothesis, namely

$$(x + y)^n = \sum_{i=0}^{n} \binom{n}{i} x^{n-i} y^i.$$

Consider

$$(x + y)^{n+1} = (x + y)(x + y)^n = (x + y) \sum_{i=0}^{n} \binom{n}{i} x^{n-i} y^i.$$

From the Distributive Law, and the properties for summation, this equals

$$= \sum_{i=0}^{n} \binom{n}{i} x^{n+1-i} y^i + \sum_{i=0}^{n} \binom{n}{i} x^{n-i} y^{i+1},$$

and, after setting $j = i + 1$ in the second summand, this equals

$$\sum_{i=0}^{n} \binom{n}{i} x^{n+1-i} y^i + \sum_{j=1}^{n+1} \binom{n}{j-1} x^{n+1-j} y^j,$$

and, by taking the summand for $i = 0$ out of the first summation symbol, taking the summand for $j = n + 1$ out of the second summation symbol, introducing a new index of summation k, and applying the properties for summation, this equals

$$x^{n+1} + \sum_{k=1}^{n} \left[\binom{n}{k} + \binom{n}{k-1} \right] x^{n+1-k} y^k + y^{n+1}.$$

By Pascal's Identity in Exercise 1.14 on page 14, the latter equals

$$x^{n+1} + \sum_{k=1}^{n} \binom{n+1}{k} x^{n+1-k} y^k + y^{n+1} = \sum_{k=0}^{n+1} \binom{n+1}{k} x^{n+1-k} y^k.$$

In other words,

$$(x + y)^{n+1} = \sum_{k=0}^{n+1} \binom{n+1}{k} x^{n+1-k} y^k.$$

The Principle of Mathematical Induction yields the result for all $n \in \mathbb{N}$. $\square$

Exercises

1.1. Prove that $\mathfrak{g}^2 = \mathfrak{g} + 1$, where $\mathfrak{g}$ is the golden ratio introduced in Equation (1.2) on page 4.

1.2. There is another version of the PMI, introduced on page 1, namely

The Principle of Mathematical Induction, Second Form — PMI2:

Suppose that $\mathcal{S} \subseteq \mathbb{Z}$, and $m \in \mathbb{Z}$ with

(a) $m \in \mathcal{S}$, and

(b) If $m < n$ and $\{m, m+1, \ldots, n-1\} \subseteq \mathcal{S}$, then $n \in \mathcal{S}$.

Then $k \in \mathcal{S}$ for all $k \in \mathbb{Z}$ such that $k \geq m$.

Use the PMI2 to prove that for any $n \in \mathbb{N}$, $\mathfrak{g}^{n-1} \geq F_n \geq \mathfrak{g}^{n-2}$, where F_n is the n-th Fibonacci number introduced on page 3 and $\mathfrak{g}$ is the golden ratio.

1.3. It can be proved that PMI and PMI2, introduced above, are logically equivalent (see [34, Theorem 1.8, p. 12], for instance). Thus, we may refer to PMI or PMI2 simply as *proof by induction*. Prove that induction is equivalent to the following. *Every nonempty subset of* $\mathbb{N}$ *contains a least element*, called **The Well-Ordering Principle**.

1.4. Prove that if $b, n \in \mathbb{N}$ with $b > 1$, then $n < b^n$. Use this fact and Exercise 1.2 to prove that for any $n \in \mathbb{N}$, there exists a $k \in \mathbb{N}$ such that $F_k > n$.

(*Hint: Use Theorem 1.2.*)

1.5. Prove that the golden ratio has an alternative representation given by

$$\mathfrak{g} = \sqrt{1 + \sqrt{1 + \sqrt{1 + \cdots}}}.$$

(*Hint: Use Exercise 1.1.*)

1.6. Prove that $\sum_{j=1}^{n} j^3 = \left(\sum_{j=1}^{n} j \right)^2$.

(*Hint: Use Theorem 1.1.*)

1.7. Prove the *sum of squares* result for the Fibonacci numbers,

$$\sum_{j=1}^{n} F_j^2 = F_n F_{n+1}.$$

1.8. Prove the following *running sum* result for the Fibonacci numbers, $\sum_{j=1}^{n} F_j = F_{n+2} - 1$.

1.9. Prove the following *Cassini formula* for the Fibonacci numbers,

$$F_{n-1}F_{n+1} - F_n^2 = (-1)^n$$

for any integer $n > 1$.

1.10. Prove the following *convolution formula* for the Fibonacci numbers,

$$F_{m+n} = F_{m-1}F_n + F_m F_{n+1}$$

for any $m, n \in \mathbb{N}$ with $m > 1$.

1.11. Fix nonzero values $a, b \in \mathbb{R}$, $b \neq 0$. Select values $R_1, R_2 \in \mathbb{R}$, and define, for any $n \in \mathbb{N}$,

$$R_{n+2} = aR_{n+1} + bR_n, \tag{1.4}$$

called a *two-term recurrence sequence*. Prove that if $aR_1 = R_2$, then

$$\sum_{j=1}^{n} b^{n-j} R_j^2 = \frac{R_n R_{n+1}}{a},$$

and observe that this is a generalization of Exercise 1.7 since the Fibonacci sequence is the case where $a = b = 1 = R_1 = R_2$. Similarly, Exercises 1.8–1.10 may be generalized to these two-term sequences.

1.12. If $u, v \in \mathbb{R}$ are nonzero and $f(x) = x^2 - ux - v$, with $x = \alpha$ and $x = \beta$ as the distinct roots of $f(x)$, where $|\alpha| > |\beta|$. Select real values R_1 and R_2. Let

$$a = \frac{R_2 - \beta R_1}{\alpha - \beta}, \text{ and } b = \frac{\alpha R_1 - R_2}{\alpha - \beta}.$$

Define R_n for any $n \in \mathbb{N}$ by

$$R_n = a\alpha^{n-1} + b\beta^{n-1} \tag{1.5}$$

(a) Prove that

$$\lim_{n \to \infty} \frac{R_{n+1}}{R_n} = \alpha.$$

(Observe that this is a generalization of the asymptotic result for the Fibonacci sequence given in Corollary 1.1, where, in that case, $u = v = 1$, $R_1 = 0$, $R_2 = 1$,

$$\alpha = (1 + \sqrt{5})/2, \quad \beta = (1 - \sqrt{5})/2,$$

$$a = 1/(\alpha - \beta) = 1/\sqrt{5}, \text{ and } b = -1/(\alpha - \beta) = -1/\sqrt{5}.$$

Notice, as well, that this also achieves Binet's formula given in Theorem 1.3 on page 4. It also readily gives the Binet formulation for the Lucas sequence given in Equation (1.3), where, in this case, $R_1 = 1$, $R_2 = 3$, $a = \alpha = (1 + \sqrt{5})/2$, and $b = \beta = (1 - \sqrt{5})/2$.)

(b) Let
$$U_n = (\alpha^n - \beta^n)/(\alpha - \beta) \text{ and } V_n = \alpha^n + \beta^n.$$
Prove that if $D = u^2 + 4v$, then

$$V_n^2 - DU_n^2 = 4(-v)^n.$$

(The values U_n and V_n are known as the *Lucas functions*, which are a distinguished sequence of numbers. For instance, in the case of the Lucas and Fibonacci sequences $u = v = 1$, and

$$L_n^2 - 5F_n^2 = 4(-1)^n.$$

These types of quadratic equations are called *norm-form* equations that we will study in Chapter 7. See [29] for an advanced perspective on the Lucas functions and their applications.)

1.13. With reference to the notation and definitions in Exercise 1.11, prove the following matrix equations hold for any $n \in \mathbb{N}$.

(a)
$$\begin{pmatrix} 0 & 1 \\ b & a \end{pmatrix}^n \begin{pmatrix} R_1 \\ R_2 \end{pmatrix} = \begin{pmatrix} R_{n+1} \\ R_{n+2} \end{pmatrix}$$

(b)
$$\begin{pmatrix} 1 & 0 \end{pmatrix} \begin{pmatrix} 0 & 1 \\ b & a \end{pmatrix}^{n+1} \begin{pmatrix} R_1 \\ R_2 \end{pmatrix} = R_{n+2},$$

which yields a matrix representation for the two-term recurrence sequence defined in Equation (1.4) on page 12.

(c) If F_j denotes the j-th Fibonacci number then,

$$\begin{pmatrix} 0 & 1 \\ 1 & 1 \end{pmatrix}^{n+1} = \begin{pmatrix} F_n & F_{n+1} \\ F_{n+1} & F_{n+2} \end{pmatrix}$$

(d) Use part (c) to establish Cassini's formula presented in Exercise 1.9.

(e) Use part (c) to establish the following *sums of squares* result for Fibonacci numbers,

$$F_{2n+1} = F_n^2 + F_{n+1}^2 \text{ for any } n \in \mathbb{N}.$$

(f) Use part (e) to prove the *Pythagorean triples* result for the Fibonacci numbers:

$$(F_{n+1}^2 - F_n^2)^2 + (2F_n F_{n+1})^2 = (F_{n+1}^2 + F_n^2)^2.$$

In Exercises 1.14–1.17, assume that $n, r \in \mathbb{N}$ with $n \geq r$.

1.14. Prove that for $n \geq r \geq 1$,

$$\binom{n+1}{r} = \binom{n}{r-1} + \binom{n}{r}.$$

(This is known as *Pascal's Identity* — see Biography 1.3.)

1.15. Prove that

$$\binom{n}{n-r} = \binom{n}{r}.$$

(This is known as the *Symmetry Property* for binomial coefficients.)

1.16. Prove that

$$\sum_{i=0}^{n}(-1)^i\binom{n}{i} = 0.$$

(This is known as the *Null Summation Property* for the binomial coefficient.)

(*Hint: Use the Binomial Theorem.*)

1.17. Prove that

$$\sum_{i=0}^{n}\binom{n}{i} = 2^n.$$

(This is the *Full Summation Property* for the binomial coefficient.)

(*Hint: Use the Binomial Theorem.*)

Biography 1.3 Blaise Pascal (1623–1662) *with his contemporaries* René Descartes (1596–1650) *and* Pierre de Fermat (1601–1665) *among others, made France the center of mathematics in the second third of the seventeenth century. When Pascal was only sixteen years old, he published a paper, which was only one page long, and has become known as Pascal's Theorem, which says that opposite sides of a hexagon, inscribed in a conic, intersect in three collinear points. In 1654, Pascal became interested in what we now call probability theory. His correspondence with Fermat on this topic might be considered the genesis of that theory. Pascal is most remembered for his connections between the study of probability and the arithmetic triangle. Although this triangle had been around for centuries before, Pascal made new and fascinating discoveries about it. Therefore, it is now called Pascal's triangle, which we will discuss later. On November 23, 1654 Pascal had an intense religious experience which caused him to abandon mathematics. However, one night in 1658, he was kept awake by a toothache, and began to distract himself by thinking about the properties of the cycloid. Suddenly, the toothache disappeared, which he took as divine intervention, and returned to mathematics. He died in his thirty-ninth year on August 19, 1662.*

1.18. This problem is intended to illustrate some properties of the number 9 in our base 10 system.

    (a) Let $n \in \mathbb{N}$ be any three digit, base ten, number with different first and last digits. Reverse the order of the digits to get a new number, and subtract the smaller from the larger one. Prove that the middle digit of the answer is always 9, and the first and last digits sum to 9.

    (b) Find a formula for the following pattern.

$$9 \cdot 0 + 8 = 8$$
$$9 \cdot 9 + 7 = 88$$
$$9 \cdot 98 + 6 = 888$$
$$9 \cdot 987 + 5 = 8888$$
$$9 \cdot 9876 + 4 = 88888$$
$$9 \cdot 98765 + 3 = 888888$$
$$9 \cdot 987654 + 2 = 8888888$$
$$9 \cdot 9876543 + 1 = 88888888$$
$$9 \cdot 98765432 + 0 = 888888888$$
$$9 \cdot 987654321 - 1 = 8888888888$$
$$9 \cdot 9876543210 - 2 = 88888888888$$

    (*Hint*: Let

$$9y_n - n + 8 = 8 \sum_{i=0}^{n} 10^i,$$

and use Theorem 1.2 on page 2 with induction.)

    (c) Prove that if the sum of the digits of a base 10 natural number n is divisible by 9, then $9 | n$.

Let $R_{(n,b)} = (1, 1, \ldots, 1)_b$ be the representation of a number to base $b > 1$ having n ones. Then $R_{(n,b)}$ is called a *repunit*. Answer the remaining questions on repunits.

1.19. Let p be a prime and b be a base.

    (a) Let $n \in \mathbb{N}$. Prove that $R_{(n,b)} = (b^n - 1)/(b - 1)$.

    (b) Prove that p divides $R_{(10,p)}$.

1.2   Division

> *Multiplication is vexation, division is as bad, the rule of three doth puzzle me, and practice drives me mad.*
>
> **Anonymous — 16th century**

1.1

The notions surrounding division are traceable to antiquity, at least to the time of Euclid (see Biography 1.4 on the facing page). We develop these notions in this section.

Definition 1.3 | Division |

If $a, b \in \mathbb{Z}$, $b \neq 0$, then to say that b divides a, or a is divisible by b, denoted by $b \mid a$, means that $a = bx$ for a unique $x \in \mathbb{Z}$, denoted by $x = a/b$. (Note that the existence and uniqueness of x implies that b cannot be 0, so we say that division by zero is undefined.) If b does not divide a, then we write $b \nmid a$ and say that a is not divisible by b. Any divisor $b \neq a$ of a is called a proper divisor of a. When $b^n \mid a$ for some $n \in \mathbb{N}$ and $b^{n+1} \nmid a$, then we denote this fact by $b^n \parallel a$, and say that a is exactly divisible by the power n of b.

We may classify integers according to whether they are divisible by 2, as follows.

Definition 1.4 | Parity |

If $a \in \mathbb{Z}$, and $a/2 \in \mathbb{Z}$, then we say that a is an even *integer. In other words, an even integer is one which is divisible by 2. If $a/2 \notin \mathbb{Z}$, then we say that a is an* odd *integer. In other words, an odd integer is one which is not divisible by 2. If two integers are either both even or both odd, then they are said to have the* same *parity. Otherwise they are said to have* opposite *or* different *parity.*

Theorem 1.7 | The Division Algorithm |

If $a \in \mathbb{N}$ and $b \in \mathbb{Z}$, then there exist unique integers $q, r \in \mathbb{Z}$ with $0 \leq r < a$, and $b = aq + r$.

[1.1]The rule of three is a centuries-old adage having many manifestations. In mathematics it is a method of finding the fourth term of a mathematical proportion when three terms are known. There is also the *double rule of three*, where five terms are known and used to find a sixth. Lewis Carroll mentions the latter in his *Mad Gardener's Song*: '*He thought he saw a Garden-Door that opened with a key: He looked again, and found it was a Double Rule of Three: "And all its mystery," he said, "Is clear as day to me!"* Also, Abraham Lincoln is known to have said that he learned to "read, write, and cipher to the rule of 3." It is also used in presentations, such as in speeches, for instance, where three things are easily remembered by an audience. For instance, "Friends, Romans, Countrymen" - William Shakespeare in Julius Caesar. In religion we find the trinity, "Father, Son and Holy Spirit", or everyday common sense, "Stop, look and listen", even in Hollywood, "The good, the bad and the ugly", etc. Succinctly, those ideas best remembered are those in a list of three.

Proof. First assume that $b \in \mathbb{N}$. If $a = 1$, then $r = 0$, so $b = q$. If $a > 1$, then by Theorem 1.5, the Base Representation Theorem, on page 8, b has a unique representation to base a,

$$b = b_0 + \sum_{j=1}^{m} b_j a^j, \text{ where } 0 \le b_j < a \text{ for each } j = 0, 1, \ldots, m.$$

Thus,

$$b = b_0 + a \sum_{j=1}^{m} b_j a^{j-1} = aq + r, \tag{1.6}$$

so $0 \le r = b_0 < a$. To show that q and r must indeed be unique, we use the above representation. If $b = aq_1 + r_1$, then q_1 has a unique representation to base a,

$$q_1 = c_0 + \sum_{j=1}^{m_1} c_j a^j, \text{ where } 0 \le c_j < a \text{ for each } j = 0, 1, \ldots, m_1,$$

so

$$b = aq_1 + r_1 = \sum_{j=0}^{m_1} c_j a^{j+1} + r_1 = b_0 + \sum_{j=1}^{m} b_j a^j,$$

and by uniqueness of representation, we may infer that $m_1 = m - 1$, $b_j = c_{j-1}$ for $j = 1, 2, \ldots, m$, and $r_1 = b_0 = r$. Hence,

$$q_1 = \sum_{j=0}^{m_1} c_j a^j = \sum_{j=1}^{m} b_j a^{j-1} = q,$$

where the last equality comes from Equation (1.6). The uniqueness of r is now apparent. We have succeeded in proving the result for all $b \in \mathbb{N}$. However, for $b = 0$, the unique result is $q = r = 0$, and for $b < 0$, the uniqueness follows from the uniqueness of the representation of $-b > 0$. $\square$

Biography 1.4 Euclid of Alexandria (ca. 300 B.C.) *is the author of the Elements.* Next to the Bible, the *Elements* is the most reproduced book in recorded history. Little is known about Euclid's life, other than that he lived and taught in Alexandria. However, the folklore is rich with quotes attributed to Euclid. For instance, he is purported to have been a teacher of the ruler Ptolemy I, who reigned from 306 to 283 B.C. When Ptolemy asked if there were an easier way to learn geometry, Euclid ostensibly responded that there is no royal road to geometry. His nature as a purist is displayed by another quotation. A student asked Euclid what use could be made of geometry, to which Euclid responded by having the student handed some coins, saying that the student had to make gain from what he learns.

Example 1.1 *If $a = 8$, $b = 17$, then $17 = 2 \cdot 8 + 1$ with $q = 2$, and $r = 1$.*

Example 1.2 *If $a = 4$, $b = -15$, then $-15 = 4(-4) + 1$ with $q = -4$, and $r = 1$.*

Example 1.3 *If $a = 19$, $b = 13$, then $13 = 0 \cdot 19 + 13$ with $q = 0$, and $r = 13$.*

Example 1.4 *If $n, a, b \in \mathbb{Z}$, where $n \mid a$, and $n \mid b$, then n is called a* common divisor *of a and b. Therefore, there are integers c, d such that $a = nc$ and $b = nd$. Thus, for any integers x, y, $xa + yb = xnc + ynd = n(xc + yd)$, so $n \mid (xa + yb)$, where $xa + yb$ is a* linear combination *of a and b. Thus, we have demonstrated one of the fundamental properties of common divisors, namely that any common divisor of a and b divides all linear combinations thereof.*

There is a special kind of common divisor that deserves singular recognition.

Definition 1.5 $\boxed{\textbf{The Greatest Common Divisor}}$

If $a, b \in \mathbb{Z}$ are not both zero, then the greatest common divisor *or* gcd *of a and b is the natural number g such that g is a common divisor of a and b, and any common divisor of a and b is also a divisor of g. We denote this by $g = \gcd(a, b)$.*

Another important property concerning divisibility is due to Euclid, and based on the gcd, is given as follows.

Lemma 1.1 $\boxed{\textbf{Euclid's Lemma}}$

Suppose that $a, b \in \mathbb{Z}$, not both zero, and $c \in \mathbb{Z}$ such that $c \mid ab$, with $\gcd(b, c) = 1$. Then $c \mid a$.

Proof. By Exercise 1.24, $\gcd(ab, ac) = |a| \gcd(b, c) = |a|$. Since $c \mid ab$, then c is a common divisor of ab and ac, so $c \mid |a| = \gcd(ab, ac)$, namely $c \mid a$.

□

Definition 1.5, by its very phraseology, *the* gcd, assumes that the *gcd* is unique. We establish this and more in what follows.

Theorem 1.8 $\boxed{\textbf{Existence and Uniqueness of the GCD}}$

If $a, b \in \mathbb{Z}$, where $a \neq 0$ or $b \neq 0$, then $\gcd(a, b)$ exists and is unique.

Proof. If $ab = 0$, then $\gcd(a, b)$ is the absolute value of the nonzero integer. If either a or b is negative, then the $\gcd(a, b) = \gcd(|a|, |b|)$, so we may assume that both a and b are positive. In fact, without loss of generality, assume that $b \geq a > 0$. If $a \mid b$, then $\gcd(a, b) = a$, so we may assume that $a \nmid b$.

We begin by exploiting Theorem 1.7 to fit our needs.

Claim 1.2 *Repeated application of the division algorithm produces the sequence of r_j values which is strictly decreasing and bounded below by zero, and if $r_n = 0$, then $r_{n-1} = \gcd(a, b)$.*

If we set $a = a_0$ and $b = b_0$, then by Theorem 1.7 there exists unique integers q_0, r_0 such that $b_0 = a_0 q_0 + r_0$ where $0 \le r_0 < a_0$. Then by repeated application of that division algorithm, there exist unique integers q_j, r_j for any $j \in \mathbb{N}$ with

$$b_j = a_j q_j + r_j, \text{ with } 0 \le r_j < a_j, \qquad (1.7)$$

where $b_j = a_{j-1}$ and $a_j = r_{j-1}$. For a given $j \in \mathbb{N}$, $0 \le r_{j-1} < a_{j-1} = r_{j-2} < a_{j-2} < \cdots < a_0$, so by induction, $0 \le r_{j-1} \le a - j$, which tells us that $r_n = 0$ for some $0 < n < a$. Note that $n > 0$ since $r_0 \ne 0$ given our assumption that $a \nmid b$. Since $b_n = a_n q_n + r_n = a_n q_n$, then $r_{n-1} = a_n \mid b_n = a_{n-1} = r_{n-2}$, and similarly $r_{n-2} \mid r_{n-3}$. Continuing in this fashion, we see that $r_{n-j} \mid r_{n-j-1}$ for each natural number $j < n$, so $r_{n-1} \mid r_1 \mid r_0 = a_1$. Therefore, by Equation (1.7), with $j = 1$, $r_{n-1} \mid b_1 = a_0 = a$, and by Equation (1.7), with $j = 0$, $r_{n-1} \mid b$. Therefore, r_{n-1} is a common divisor of a and b. Moreover, if d is a common divisor of a and b, then $d \mid r_0$, but $a = a_0 = b_1 = q_1 a_1 + r_1 = q_1 r_0 + r_1$, so $d \mid r_1$. Continuing in this fashion we see that $d \mid r_j$ for all $j < n$. Hence, r_{n-1} satisfies Definition 1.5. In other words, $\gcd(a, b) = r_{n-1}$. This is Claim 1.2. In fact, we have shown something somewhat stronger, namely that

$$\gcd(a, b) = \gcd(r_j, r_{j+1}) \text{ for any integer } j \text{ with } 0 \le j < n.$$

We have shown existence. It remains to demonstrate uniqueness. If g_1 and g_2 are greatest common divisors, then they must divide each other by Definition 1.5, so there are $x, y \in \mathbb{N}$ such that $g_1 = xg_2$ and $g_2 = yg_1$. Therefore, $g_2 = yxg_2$, so $xy = 1$ forcing $x = y = 1$. This is uniqueness and secures the proof. $\quad\square$

The following is immediate from the above proof.

Corollary 1.3 *Suppose that $a, b \in \mathbb{Z}$ not both zero, and neither a nor b is a divisor of the other. If n is the smallest integer such that $r_n = 0$ in the application of the division algorithm in Theorem 1.7, then $r_{n-1} = \gcd(a, b)$.*

What is implicit in the above is the following:

◆ **The Euclidean Algorithm (EA)**

If we have integers a, b with $a > 0$, then the *Euclidean algorithm* is obtained by repeated application of the division algorithm, Theorem 1.7, to yield a set of equations

$$b = aq_1 + r_1, \text{ with } 0 < r_1 < a,$$
$$a = r_1 q_2 + r_2, \text{ with } 0 < r_2 < r_1,$$
$$r_1 = r_2 q_3 + r_3, \text{ with } 0 < r_3 < r_2,$$
$$\vdots \qquad\qquad \vdots$$
$$r_{n-3} = r_{n-2} q_{n-1} + r_{n-1}, \text{ with } 0 < r_{n-1} < r_{n-2},$$
$$r_{n-2} = r_{n-1} q_n.$$

where $r_n = 0$, and $\gcd(a, b) = r_{n-1}$.

Example 1.5 *Let $a = 210$ and $b = 1001$. Using Theorem 1.7, we get*

$$1001 = b = aq + r = 210 \cdot 4 + 161, \tag{1.8}$$

so any common divisor of 1001 and 210 is also a divisor of 161, which is a linear combination $b - aq = 1001 - 210 \cdot 4$. Thus, repeatedly using Theorem 1.7,

$$210 = b_1 = a_1 q_1 + r_1 = 161 \cdot 1 + 49, \tag{1.9}$$

$$161 = b_2 = a_2 q_2 + r_2 = 49 \cdot 3 + 14, \tag{1.10}$$

$$49 = b_3 = a_3 q_3 + r_3 = 14 \cdot 3 + 7, \tag{1.11}$$

$$14 = b_4 = a_4 q_4 + r_4 = 7 \cdot 2 + 0.$$

Now we see that in the last calculation where $r_4 = 0$, $7 = a_4$ divides both $a = 7 \cdot 30$ and $b = 7 \cdot 143$. Moreover, if d is a common divisor of a and b, then $d \mid 161$ by Equation (1.8), $d \mid 49$ by Equation (1.9), $d \mid 14$ by Equation (1.10), and $d \mid 7$ by Equation (1.11). Hence, Definition 1.5 is satisfied and $\gcd(210, 1001) = 7$.

One might wonder about the number of iterations or divisions, namely the value of n, required in the EA in order to find the *gcd* in general. It may seem surprising, but it can be shown that for any $n \in \mathbb{N}$ there exist $a, b \in \mathbb{N}$ such that it takes n iterations in the EA to find $\gcd(a, b)$, and the following example not only demonstrates this, but also motivates a result that determines when such maximum iterations occur.

Example 1.6 *We seek $\gcd(F_{n+2}, F_{n+1})$ where F_j is the jth Fibonacci number defined on page 3. We have the following sequence,*

$$F_{n+2} = 1 \cdot F_{n+1} + F_n,$$

$$F_{n+1} = 1 \cdot F_n + F_{n-1},$$

$$\vdots \quad \vdots$$

$$F_4 = 1 \cdot F_3 + F_2,$$

$$F_3 = 2F_2,$$

showing that n iterations are required to find that the gcd of F_{n+2} and F_{n+1} is 1, a fact that we know, without the Euclidean algorithm, from Exercise 1.9 on page 12, for instance. Indeed, the following result, which this example illustrates, provides us with the answer to the query: What are the least positive integers requiring exactly n divisions to find their gcd via the EA?

Theorem 1.9 $\boxed{\textbf{EA — Least Values for Maximum Iterations}}$

If $a > b > 0$ are integers and the application of the Euclidean algorithm requires n iterations to find the $\gcd(a, b)$, then $a \geq F_{n+2}$ and $b \geq F_{n+1}$.

Proof. We use induction on n to verify the result. If $n = 1$, then the application of the EA is one iteration, namely $a = q_0 b$, and since $a > b$, to find the least such a and b, we set $b = 1 = F_2$ and $a = 2 = F_3$, which is the induction step. The induction hypothesis is that for any $j < n$, the result holds. We now prove it holds for $j = n$. Since $a = a_0 = q_0 a_1 + r_1$ is the first iteration with $a_1 = b$, and it takes $n - 1$ iterations to find $\gcd(a_1, r_1)$, then $b = a_1 \geq F_{n+1}$ and $r_1 \geq F_n$ by the induction hypothesis. Hence, $a = a_0 \geq a_1 + r_1 \geq F_{n+1} + F_n = F_{n+2}$. $\square$

Biography 1.5 Gabriel Lamé *was born in Tours, France on July 22, 1795. He was educated at the École Polytechnique after which he studied engineering at the École des Mines in Paris from which he graduated in 1820. In that year he went to Russia where he was appointed professor and engineer at the Institut et Corps du Genie des Voies de Communication in St. Petersburg where he taught various subjects. He also published papers in both Russian and French journals in his 12 years there. He returned to Paris in 1832, where he was appointed the chair of physics at École Polytechnique. By 1836, he was appointed chief engineer of mines, and in that capacity was associated with the construction of the railways from Paris to Versailles, and from Paris to St. Germain, completed in 1837. He was elected to the Académie des Sciences in 1843, and a year later vacated his chair at École Polytechnique to accept a position at the Sorbonne in physics, and by 1851, he was appointed chair of physics and probability there.*

From the above, we see that Lamé was essentially a mathematical physicist. Other than the contribution to which his name is attached in Corollary 1.4, he provided the first proof of Fermat's Last Theorem for the exponent 7. He actually believed he had a proof for the general case, but he made an error similar to that which Kummer made, assuming unique factorization in rings of cyclotomic integers. He died on May 1, 1870 in Paris.

Corollary 1.4 | **Lamé's Theorem** |

If $a > b > 0$ are integers and it takes $n + 1$ iterations (divisions) *to find* $\gcd(a, b)$ *via the EA, then $n < \log_{\mathfrak{g}}(b)$, where $\mathfrak{g}$ is the golden ratio defined on page 4.*

Proof. By Theorem 1.9, $b \geq F_{n+2}$, and by Exercise 1.2 on page 11, $F_{n+2} \geq \mathfrak{g}^n$. Thus, $b \geq \mathfrak{g}^n$, so $n < \log_{\mathfrak{g}}(b)$. $\square$

Another formulation of Lamé's theorem is the following, which is the one most often cited.

Corollary 1.5 *If $a > b > 0$ are integers and it takes $n + 1$ iterations* (divisions) *to find* $\gcd(a, b)$ *via the EA, then $n < 5\log_{10}(b)$.*

Proof. It is straightforward to verify that $\log_{10}\mathfrak{g} > 1/5$. Therefore, by Lamé's Theorem, $n/5 < \log_{\mathfrak{g}}(b)\log_{10}(\mathfrak{g}) = \log_{10}(b)$. Hence, $n < 5\log_{10}(b)$.   □

Remark 1.1 *Succinctly, what Corollary 1.5 says is that the number of iterations required to find the* $\gcd(a, b)$ *is less than five times the number of decimal digits in the smaller value* b. *Indeed, suppose that* b *has* s *decimal digits, so* $b < 10^s$, *namely* $\log_{10}(b) < s$. *Therefore,* $5s > n$ *by Corollary 1.5, so* $5s \geq n+1$, *which is the number of iterations required to find the* $\gcd(a, b)$. *This result is often used to find the* computational complexity *of the gcd, which is* $O(\log_2^3(a))$. (*See Appendix B for a discussion of computational complexity.*)

There is another important characterization of the gcd given in what follows.

Theorem 1.10 │ **The GCD as a Linear Combination** │

If $g = \gcd(a, b)$ *where* $a, b \in \mathbb{Z}$ *are not both zero, then* g *is the least positive value of* $ax + by$ *where* x *and* y *range over all integers.*

Proof. Let $\mathcal{S} = \{ax + by : x, y \in \mathbb{Z}\}$. By Exercise 1.3 on page 11, there is a least positive element in $\mathcal{S}$, which we will denote by $\ell = ax_0 + by_0$. Now we show that $\ell \mid a$ by assuming it does not (proof by contradiction). Thus, by Theorem 1.7, there exist unique integers q, r with $0 < r < \ell$ such that $a = \ell q + r$. Therefore, $r = a - \ell q = a - q(ax_0 + by_0) = a(1 - qx_0) - qy_0 b$, so $r \in \mathcal{S}$, contradicting the minimality of ℓ. Thus $\ell \mid a$. The same argument shows that $\ell \mid b$, so ℓ is a common divisor of a and b. Since g is also a divisor, then there exist $c, d \in \mathbb{Z}$ such that $a = gc$ and $b = gd$. Thus,

$$\ell = ax_0 + by_0 - gcx_0 + gdy_0 = g(cx_0 + dy_0),$$

so $g \mid \ell$. Hence, $\ell = gf$ for some $f \in \mathbb{N}$. However, if $f > 1$, then g is not the greatest common divisor. Therefore, $f = 1$ and this gives us the desired result that $\ell = g$.   □

The following consequence of Theorem 1.10 will be useful for later developments in this section and elsewhere.

Corollary 1.6 *If* $a, b \in \mathbb{N}$ *with* $\gcd(a, b) = 1$, *then there exist natural numbers* z, w *such that* $az - bw = 1$.

Proof. By Theorem 1.10 there integers x, y such that $ax + by = 1$. Select an integer x_0 such that $x_0 > -x/b$ and $x_0 > y/a$, and set $z = x + bx_0 \in \mathbb{N}$ and $w = ax_0 - y \in \mathbb{N}$. Then $az - bw = ax + by = 1$, as required.   □

Remark 1.2 *It is straightforward to extrapolate from the gcd of two integers to any finite number. For instance, if* $a, b, c \in \mathbb{Z}$ *not all zero, then* $\gcd(a, b, c) = \gcd(\gcd(a, b), c)$. *Continuing in this fashion for any* $n \geq 3$, *we get a well-defined gcd. Thus, if* $x_1, x_2, \ldots, x_n$ *are integers not all zero, then*

$$\gcd(x_1, x_2, \ldots, x_n) = \gcd(\gcd(x_1, x_2, \ldots, x_{n-1}), x_n).$$

Some further basic properties of the *gcd* are developed in the exercises at the end of this section.

The EA leads naturally to a topic that we will cover in depth in Chapter 5. By viewing Example 1.5 from a different perspective, we get the following

Example 1.7 $b/a = 1001/210 = 4 + 161/210 = 4 + \frac{1}{210/161}$, *but*

$$\frac{210}{161} = 1 + \frac{1}{161/49},$$

so

$$\frac{1001}{210} = 4 + \cfrac{1}{1 + \cfrac{1}{161/49}}.$$

Given that

$$\frac{161}{49} = 3 + 14/49 = 3 + \frac{1}{49/14},$$

then

$$\frac{1001}{210} = 4 + \cfrac{1}{1 + \cfrac{1}{3 + \cfrac{1}{49/14}}}.$$

However,

$$49/14 = 3 + 7/14 = 3 + 1/2,$$

so

$$\frac{1001}{210} = 4 + \cfrac{1}{1 + \cfrac{1}{3 + \cfrac{1}{3 + \cfrac{1}{2}}}}.$$

Example 1.7 illustrates a very important general fact, namely that *all* rational numbers can be expressed in this way. In order to formalize this statement, we need some terminology, and we need some convenient notation rather than the cumbersome notation displayed in Example 1.7.

Definition 1.6 Finite Continued Fractions

If $q_j \in \mathbb{R}$ for $j = 0, 1, \ldots, \ell$ where $\ell \in \mathbb{Z}$ is nonnegative and $q_j \in \mathbb{R}^+$ for $j > 0$, then an expression of the form

$$\alpha = q_0 + \cfrac{1}{q_1 + \cfrac{1}{q_2 +}}$$

$$\ddots$$

$$+ \cfrac{1}{q_{\ell-1} + \cfrac{1}{q_\ell}}$$

denoted by $\langle q_0; q_1, \ldots, q_\ell \rangle$, *is a* finite continued fraction *of length* ℓ. *A finite continued fraction is said to be* simple *if* $q_j \in \mathbb{Z}$ *for all* $j = 0, 1, 2, \ldots, \ell$. *The values* q_j *are the* partial quotients. *The semi-colon is used after the first partial quotient* q_0 *to separate the* integer value *of* α *from the rest of the partial quotients, namely* $q_0 = \lfloor \alpha \rfloor$, *the floor function (see Definition 2.15 on page 108). The reader may verify that this follows from the Euclidean algorithm.*

Example 1.8 *From Example 1.7,* $1001/210 = \langle 4; 1, 3, 3, 2 \rangle$, *a much more compact and useful notation than the one in that example.*

Another celebrated example is derived from Example 1.6 on page 20 for continued fractions.

Example 1.9 *If* F_n *denotes the* n-*th Fibonacci number, then as a simple continued fraction,*

$$F_{n+1}/F_n = \underbrace{\langle 1; 1, 1, \ldots, 1 \rangle}_{n \text{ copies of } 1} = \underbrace{\langle 1; 1, \ldots, 1, 2 \rangle}_{n-2 \text{ copies of } 1},$$

(Exercise 1.32 on page 29 shows that all finite simple continued fractions have two representations, with the Fibonacci sequence the simplest such.)

Now we can prove what we asserted above.

Theorem 1.11 | **Finite Simple Continued Fractions are Rational** |

Let $\alpha \in \mathbb{R}$. *Then* $\alpha \in \mathbb{Q}$ *if and only if* α *can be written as a finite simple continued fraction.*

Proof. If $\alpha = \langle q_0; q_1, \ldots, q_\ell \rangle$ with $q_i \in \mathbb{Z}$, then we use induction on ℓ. If $\ell = 1$, then

$$\alpha = q_0 + \frac{1}{q_1} = \frac{q_0 q_1 + 1}{q_1} \in \mathbb{Q}.$$

Assume that all simple continued fractions of length less than ℓ are in $\mathbb{Q}$. Since

$$\langle q_0; q_1, \ldots, q_\ell \rangle = q_0 + \frac{1}{\langle q_1; \ldots, q_\ell \rangle},$$

then by the induction hypothesis $\langle q_0; q_1, \ldots, q_\ell \rangle \in \mathbb{Q}$.

Conversely, assume that $b/a \in \mathbb{Q}$ with $a \in \mathbb{N}$ and $b \in \mathbb{Z}$. Then we may set $a = r_0$, $b = r_{-1}$ and invoke the EA to get the recursive relation $r_{j-1} = r_j q_j + r_{j+1}$ where $0 < r_{j+1} < r_j$, for $j = 0, 1, \ldots, n$, $r_n = 0$ and $r_{n-1} = \gcd(a, b)$. Also, if $\alpha_{j-1} = r_{j-1}/r_j$, then $\alpha_{j-1} = q_j + 1/\alpha_j$ for $j = 0, 1, \ldots, n-1$. Thus, $b/a = \alpha_{-1} = \langle q_0; q_1, \ldots, q_{n-1}, \alpha_{n-1} \rangle = \langle q_0; q_1, \ldots, q_n \rangle$, and the result is complete. $\square$

The inductive process in the proof of Theorem 1.11 contains the seeds of some new and valuable information.

Example 1.10 *In Example 1.8, we have* $1001/210 = \langle 4; 1, 3, 3, 2 \rangle = \langle q_0; q_1, q_2, q_3, q_4 \rangle$. *Consider* $C_0 = \langle 4 \rangle = 4 = q_0$, *and*

$$C_1 = \langle 4; 1 \rangle = 5 = q_0 + 1/q_1 = (q_0 q_1 + 1)/(q_1).$$

Similarly the reader may verify that:

$$C_2 = \langle 4; 1, 3 \rangle = \frac{19}{4} = \frac{(q_0 q_1 + 1)q_2 + q_0}{q_1 q_2 + 1},$$

and

$$C_3 = \langle 4; 1, 3, 3 \rangle = \frac{62}{13} = \frac{((q_0 q_1 + 1)q_2 + q_0)q_3 + (q_0 q_1 + 1)}{(q_1 q_2 + 1)q_3 + q_1}.$$

The illustration in Example 1.10 is a special case of a general phenomenon motivated by the proof of Theorem 1.11.

Definition 1.7 | **Convergents**

Let $n \in \mathbb{N}$ *and let* α *have finite continued fraction expansion* $\langle q_0; q_1, \ldots, q_\ell \rangle$ *for* $q_j \in \mathbb{R}^+$ *when* $j > 0$. *Then* $C_k = \langle q_0; q_1, \ldots, q_k \rangle$ *is the* k^{th} *convergent of* α *for any nonnegative integer* $k \leq n$.

Theorem 1.12 | **Representation of Convergents**

Let $\alpha = \langle q_0; q_1, \ldots, q_\ell \rangle$ *for* $\ell \in \mathbb{N}$ *be a finite continued fraction expansion. Define two sequences for* $k \in \mathbb{Z}$ *nonnegative:*

$$A_{-2} = 0, A_{-1} = 1, A_k = q_k A_{k-1} + A_{k-2},$$

and

$$B_{-2} = 1, B_{-1} = 0, B_k = q_k B_{k-1} + B_{k-2}.$$

Then

$$C_k = A_k/B_k = \frac{q_k A_{k-1} + A_{k-2}}{q_k B_{k-1} + B_{k-2}},$$

is the kth *convergent of* α *for any nonnegative integer* $k \leq \ell$.

Proof. We use induction on k. If $k = 0$, then

$$C_0 = q_0 = A_0/B_0 = \frac{q_0 A_{-1} + A_{-2}}{q_0 B_{-1} + B_{-2}}.$$

Assume that

$$C_k = A_k/B_k = \frac{q_k A_{k-1} + A_{k-2}}{q_k B_{k-1} + B_{k-2}}$$

for k, and prove the result for $k + 1$.

$$C_{k+1} = \langle q_0; q_1, \ldots, q_{k+1} \rangle = \langle q_0; q_1, \ldots, q_{k-1}, q_k + 1/q_{k+1} \rangle,$$

1. *Arithmetic of the Integers*

so we may use the induction hypothesis in C_{k+1} since it is of length k in the last representation. Thus,

$$C_{k+1} = \frac{(q_k + 1/q_{k+1})A_{k-1} + A_{k-2}}{(q_k + 1/q_{k+1})B_{k-1} + B_{k-2}} = \frac{(q_k q_{k+1} + 1)A_{k-1} + q_{k+1}A_{k-2}}{(q_k q_{k+1} + 1)B_{k-1} + q_{k+1}B_{k-2}} =$$

$$\frac{A_{k-1} + q_{k+1}(q_k A_{k-1} + A_{k-2})}{B_{k-1} + q_{k+1}(q_k B_{k-1} + B_{k-2})} = \frac{A_{k-1} + q_{k+1}A_k}{B_{k-1} + q_{k+1}B_k} = \frac{A_{k+1}}{B_{k+1}},$$

so by induction we have the result. □

Remark 1.3 *Exercise 1.33 on page 29 provides a relationship between the A_k and B_k that allows one to conclude that $\gcd(A_k, B_k) = 1$. It turns out that these sequences will be a key element in solving Diophantine equations, such as the classic Pell equation, $x^2 - DY^2 = \pm 1$. We will return to this topic in Chapter 5 — see Theroem 5.15 on page 234 — where we cover infinite as well as finite simple continued fractions. This, however, was an appropriate juncture to introduce the notion. Now we turn to a different topic related to the gcd that deserves special mention.*

Definition 1.8 | The Least Common Multiple

If $a, b \in \mathbb{Z}$, then the smallest natural number which is a multiple of both a and b is the least common multiple *of a and b, denoted by $\mathrm{lcm}(a, b)$.*

The following are illustrations of basic facts about the lcm established in exercises at the end of this section.

Example 1.11 *If $a = 15$, $b = 21$, then $\mathrm{lcm}(a, b) = 105 = 3 \cdot \mathrm{lcm}(5, 7)$.*
If $a = 6$, $b = 15$, then $\mathrm{lcm}(2, 5) = 10 = \mathrm{lcm}(a/3, b/3) = \mathrm{lcm}(a, b)/3$.

Theorem 1.13 | Relative Properties of the gcd and lcm

Let $a, b \in \mathbb{N}$, $\ell = \mathrm{lcm}(a, b)$, and $g = \gcd(a, b)$.

(a) If $g = 1$, then $\ell = ab$.

(b) $\ell g = ab$.

Proof. Since $b | \ell$, $\ell = bn$ for some $n \in \mathbb{Z}$. Also, since $a | \ell$, and $g = 1$, in part (a), then by Lemma 1.1 on page 18, $a | n$. However, since $\ell \mid ab$, then $\ell \leq ab$. Therefore, $ab \geq \ell = ab(n/a) = bn \geq ba$. In other words, $n = a$, so $\ell = ab$.
For part (b), we need the following.

Claim 1.3 $\gcd(a/g, b/g) = 1$

If c is a common divisor of a/g and b/g, then gc is a common divisor of a and of b. However, g is the *greatest* common divisor, so $c = 1$. In other words, $\gcd(a/g, b/g) = 1$.

Using Claim 1.3 and part (a), we get that $\operatorname{lcm}(a/g, b/g) = ab/g^2$. Therefore, $ab = g^2 \operatorname{lcm}(a/g, b/g) = g \cdot \operatorname{lcm}(g(a/g), g(b/g)) = g\ell$, where the penultimate equality follows from Exercise 1.35. Thus, $ab = g\ell$.   □

Example 1.12 *If $a = 21$, $b = 55$, then $\operatorname{lcm}(a, b) = 1155 = ab$, and $\gcd(a, b) = 1$. If $a = 21$ and $b = 33$, then $\ell = \operatorname{lcm}(a, b) = \operatorname{lcm}(21, 33) = 231$ and $g = \gcd(a, b) = \gcd(21, 33) = 3$. Thus, in both cases, $\ell g = ab$.*

Just as we could extend the notion of the gcd of two numbers to any finite number, we can do this with the lcm as well by using the same reasoning as in Remark 1.2 on page 22.

Another illustration of properties of the gcd are given in the following.

Example 1.13 *If $a, b \in \mathbb{N}$ and $c \in \mathbb{Z}$, $c \neq 0$, with $b \mid c$, then it is straightforward to check that*

$$(a^c - 1) = (a^b - 1)(a^{c-b} + a^{c-b-1} + \cdots + a + 1).$$

In particular, this holds if $b = \gcd(c, d)$ for any $d \in \mathbb{Z}$. We will use this in the following illustration.

Example 1.14 *We will show that if $a > 1$ and $m, n \in \mathbb{N}$, then*

$$\gcd(a^m - 1, a^n - 1) = a^{\gcd(m,n)} - 1.$$

Let $g = \gcd(m, n)$ and $g' = \gcd(a^m - 1, a^n - 1)$. By Example 1.13, $a^g - 1$ divides both $a^n - 1$ and $a^m - 1$. Hence,

$$(a^g - 1) \mid g'. \tag{1.12}$$

Since $m = gm_1$ and $n = gn_1$ where $\gcd(m_1, n_1) = 1$, then By Corollary 1.6 on page 22, there exist $z, w \in \mathbb{N}$ such that $m_1 z - n_1 w = 1$. Therefore,

$$gm_1 z - gn_1 w = mz - nw = g. \tag{1.13}$$

Also, since $g' \mid (a^m - 1)$, then by Example 1.13, $g' \mid (a^{mz} - 1)$ and $g' \mid (a^{nw} - 1)$. It follows that

$$g' \mid (a^{mz} - a^{nw}) = a^{nw}(a^{mz-nw} - 1) = a^{nw}(a^g - 1), \tag{1.14}$$

where the last equality follows from (1.13). Since $g' \mid (a^m - 1)$, then $\gcd(g', a) = 1$. Thus, by Lemma 1.1 on page 18, (1.14) tells us that $g' \mid (a^g - 1)$. This, coupled with (1.12), shows that

$$\gcd(a^m - 1, a^n - 1) = a^{\gcd(m,n)} - 1.$$

The last topic we consider in this section is one that we will study in much detail later.

Example 1.15 *Let* $a, b, c, x, y \in \mathbb{Z}$ *and set*

$$ax + by = c, \tag{1.15}$$

which is a linear combination of a, b *introduced in Example 1.4 on page 18. Equation (1.15) is also called a* linear Diophantine equation. (*We will study general Diophantine equations in Chapter 7. Also, see Biography 1.15 on page 48.*)

We now demonstrate that Equation (1.15) has solutions $x, y \in \mathbb{Z}$ *for given* $a, b, c \in \mathbb{Z}$ *if and only if* $\gcd(a, b) \mid c$.

If $g = \gcd(a, b) \nmid c$, *then clearly there are no solutions. If* $g \mid c$, *then by Theorem 1.10, there exist integers* x, y *such that* $ax + by = g$. *Therefore, a solution to Equation (1.15) is given by* $(x_0, y_0) = (cx/g, cy/g)$. *This proves the assertion.*

Now we show that one solution (x_0, y_0) *leads to infinitely many that are given by*

$$(x, y) = \left(x_0 + \frac{bz}{g}, y_0 - \frac{az}{g} \right) \text{ where } z \text{ ranges over the integers.}$$

Given a solution (x_0, y_0) *to Equation (1.15), it is a straightforward check that as* z *ranges over all integers,*

$$(X, Y) = \left(x_0 + \frac{bz}{g}, y_0 - \frac{az}{g} \right) \tag{1.16}$$

are also solutions to $aX + bY = c$. *Conversely, if* $(x, y) = (s, t)$ *is a solution of Equation (1.15), then* $a(s - x_0) + b(t - y_0) = c - c = 0$, *which implies that*

$$\frac{a/g}{b/g} = \frac{-(t - y_0)}{s - x_0}.$$

Thus, there must exist an integer z *such that*

$$s - x_0 = bz/g \text{ and } t - y_0 = -az/g,$$

which is of the form in Equation (1.16), which must therefore represent the general solution set for Equation (1.15).

The notion of divisibility leads us naturally into a discussion of those integers $p > 1$ which are *not* divisible by any positive integers other than p itself or 1. These are the atoms of the theory of numbers, **prime** numbers, the topic of the next section.

Exercises

1.20. For integers a, b prove that $a \mid b$ implies $a \mid bc$ for any integer c.

1.21. If $a, b, c \in \mathbb{Z}$, prove that $a \mid b$ and $b \mid c$ imply that $a \mid c$, called the *transitivity* property for division.

1.22. If $a, b \in \mathbb{Z}$ with $a \mid b$ and $b \mid a$, prove that $a = \pm b$.

1.23. If $a, b, c \in \mathbb{Z}$ with $c \neq 0$, prove that $a \mid b$ if and only if $ca \mid cb$.

1.24. Given $m \in \mathbb{N}$ and $a, b \in \mathbb{Z}$, not both zero, prove that $\gcd(ma, mb) = m \gcd(a, b)$.

1.25. Given $a, b \in \mathbb{Z}$, prove that $\gcd(a, b) = |a|$ if and only if $|a| \mid b$.

1.26. (a) Prove that if c is a common divisor of integers a and b, not both zero, then $\gcd(a/c, b/c) = g/c$.

(b) Prove that for any $c \in \mathbb{Z}$, and any $a, b \in \mathbb{Z}$, not both zero, $\gcd(a, b) = \gcd(b, a) = \gcd(a, -b) = \gcd(a, b + ac)$.

1.27. Prove that for any integer $n \geq 0$, $57 \mid (10^{36n+7} + 23)$.

1.28. Find all integers x, y such that $3x + 7y = 1$.

1.29. If $n \in \mathbb{N}$ with $n \geq 2$, prove that the linear Diophantine equation in n variables, $a_1 x_1 + a_2 x_2 + \cdots + a_n x_n = c$ has a solution if and only if $\gcd(a_1, a_2, \ldots, a_n) \mid c$.

(*Hint: See Remark 1.2 and use induction on Example 1.15.*)

1.30. Find the gcd of each of the following pairs.
    (a) $a = 22$, $b = 55$.   (b) $a = 15$, $b = 113$.

☆ 1.31. Prove that if $m, n \in \mathbb{N}$ with $n \mid m$, then $F_n \mid F_m$, where the latter are the Fibonacci numbers defined on page 3. Use this fact to prove that for *any* $m, n \in \mathbb{N}$, $\gcd(F_m, F_n) = F_{\gcd(m,n)}$.

1.32. Prove that a finite simple continued fraction has two representations, one of which ends with a partial quotient equal to 1. If we seek uniqueness of representations of finite simple continued fractions, then we may stipulate that the last partial quotient is greater than 1.

1.33. Let A_k/B_k be the k-th convergent, identified in Theorem 1.12 on page 25. Prove that for any $k \in \mathbb{N}$, $A_k B_{k-1} - A_{k-1} B_k = (-1)^{k-1}$. Conclude that $\gcd(A_k, B_k) = 1$.

1.34. If $a \in \mathbb{Z}$, $b \in \mathbb{N}$, and $\ell = \mathrm{lcm}(a, b)$, prove that $\ell = b$ if and only if $a \mid b$.

1.35. (a) If $a, b \in \mathbb{Z}$ and $\ell = \mathrm{lcm}(a, b)$, prove that $n\ell = \mathrm{lcm}(an, bn)$ for any $n \in \mathbb{N}$.

(b) If $a, b \in \mathbb{Z}$, $c \in \mathbb{N}$ where $c|a$ and $c|b$, prove that $\mathrm{lcm}(a/c, b/c) = \ell/c$.

1.36. Find the least common multiple (lcm) of the following pairs.
    (a) $a = 15$, $b = 385$.   (b) $a = 28$, $b = 577$.
    (c) $a = 73$, $b = 561$.   (d) $a = 110$, $b = 5005$.

1.3   Primes

> *From the intrinsic evidence of His creation, the Great Architect of the Universe now begins to appear as a pure mathematician.*
>   From Chapter 5 of **The Mysterious Universe (1930)**
>   **James Jeans (1877–1946), English astronomer, physicist, and**
> **mathematician**

Although we have discussed many fundamentals, we have not yet defined the building bricks of arithmetic, prime numbers.

Definition 1.9 ┃ Primes

If $p > 1$ is an integer having no positive divisors, other than itself and 1, then p is called a prime number, *or simply a* prime. *If $n > 1$ is an integer that is not a prime, then n is said to be* composite.

In Definition 1.3 on page 16, we introduced the notion of *exact* divisibility. Now that we have the notion of primes, we may extend that notion.

Definition 1.10 ┃ Squarefree and p-Components

If p is a prime and $p^m \| a \in \mathbb{Z}$, then we say that m is the p-component of a. Also, if $p \| a$ for all primes p dividing a, then we say that a is squarefree *— see Exercises 1.39–1.42 on page 39.*

In Definition 1.5 on page 18, we introduced the notion of the *greatest common divisor.* When the gcd is equal to one, there is a special name for such an occurrence.

Definition 1.11 ┃ Relative Primality

If $a, b \in \mathbb{Z}$ with $\gcd(a, b) = 1$, then a and b are said to be relatively prime.

Remark 1.4 *The notion in Definition 1.11 is derived from the fact that $\gcd(a, b) = 1$ if and only if the only common divisor of a and b is 1. In other words, relative to each other, they exhibit the property of a prime. This notion may be generalized as follows.*

If $a_i \in \mathbb{Z}, a_i \neq 0$ for $1 \leq i \leq n$, then $a_1, a_2, \ldots, a_n$ are pairwise relatively prime *if $\gcd(a_i, a_j) = 1$ for all $i \neq j$ with $1 \leq i, j \leq n$.*

For instance, if $a_1 = 15, a_2 = 77, a_3 = 26$, then $\gcd(15, 77) = \gcd(15, 26) = \gcd(77, 26) = 1$. Therefore, the a_i are pairwise relatively prime.

The following result is required to establish some key results in this section.

Theorem 1.14 ⟨Prime Factors⟩

Every composite integer has a prime factor.

Proof. Let $n = n_1 n_2$ where $n > n_1, n_2$. If either n_1 or n_2 is prime, we are done. If both n_1 and n_2 are composite, then $n_1 = n_3 n_4$, where $n_1 > n_3, n_4 > 1$. If n_3 and n_4 are not prime we may write $n_3 = n_5 n_6$, where $n_3 > n_5, n_6 > 1$. Continuing in this fashion, we write $n_{2k-1} = n_{2k+1} n_{2k+2}$ for any $k \in \mathbb{N}$ with $n_{2k+1}, n_{2k+2} < n_{2k-1}$. This process must terminate since $n > n_1 > n_3 > \cdots > n_{2k-1} > 0$. In other words, n_{2k-1} is prime for some k. □

Corollary 1.7 *If n is composite, then n has a prime divisor p such that $p \leq \sqrt{n}$.*

Proof. Suppose that all prime divisors are larger than $\sqrt{n}$. Let p_1 and p_2 be two of them. Then $n \geq p_1 p_2 > \sqrt{n}\sqrt{n} = n$, a contradiction. □

Now the above may be used to prove the following result with a proof attributed to Euclid — see Biography 1.4 on page 17.

Theorem 1.15 ⟨The Infinitude of Primes⟩

There are infinitely many prime numbers.

Proof. Assume that there are only finitely many primes. Let that set be

$$\mathcal{S} = \{p_1, p_2, \ldots, p_n\},$$

$n \in \mathbb{N}$, and let $N = 1 + \prod_{i=1}^{n} p_i$. By Theorem 1.14, there is a prime q which divides N. However, $q \in \mathcal{S}$. Therefore, $q \mid \prod_{i=1}^{n} p_i$. In other words, q is a common divisor of N and $\prod_{i=1}^{n} p_i$, so $q | N - \prod_{i=1}^{n} p_i$. In other words, $q|1$, a contradiction. Hence, there are infinitely many primes. □

Corollary 1.7 tells us that any composite n has a prime divisor less than $\sqrt{n}$. How do we find all primes less than n? The following gives an illustration of how to do this.

Example 1.16 *Suppose that we want all primes less than* 40. *First, we write down all numbers less than* 40 *and bigger than* 1, *and cross out all numbers (bigger than* 2*) which are multiples of* 2, *the smallest prime.*

$$\{2, 3, \cancel{4}, 5, \cancel{6}, 7, \cancel{8}, 9, \cancel{10}, 11, \cancel{12}, 13, \cancel{14}, 15, \cancel{16}, 17, \cancel{18}, 19, \cancel{20}, 21, \cancel{22},$$
$$23, \cancel{24}, 25, \cancel{26}, 27, \cancel{28}, 29, \cancel{30}, 31, \cancel{32}, 33, \cancel{34}, 35, \cancel{36}, 37, \cancel{38}, 39\}.$$

Next, we cross out all numbers (bigger than 3*) which are multiples of* 3, *the next prime.*

$$\{2, 3, 5, 7, \cancel{9}, 11, 13, \cancel{15}, 17, 19, \cancel{21}, 23, 25, \cancel{27}, 29, 31, \cancel{33}, 35, 37, \cancel{39}\}.$$

Then we cross out all numbers (bigger than 5) which are multiples of 5, the next prime.

$$\{2,3,5,7,11,13,17,19,23,\cancel{25},29,31,\cancel{35},37\}.$$

What we have left is the set of primes less than 40.

$$\{2,3,5,7,11,13,17,19,23,29,31,37\}.$$

Example 1.16 illustrates the *Sieve of Eratosthenes.*

Biography 1.6 Eratosthenes of Cyrene (ca. 274–194 B.C.), *librarian of the renowned library in Alexandria, was a Greek astronomer and mathematician. He wrote* On the Measurement of the Earth, *in which he gave a remarkably accurate measurement of the diameter of the Earth, accomplished by taking measurements of the Sun's angle at two distinct locations with a given distance apart. Also included in this distinguished volume was a measurement of the tilt of the Earth's axis. However, mathematicians know him best for his sieve.*

Sieving is a process whereby we find numbers by searching up to a prescribed bound, and eliminate candidates as we proceed, leaving only the solution. By Corollary 1.7 on the page before, we see that, in order to determine whether a particular $n \in \mathbb{N}$ is prime, we need only check for divisibility by all primes less than $\sqrt{n}$. For instance, if $n = 37$, we need only check for divisibility by the primes $2, 3, 5$. Nevertheless, this is a highly inefficient method. Later, we will learn about several more efficient methods — see §1.8 and §2.7.

In anticipation of proving the Fundamental Theorem of Arithmetic, another main goal of this section, we need the following basic result.

Lemma 1.2 | **Prime Divisibility on Products**

If p is a prime and $a, b \in \mathbb{N}$ such that $p \mid ab$, then $p \mid a$ or $p \mid b$.

Proof. If $p \nmid a$, then $\gcd(p, a) = 1$, so by Euclid's Lemma 1.1 on page 18, $p \mid b$.□

Now we are ready for the following fundamental result which tells us why the primes are the building bricks of arithmetic.

Theorem 1.16 | **The Fundamental Theorem of Arithmetic**

Every $n \in \mathbb{N}$ with $n > 1$ has a unique factorization

$$n = \prod_{j=1}^{k} p_j^{a_j},$$

where the $p_1 < p_2 < \cdots < p_k$ for $j = 1, 2, \ldots k \in \mathbb{N}$ are primes and $a_j \in \mathbb{N}$. This is called the canonical prime factorization *of n.*

Proof. First we demonstrate existence of such factorizations. If n is prime, then we are done with $n = p_k$ where $k = 1 = a_k$. If n is not prime, with $n = n_1 n_2$, we apply the same argument to n_1 and n_2. The process must terminate since each subsequent factorization leads to positive integers less than the former, so ultimately all of these natural numbers are prime.

Now we establish the uniqueness of such factorizations. We use proof by contradiction to establish it. Let $n > 1$, and

$$n = \prod_{i=1}^{r} p_i^{a_i} = \prod_{i=1}^{s} q_i^{b_j}$$

be the smallest natural number (bigger than 1) which does not have unique factorization, where $p_1 < p_2 < \cdots < p_r$, and $q_1 < q_2 < \cdots < q_s$ with $a_i, b_j \in \mathbb{N}$. Suppose that $p_u = q_v$ for some u, v with $1 \le u \le r$ and $1 \le v \le s$. If $n = p_u$, then we are done, so assume that $n > p_u$. Since $1 < n/p_u < n$, n/p_u has unique factorization, and so

$$n/p_u = p_1^{a_1} p_2^{a_2} \cdots p_u^{a_u-1} p_{u+1}^{a_{u+1}} \cdots p_r^{a_r} = q_1^{b_1} q_2^{b_2} \cdots q_v^{b_v-1} q_{v+1}^{b_{v+1}} \cdots q_s^{b_s}$$

with $r = s$, $p_i = q_i$, and $a_i = b_i$ for all $i = 1, 2, \ldots, r = s$. Therefore,

$$n = p_u p_1^{a_1} p_2^{a_2} \cdots p_u^{a_u-1} p_{u+1}^{a_{u+1}} \cdots p_r^{a_r} = q_v q_1^{b_1} q_2^{b_2} \cdots q_v^{b_v-1} q_{v+1}^{b_{v+1}} \cdots q_s^{b_s}$$

has unique factorization, a contradiction. Hence, $p_u \ne q_v$ for all u, v. However, by Lemma 1.2, since $p_1 | \prod_{j=1}^{s} q_j^{b_j}$, then $p_1 | q_j$ for some j. Therefore, $p_1 = q_j$, a contradiction. We have established unique factorization. $\square$

Remark 1.5 *The first formal statement and proof of Theorem 1.16 was given by Gauss in [16, Theorem 16, p. 6]. However, if one is willing to stretch attribution to the limit, then one could say that Euclid almost had it. Some results by Euclid may be deemed to be* almost *equivalent to Theorem 1.16. These are contained in book 7 of the* Elements — *see Biography 1.4 on page 17. What Euclid actually proved was that the* lcm *of a finite set of primes has no other prime divisors — see Exercise 1.37 on page 38.*

Biography 1.7 *One of the greatest mathematicians who ever lived was* Carl Friedrich Gauss *(1777–1855). At the age of eight, he astonished his teacher, Büttner, by rapidly adding the integers from 1 to 100 via the observation that the fifty pairs $(j+1, 100-j)$ for $j = 0, 1, \ldots, 49$ each sum to 101 for a total of 5050. When still a teenager, he cracked the age-old problem of dividing a circle into 17 equal parts using only straightedge and compass. The ancient Greeks had known about construction of such regular n-gons for the cases where $2 \le n \le 6$, but the case $n = 7$ eluded solution, since as Gauss showed, the only ones that could be constructed in this fashion are those derivable from Fermat primes — see page 37. By the age of fifteen, Gauss entered Brunswick Collegium Carolinum. In 1795, Gauss was accepted to Göttingen University and by the age of twenty achieved his doctorate. Gauss remained a professor at Göttingen until the early morning of February 23, 1855, when he died in his sleep.*

Theorem 1.16 is sometimes called the *Unique Factorization Theorem for Integers*. Unique factorization is a fundamental concept in number theory, about which we will have more to say later. In §1.2 we introduced our notions of the gcd and the lcm. Theorem 1.16 on page 32 may be used to link prime factorization with these notions as follows.

Theorem 1.17 | **Factorization of the GCD and the LCM** |

   Let $a = \prod_{i=1}^{r} p_i^{m_i}, b = \prod_{i=1}^{r} p_i^{n_i}$ for integers $m_i, n_i \geq 0$, and distinct primes p_i with $1 \leq i \leq r$. Then each of the following holds.

   (a) If $t_i = \min\{m_i, n_i\}$ denotes the minimum value of m_i and n_i, then

$$g = \gcd(a, b) = \prod_{i=1}^{r} p_i^{t_i}.$$

   (b) If $M_i = \max\{m_i, n_i\}$ denotes the maximum value of m_i and n_i, then

$$\ell = \operatorname{lcm}(a, b) = \prod_{i=1}^{r} p_i^{M_i}.$$

   Proof. For part (a), set $c = \prod_{i=1}^{r} p_i^{t_i}$. To prove that $g = \gcd(a, b) = c$, it suffices to show that c is divisible by any common divisor of a and b by the definition of a gcd since clearly c is a common divisor of a and b. If $d \mid a$ and $d \mid b$, then $d = \prod_{i=1}^{r} p_i^{a_i}$ where $a_i \leq m_i$ and $a_i \leq n_i$ for each i. Hence, $a_i \leq \min\{m_i, n_i\} = t_i$. Thus, $d \mid c$, so $c = g$. This is part (a).

   For part (b), since $\ell g = ab$ by part (b) of Theorem 1.13 on page 26, and since $ab = \prod_{i=1}^{r} p_i^{m_i + n_i}$, then

$$ab = \prod_{i=1}^{r} p_i^{m_i + n_i} = \prod_{i=1}^{r} p_i^{M_i + t_i} = \ell g.$$

Thus, by part (a),

$$ab = \prod_{i=1}^{r} p_i^{M_i + t_i} = \ell g = \ell \prod_{i=1}^{r} p_i^{t_i},$$

so $\ell = \prod_{i=1}^{r} p_i^{M_i}$, which secures part (b) and so the entire result.   □

Example 1.17 If $a = 77,175 = 3^2 \cdot 5^2 \cdot 7^3$ and $b = 88,935 = 3 \cdot 5 \cdot 7^2 \cdot 11^2$, then $\operatorname{lcm}(a, b) = 3^2 \cdot 5^2 \cdot 7^3 \cdot 11^2 = 9,338,175$. Also, $\gcd(a, b) = 3 \cdot 5 \cdot 7^2 = 735$.

   Given Euclid's result on the infinitude of primes, we may ask about the infinitude of certain subsets of the primes.

Theorem 1.18 *There are infinitely many primes of the form* $4n - 1, n \in \mathbb{N}$.

Proof. If there are only finitely many, namely $S = \{p_1, p_2, \ldots, p_r\}$, then we may form $N = \prod_{i=1}^{r} p_i$ and set $M = 4N - 1$. Let $p|M$ where p is prime. If p is of the form $4n - 1$, then $p|N$, so $p|(M - 4N) = -1$, a contradiction. This suffices to show that only primes of the form $4n + 1$ divide M, since all odd integers are of the form $4n + 1$ or $4n - 1 = 4(n - 1) + 3$ since division of an odd number by 4 leaves a remainder of 1 or 3. Therefore, M must be of the form $4n + 1$, it is a straightforward induction argument for the reader to show that products of integers of the form $4n + 1$ are also of that form. Thus, $4N - 1 = M = 4n + 1$ for some $n \in \mathbb{N}$, so $4(N - n) = 2$, which implies that $4|2$, an absurdity. Hence, the set of primes of the form $4n - 1$ is infinite. $\square$

Theorem 1.18 is a very special case of a celebrated theorem of Dirichlet.

Theorem 1.19 | **Dirichlet: Primes in Arithmetic Progression**

Let $a, b \in \mathbb{N}$ be given, relatively prime, and fixed. Then there exist infinitely many primes of the form $an + b$ as n ranges over values of $\mathbb{N}$.

Unfortunately, we do not have the tools at our disposal to prove this result since it is beyond the scope of a first course in number theory. Nevertheless, such elegant theorems are worth seeing for the simple fact that our work on the special cases $4n - 1$ allows us to more readily appreciate this stronger gem from the mind of one of the great number-theorists.

Biography 1.8 Peter Gustav Lejeune Dirichlet (1805–1859) *was born into a French family who lived in Cologne, Germany. He studied at the University of Paris, and held positions at the Universities of Bres- lau and Berlin. He left Berlin in 1855 to succeed Gauss at Göttingen. Dirichlet did the most to amplify Gauss's great work, the Disquisitiones Arithmeticae* [16] *through his own book,* Vorlesungen über Zahlentheo- rie. *Moreover, another of his famous results is his* Pigeonhole Princi- ple, *sometimes called Dirichlet's Box Principle, which is widely useful in number-theoretic applications, some of which we will see in this text. We describe this principle below.*

◆ **The Pigeonhole Principle**

If n objects are placed in r boxes where $r < n$, then at least one of the boxes contains more than one object.

Although this appears to be obvious, mathematically we can demonstrate this as follows. We formulate this in terms of set theory. Let S be a set with $n \in \mathbb{N}$ elements, and let $S = S_1 \cup S_2 \cup \ldots \cup S_m$, $m \in \mathbb{N}$ with $m < n$ and $S_j \cap S_k = \varnothing$ where $1 \le j, k \le m$. Since the cardinality is $|S| = \sum_{j=1}^{m} |S_j|$, $n = \sum_{j=1}^{m} |S_j|$. If $|S_j| \le 1$ for all $j = 1, 2, \ldots, m$, then $n \le m$, a contradiction. Hence, at least one of the S_j has more than one element.

The term *pigeonhole* principle comes from the notion of $n + 1$ pigeons flying into n holes. This principle, along with induction will be a very useful tool for us to use throughout the text. The following is a simple illustration.

Example 1.18 *Suppose that* $n \in \mathbb{N}$, *and* $\mathcal{S}$ *is a subset of*

$$\mathcal{R} = \{j \in \mathbb{N} : 1 \le j \le 2n\}$$

with $|\mathcal{S}| = n + 1$. *We now show that* $\mathcal{S}$ *contains two relatively prime integers. Since the cardinality of* $\mathcal{S}$ *is* $n + 1$ *and the cardinality of* $\mathcal{R}$ *is* $2n$, *then for some natural number* $j \le n$ *we must have both* $2j - 1$ *and* $2j$ *in* $\mathcal{S}$, *and* $\gcd(2j - 1, 2j) = 1$. *Here the "pigeons" are the* $n + 1$ *elements of* $\mathcal{S}$, *and the "pigeonholes" are the* n *relatively prime pairs* $(2j - 1, 2j)$ *for* $j = 1, 2, \ldots, n$.

Now we turn our attention to some special sequences of integers that will allow us to apply some of what we have learned, as well as provide a preamble for some concepts yet to be studied.

◆ Mersenne Numbers

An integer of the form $M_n = 2^n - 1$ is called a *Mersenne number.*

The search for Mersenne primes is on ongoing affair. The reader may see *http://www.mersenne.org/* for the largest Mersenne prime, which is updated on a regular basis. Later in the text we will be able to employ Mersenne numbers in primality tests (see §1.8) and other important scenarios. For now we prove the following basic result.

Biography 1.9 Marin Mersenne (1588–1648) *was a Franciscan friar and mathematician whose name became attached to these numbers primarily through a claim that he made in his book* Cognitata Physica-Mathematica (1644). *In his book, he claimed, without proof, that the only primes* $p \le 257$ *such that* M_p *is prime are* $p = 2, 3, 5, 7, 13, 17, 19, 31, 67, 127, 257$. *However, it was not until more than 300 years later that this list was resolved. We now know that Mersenne made five mistakes. In 1947, it was shown that* M_{67} *and* M_{257} *are composite, and that* M_{61}, M_{89}, *and* M_{107} *are all primes.*

Mersenne studied theology at the Sorbonne, then joined the Franciscan Order of Minims. In 1619, he entered the Minim Convent de l'Annociade near Place Royal, which became home base for the rest of his life. He fostered a learned circle of mathematicians and scientists who would meet at the Minim convent to discuss ideas of mutual interest. Mersenne was also an important conduit for communication with his contemporaries, among whom were Descartes, Fermat, Galileo, and Pascal. Indeed, after his death, letters for nearly 80 writers were found in his quarters, among whom were the above, as well as Huygens, Torricelli, and Hobbes. Although Mersenne himself did not contribute significantly, his questions and conjectures were inspiration to others.

Theorem 1.20 | **Mersenne Prime Exponents**

If $M_n = 2^n - 1$ *is prime, then* n *is prime.*

Proof. First we establish the following.

Claim 1.4 *If $x^n - 1$ is prime where $x, n \in \mathbb{N}$, with $n > 1$, then $x = 2$.*

Since
$$x^n - 1 = (x - 1)(x^{n-1} + x^{n-2} + \cdots + x + 1),$$
then if $x^n - 1$ is prime, we must have that $x - 1 = 1$, since $n > 1$, which is the Claim 1.4.

Now we use Claim 1.4 to complete the proof. If M_n is prime and $n = ab$, then
$$M_n = 2^{ab} - 1 = (2^a)^b - 1,$$
so by Claim 1.4, either $2^a = 2$, or $b = 1$, namely $a = 1$ or $b = 1$, so n must be prime. $\qquad\square$

Another celebrated sequence is the following.

◆ Fermat Numbers

An integer of the form $\mathcal{F}_n = 2^{2^n} + 1$ is called a *Fermat number* — see Biography 1.10 on the next page.

Later we will use Fermat numbers in several applications. For now we prove the following, which is Theorem 1.15 on page 31 from the perspective of Fermat numbers. The proof is taken from [2].

Theorem 1.21 | The Infinitude of Primes

There are infinitely many prime numbers.

Proof. | Fermat Number Perspective |: If we prove that any two distinct Fermat numbers are relatively prime, then the result will be immediate since there are Fermat numbers $\mathcal{F}_n$ for all $n \in \mathbb{N}$. First we show the following.

Claim 1.5 $\prod_{j=0}^{n-1} \mathcal{F}_j = \mathcal{F}_n - 2$.

We use induction on n. If $n = 1$, then
$$\prod_{j=0}^{n-1} \mathcal{F}_j = \mathcal{F}_0 = 3 = \mathcal{F}_1 - 2,$$
which is the induction step. Assume that
$$\prod_{j=0}^{n-1} \mathcal{F}_j = \mathcal{F}_n - 2$$
and consider
$$\prod_{j=0}^{n} \mathcal{F}_j = \mathcal{F}_n \prod_{j=0}^{n-1} \mathcal{F}_j = \mathcal{F}_n(\mathcal{F}_n - 2) = (2^{2^n} + 1)(2^{2^n} - 1) =$$

$$2^{2^{n+1}} - 1 = \mathcal{F}_{n+1} - 2,$$

which secures the claim.

Now, by Claim 1.5, if $0 \leq m < n \in \mathbb{N}$, and p is a prime dividing $\mathcal{F}_m$ and $\mathcal{F}_n$, then

$$p \mid \left(\prod_{j=0}^{n-1} \mathcal{F}_j - \mathcal{F}_n \right) = 2.$$

However, this is not possible since Fermat numbers are odd. □

Biography 1.10 Pierre Fermat (1607–1665) *is most often listed in the historical literature as having been born on August 17, 1601, which was actually the baptismal date of an elder brother, also named Pierre Fermat, born to Fermat's father's first wife, who died shortly thereafter. Fermat, the mathematician, was a son of Fermat's father's second wife. Note also that Fermat's son gave Fermat's age as fifty-seven on his tombstone —* see http://library.thinkquest.org/27694/Pierre%20de%20Fermat.htm, *for instance. Fermat attended the University of Toulouse and later studied law at the University of Orléans where he received his degree in civil law. By 1631, Fermat was a lawyer as well as a government official in Toulouse. This entitled him to change his name to Pierre de Fermat. He was ultimately promoted to the highest chamber of the criminal court in 1652. Throughout his life Fermat had a deep interest in number theory and incisive ability with mathematics. There is little doubt that he is best remembered for Fermat's Last Theorem (FLT). FLT says that*

$$x^n + y^n = z^n$$

has no solutions $x, y, z, n \in \mathbb{N}$ for $n > 2$. This was solved in 1995, after more than 300 years of struggle, by Andrew Wiles and Richard Taylor, a former student of Wiles. The original result by Wiles, announced in 1993, had a gap in it. However, Fermat published none of his discoveries. It was only after Fermat's son Samuel published an edition of Bachet's translation of Diophantus's Arithmetica *in 1670 that his father's margin notes, claiming to have had a proof, came to light. Fermat died on January 12, 1665, in Castres, France.*

Exercises

1.37. Let p_j for $j = 1, 2, \ldots, k$ be distinct primes. Prove that if

$$p \mid \text{lcm}(p_1, p_2, \ldots, p_k),$$

then $p = p_j$ for some natural number $j \leq k$.

1.38. Prove that any product of $k \in \mathbb{N}$ consecutive integers is divisible by $k!$.

(*Hint: Use properties of $\binom{n}{k}$, the binomial coefficient — see Definition 1.2 on page 9.*)

In Exercises 1.39–1.42, we assume that p is prime, $a, b, m, n \in \mathbb{N}$.

1.39. Prove that if $p^m || a$, $p^n || b$, then $p^{m+n} || ab$.

1.40. Prove that, if $p > 2$, $m \leq n$, $p^n || (a - 1)$, and $p^{m+n} || (a^t - 1)$, then $p^m || t$. (*Hint: Use the Binomial Theorem 1.6 on page 9.*)

1.41. Prove that $p^n || a$ implies that $p^{mn} || a^m$.

1.42. Prove that if $a \neq b$, $m \neq n$, $p^m || a$, and $p^n || b$, then

$$p^{\min(m,n)} || (a + b).$$

1.43. Prove that if p is a prime dividing a^n for some $a \in \mathbb{Z}$ and $n \in \mathbb{N}$, then $p^n \mid a^n$.

An integer n is said to be *powerful* if the property holds that whenever a prime p divides n, then $p^2 \mid n$. Prove each of the Exercises 1.44–1.46 below.

1.44. If n is powerful, then there exist $r, s \in \mathbb{Z}$ such that $n = r^2 s^3$.

1.45. If a, $a + 1$, and $a + 2$ are all powerful, then there exist $n \in \mathbb{N}$ such that $a = 4n - 1$.

1.46. Prove that three consecutive powerful numbers exist if and only if there exist powerful numbers P, Q such that P is even, Q is odd, and $P^2 - Q = 1$.

1.47. Prove that $1 + 2^a$ cannot be a square for any integer $a > 3$.

(This was proved by Frénicle de Bessy in 1657. He also proved that $p^a + 1$ is not a square for any odd prime p and any integer $a > 1$. Perhaps de Bessy is best known for a letter written to him by Pierre de Fermat, dated October 18, 1640, in which *Fermat's Last Theorem* made its first recorded appearance.)

1.48. Prove that if a and b are relatively prime, $c \in \mathbb{Z}$ with $c \mid (a + b)$, then $\gcd(a, c) = 1 = \gcd(b, c)$.

1.49. Prove that if $a, b, c \in \mathbb{Z}$ with $a \mid c$, $b \mid c$, and $\gcd(a, b) = 1$, then $ab \mid c$.

☆ 1.50. Let a and b be relatively prime integers and let c be any natural number. Prove that there are infinitely many natural numbers of the form $a + bn$ with $n \in \mathbb{N}$ such that $\gcd(c, a + bn) = 1$. (*Hint: Use both Exercises 1.48–1.49.*)

1.51. Prove that for any integer $n > 2$, there exists at least one prime between n and $n!$.

1.52. Prove that for any $n \in \mathbb{N}$, $8^n + 1$ is composite.

1.4   The Chinese Remainder Theorem

> *All things began in order, so shall they end, and so shall they begin again;*
> *according to the ordainer of order and mystical mathematics of the city of*
> *heaven.*
>
> From Chapter Five of **The Garden of Cyrus (1658)**
> **Sir Thomas Browne (1605–1682), English writer and physician**

Example 1.15 on page 28 and Exercise 1.29 on page 29 taught us how to solve linear Diophantine equations. However these involved single linear equations for which we sought integral solutions. Now we seek *simultaneous* integral solutions of several such equations. The problem goes back to the first century A.D., when Sun Tsŭ, in a Chinese work on arithmetic *Suang-ching*, provided a method for determining integers having remainders $2, 3, 2$ when divided by $3, 5, 7$, respectively. In the modern terminology, he determined how to solve the simultaneous linear equations for natural numbers n_1, n_2, n_3.

$$x = 3n_1 + 2, \qquad x = 5n_2 + 3, \text{ and} \qquad x = 7n_3 + 2.$$

He calculated that $x = 233 = 3 \cdot 77 + 2 = 5 \cdot 46 + 3 = 7 \cdot 33 + 2$ is a solution to the above simultaneous linear equations. However, 233 is not the smallest such solution since we may remove multiples of $3 \cdot 5 \cdot 7 = 105$ from it to get $x = 233 - 2 \cdot 105 = 23$ as the unique smallest positive integer. This is formalized in the following, which takes its name from the work of Sun Tsŭ and other Chinese mathematicians of antiquity.

Theorem 1.22 | Chinese Remainder Theorem (CRT)

Suppose that $k > 1$ is an integer, $n_i \in \mathbb{N}$ for natural numbers $i \le k$ are pairwise relatively prime, and $r_i \in \mathbb{Z}$ for $i \le k$ are arbitrary. Then there exist integers x_i for $1 \le i \le k$ such that

$$n_1 x_1 + r_1 = n_2 x_2 + r_2 = \cdots = n_k x_k + r_k. \tag{1.17}$$

Proof. We use induction on k. If $k = 2$, the result holds since when $\gcd(n_1, n_2) = 1$, then $n_1 x - n_2 y = r_2 - r_1$ has a solution by Example 1.15 on page 28. Now assume that the result holds for $k \ge 2$, the induction hypothesis, and we prove it for $k + 1$. Let $n_1, n_2, \ldots, n_{k+1} \in \mathbb{N}$ be pairwise relatively prime and let $r_1, r_2, \ldots, r_{k+1} \in \mathbb{Z}$ be arbitrarily chosen. By the induction hypothesis, there exist integers $x_1, x_2, \ldots, x_k \in \mathbb{Z}$ satisfying Equation (1.17). The relative primality assumption implies that $\gcd(n_1 n_2 \cdots n_k, n_{k+1}) = 1$ so by Example 1.15, again, there exist $X, Y \in \mathbb{Z}$ such that $n_1 n_2 \cdots n_k X - n_{k+1} Y = r_{k+1} - n_1 x_1 - r_1$. Set

$$X_j = \frac{n_1 n_2 \cdots n_k X}{n_j} + x_j \in \mathbb{Z}, \text{ for } 1 \le j \le k \text{ and } X_{k+1} = Y.$$

Thus, $n_1 X_1 + r_1 = n_2 X_2 + r_2 = \cdots = n_{k+1} X_{k+1} + r_{k+1}$, so we have the result by induction. $\square$

We see that the CRT generalizes Sun Tsŭ's problem since it says that given any pairwise relatively prime integers $n_1, n_2, \ldots, n_k$ for $k \geq 2$ and arbitrary integers $r_1, r_2, \ldots, r_k$, there exists an integer x such that dividing x by $n_1, n_2, \ldots, n_k$ leaves remainders $r_1, r_2, \ldots, r_k$, respectively, whence the term *remainder* theorem.

The CRT can be applied to a variety of problems. We begin illustrations with the following example employing Fermat numbers introduced on page 37.

Example 1.19 *We show that for any $n, k \in \mathbb{N}$, with $k \geq 2$, there exist $x, y_j \in \mathbb{N}$ such that $y_j > 1$ for $j = 1, 2, \ldots, k$, and $y_j^n \mid (x + j)$ for $j = 1, 2, \ldots, k$.*

By Claim 1.5 in the proof of Theorem 1.21 on page 37 any two distinct Fermat numbers are relatively prime and we know that $\mathcal{F}_j > 1$ for all $j \in \mathbb{N}$. Thus, set $\mathcal{F}_j^n = n_j$ and $r_j = -j$ in the notation of Theorem 1.22, which therefore assures us that there are integers x_j such that $n_1 x_1 + r_1 = \cdots = n_k x_k + r_k$, so $\mathcal{F}_j^n x_j = n_1 x_1 + r_1 - r_j$. Hence, $\mathcal{F}_j^n \mid (x + j)$ for all $j = 1, 2, \ldots, k$ where $x = n_1 x_1 + r_1$.

Next we look at a well-known problem with an intriguing name.

♦ The Coconut Problem

Three sailors and a monkey are shipwrecked on an island. The sailors pick n coconuts as a food supply, and place them in a pile. During the night, one of the sailors wakes up and goes to the pile to get his *fair share*. He divides the pile into three, and there is a coconut left over, which he gives to the monkey. He then hides his third and goes back to sleep. Each of the other two sailors does the exact same thing, by dividing the remaining pile into three, giving the leftover coconut to the monkey and hiding his third. In the morning, the sailors divide the (much diminished) remaining pile into three and give the monkey its fourth coconut. What is the minimum number of coconuts that could have been in the original pile?

We begin by observing that the first sailor began with a pile of $y = 3n_1 + 1$ coconuts. The second sailor began with a pile of

$$y_1 = \frac{2(y-1)}{3} = 3n_2 + 1$$

coconuts, and the third sailor began with a pile of

$$y_2 = \frac{2(y_1-1)}{3} = 3n_3 + 1$$

coconuts, after which the three of them divided up the remaining pile of

$$y_3 = \frac{2(y_2-1)}{3} = 3n_4 + 1$$

coconuts. We calculate y_3 from the above equations and get

$$y_3 = \frac{8}{27}y - \frac{38}{27} = 3n_4 + 1$$

We now solve for y by multiplying through both sides of the right-hand equality by 27, then simplifying to get $8y = 81n_4 + 65$. (Note that each of y, $y_1 = 2(y-1)/3$, $y_2 = 2(y_1 - 1)/3$, and y_3 must be natural numbers.) In order that $81n_4 + 65$ be divisible by 8, (since $80n_4$ certainly is) we must have that $n_4 + 65$ is divisible by 8, and the smallest positive value of n_4 for which this occurs is $n_4 = 7$, so $y = 79$ is the smallest solution to the problem.

Next we look at another problem from antiquity that actually generalizes the CRT.

Example 1.20 *In 717 AD, a priest named Yih-hing generalized Theorem 1.22 in his book T'ai-yen-lei-shu, as follows. Suppose that $n_1, n_2, \ldots, n_k$ are (not necessarily relatively prime) natural numbers, and $r_1, r_2, \ldots, r_k$ are arbitrarily chosen integers. We now show that the system of equations*

$$x = n_1 x_1 + r_1 = n_2 x_2 + r_2 = \cdots = n_k x_k + r_k \qquad (1.18)$$

has a solution in integers x_j for $j = 1, 2, \ldots, k$ if and only if $\gcd(n_i, n_j)|(r_i - r_j)$ for all subscripts $i, j \leq k$.

If $\gcd(n_i, n_j) \nmid (r_i - r_j)$, then Equation (1.18) clearly has no solutions. Conversely, if $\gcd(n_i, n_j)|(r_i - r_j)$ for all subscripts $i, j \leq k$, we show that Equation (1.18) holds by induction on k. If $k = 2$, then the result follows from Example 1.15 on page 28. Now, assume that Equation (1.18) holds our induction hypothesis, and prove it holds for $k + 1$. Given the assumption that $\gcd(n_j, n_{k+1}) \mid (r_j - r_{k+1})$ for any $j = 1, 2, \ldots, k$, then it follows that $\gcd(\prod_{i=1}^{k} n_i, n_{k+1})$ divides $(r_{k+1} - n_1 x_1 - r_1)$, so by Example 1.15 again, there exists $X, Y \in \mathbb{Z}$ such that $\left(\prod_{i=1}^{k} n_i\right) X - n_{k+1} Y = r_{k+1} - n_1 x_1 - r_1$. By setting $X_j = (\prod_{i=1}^{k} n_i) X / n_j + x_j$, we may now proceed exactly as in the proof of the CRT, and we have the result by induction.

Another famous problem comes from the mathematician Brahmagupta— see Biography 1.11 on the facing page, and we will use Example 1.20 to illustrate.

◆ The Egg-Basket Problem

The Hindu mathematician Brahmagupta is credited with the following problem, known as the *egg-basket problem*. Suppose that a basket has n eggs in it. If the eggs are taken from the basket $2, 3, 4, 5,$ and 6 at a time, there remain $1, 2, 3, 4,$ and 5 eggs in the basket, respectively. If the eggs are removed from the basket 7 at a time, then no eggs remain in the basket. What is the smallest value of n such that the above could occur?

Translated into the notation of Example 1.20, the above becomes

$$x = 2x_1 + 1 = 3x_2 + 2 = 4x_3 + 3 = 5x_4 + 4 = 6x_5 + 5 = 7x_6. \qquad (1.19)$$

It is straightforward to check that the condition $\gcd(n_i, n_j)|(r_i - r_j)$ holds for all subscripts $i, j \leq 6$, where $n_1 = 2, n_2 = 3, n_3 = 4, n_4 = 5, n_5 = 6, n_6 = 7$

and $r_1 = 1, r_2 = 2, r_3 = 3, r_4 = 4, r_5 = 5, r_6 = 0$. Hence, by Example 1.20, Equation (1.19) has solutions. Indeed, the smallest such is $x = 119$.

Biography 1.11 Brahmagupta (598–668 A.D.) *was considered to be the greatest of the Hindu mathematicians. He was born in northwest India, and consensus is that he lived most of his life in Bhillamala in the empire of Harasha. (Bhillamala is now known as Bhinmal in what is now Rajasthan.) In 628 he wrote his masterpiece on astronomy* Brahma-sphuta-siddhanta *or* The revised system of Brahma, *which had more than four chapters devoted to mathematics. This included a method for solving the linear Diophantine equation $ax + by = c$ (see Example 1.15 on page 28). He is also credited with first studying the equation $x^2 - py^2 = 1$ for a prime p. This equation was mistakenly attributed, by Euler, to John Pell (1611–1685). (However, instances of the Pell equation can be traced back to Archimedes in his book* Liber Assumptorum *or* Book of Lemmas, *where we find the* Cattle Problem *that involves the equation $x^2 - 4729494y^2 = 1$.) He is credited with the definition of zero as the result of subtracting a number from itself. Some of the contributions he made to astronomy included calculation of the eclipses of the sun and moon, and provided a means of calculating the motion of various planets including their conjunctions.*

Exercises

In each of Exercises 1.53–1.59, find the least positive value x, which is a simultaneous solution to the equations.

1.53. $x = 3n_1 + 4 = 4n_2 + 5$.

1.54. $x = 5n_1 + 1 = 6n_2 + 2 = 7n_3 + 3$.

1.55. $x = 9n_1 + 2 = 11n_2 + 2 = 13n_3 + 3$.

1.56. $x = 15n_1 - 5 = 17n_2 - 1 = 22n_3 - 3$.

1.57. $x = 5n_1 - 4 = 23n_2 + 1 = 27n_3 - 18$.

1.58. $x = 5n_1 + 1 = 29n_2 + 12 = 31n_3 - 1 = 41n_4 + 39$.

1.59. $x = 2n_1 + 1 = 19n_2 + 2 = 33n_3 - 1 = 53n_4 - 2$.

1.60. Prove that there exist arbitrarily long blocks of consecutive natural numbers, no one of which is square-free — see Definition 1.10 on page 30.

1.5   Thue's Theorem

> *If in other sciences we should arrive at certainty without doubt and*
> *truth without error, it behooves us to place the foundation of knowledge*
> *in mathematics.*
>
> From Chapter 4 in book I of **Opus Majus**
> **Roger Bacon (ca. 1220–1292), English philosopher, scientist,**
> **and Franciscan friar**

The featured result in this section is a powerful tool in number theory deserving special attention. We first prove it, then apply it to other topics herein as illustrations. However, later in the text, when we develop the tools to look at other areas, we will see the greater applicability of this renowned theorem.

Theorem 1.23 | Thue's Theorem |

Suppose that $m \in \mathbb{N}$, $n \in \mathbb{Z}$ with $\gcd(m,n) = 1$, $m > 1$. Then there exist integers x, y such that $1 \le x < \sqrt{m}$, and $1 \le |y| < \sqrt{m}$ such that $m \mid (nx - y)$.

Proof. First we need a general result as follows.

Claim 1.6 *Let $m \in \mathbb{N}$ and let $\mathcal{S} \subseteq \mathbb{Z}$ be a set with cardinality $|\mathcal{S}| > m$. Then there exist $s_1, s_2 \in \mathcal{S}$ with $s_1 \ne s_2$ such that $m \mid (s_1 - s_2)$.*

By the division algorithm — Theorem 1.7 on page 16 — we may take each $s \in \mathcal{S}$ and write $s = q_s m + r_s$ for integers q_s, r_s with $0 \le r_s < m$. Now form the set $\mathcal{S}_r = \{r_s : s \in \mathcal{S}\}$. Since there are at most m distinct such r_s, given that $0 \le r_s < m$, then $|\mathcal{S}_r| \le m$. Hence, since $|\mathcal{S}| > m$, there must exist distinct $s_1, s_2 \in \mathcal{S}$ such that $r_{s_1} = r_{s_2}$, by the Dirichlet Pigeonhole Principle introduced on page 35. Hence, m divides $s_1 - s_2$ since,

$$s_1 - s_2 = q_{s_1} m + r_{s_1} - q_{s_2} m - r_{s_2} = m(q_{s_1} - q_{s_2}),$$

and this establishes the claim.

Let c denote the least integer larger than $\sqrt{m}$, and set

$$\mathcal{S} = \{nx - y : x = 0, 1, 2, \ldots, c - 1 \text{ and } y = 0, 1, 2, \ldots, c - 1\}.$$

Then the cardinality of $\mathcal{S}$ is $|\mathcal{S}| = c^2$. Since $c^2 > m$, then by Claim 1.6, there are at least two elements, $nx_1 - y_1 \ne nx_2 - y_2$, say, of $\mathcal{S}$, which satisfy the property that $m \mid (n(x_1 - x_2) - (y_1 - y_2))$, with $x_1 > x_2$. Note that we may assume $x_1 \ne x_2$ since if $x_1 = x_2$, then $m \mid (y_1 - y_2)$. However, $|y_1 - y_2| \le c - 1 < m$, so $y_1 = y_2$, a contradiction to $nx_1 - y_1 \ne nx_2 - y_2$. Also, if $y_1 = y_2$, and $x_1 \ne x_2$, then since $\gcd(m,n) = 1$, we must have that $m \mid (x_1 - x_2)$, by Euclid's Lemma 1.1 on page 18. Thus, since $|x_1 - x_2| \le c - 1 < m$, then $x_1 = x_2$, a contradiction. Therefore, $y_1 \ne y_2$. Set $x = (x_1 - x_2) \in \mathbb{N}$ and $y = (y_1 - y_2) \in \mathbb{Z}$. Then $m \mid (nx - y)$, as required. $\square$

We now apply Thue's Theorem to prove a result on sums of squares, a topic that we will delve into much greater detail in Chapter 6.

Biography 1.12 Axel Thue (1863–1922) *was born on February 19, 1863 in Tønsberg, Norway. He studied under S. Lie. In 1909, he produced an important paper on algebraic numbers showing that, for instance,* $y^3 - 2x^2 = 1$ *has only finitely many integer solutions. His work was extended by Siegel in 1920, and again by Roth in 1958. In 1922, Landau described Thue's work as* the most important discovery in elementary number theory that I know. *Thue's celebrated Theorem says: If* $f(x,y)$ *is a homogeneous integral polynomial, irreducible over* $\mathbb{Q}$ *and of degree bigger than two, then* $f(x,y) = c$ *has only finitely many solutions for any* $c \in \mathbb{Z}$. *He died on March 7, 1922 in Oslo, Norway.*

Example 1.21 *Let* $m, n \in \mathbb{N}$, $m > 1$, *such that* $m \mid (n^2 + 1)$. *We now show that there exist unique* $a, b \in \mathbb{N}$ *such that* $m = a^2 + b^2$ *with* $\gcd(a, b) = 1$, *and* $m \mid (nb - a)$.

Since $m \mid (n^2 + 1)$, *then by Thue's Theorem, there exist* $x, y \in \mathbb{Z}$ *with* $m \mid (nx - y)$, *where* $1 \leq x < \sqrt{m}$ *and* $|y| < \sqrt{m}$. *Set* $a = x$ *and* $b = |y|$. *Therefore, there exist* $d, f \in \mathbb{Z}$ *such that* $n^2 + 1 = md$ *and* $a = nb + mf$. *It follows that,*

$$a^2 + b^2 = (nb + mf)^2 + b^2 = b^2(n^2 + 1) + 2nbfm + m^2f^2 =$$

$$b^2md + 2nbfm + m^2f^2 = mt, \tag{1.20}$$

where

$$t = b^2d + 2nbf + mf^2.$$

Thus, $m \mid (a^2 + b^2)$. *However, since* $a, b < \sqrt{m}$, *then* $a^2 + b^2 < 2m$. *Hence,* $m = a^2 + b^2$. *Thus, by Equation (1.20),* $m = a^2 + b^2 = mt$, *so* $t = 1 = b^2d + 2nbf + mf^2$. *Therefore,*

$$\gcd(b^2d + 2nbf, mf^2) = \gcd(b(bd + 2nf), f(mf)) = \gcd(b(bd + nf), f(a - nb)) = 1$$

and this forces $\gcd(a, b) = 1$. *This establishes existence of representation.*

Now we show uniqueness of representation. If $a_0^2 + b_0^2 = m$ *for some* $a_0, b_0 \in \mathbb{N}$, *with* $a_0, b_0 < \sqrt{m}$, $\gcd(a_0, b_0) = 1$, *and* $m \mid (nb_0 - a_0)$, *it follows that (see Remark 1.6 on the next page),*

$$m^2 = (a^2 + b^2)(a_0^2 + b_0^2) = (aa_0 + bb_0)^2 + (ab_0 - ba_0)^2, \tag{1.21}$$

so $0 < aa_0 + bb_0 \leq m$. *Since both* $m \mid (nb_0 - a_0)$ *and* $m \mid (nb - a)$, *then* $a_0 = nb_0 + mk_0$ *for some* $k_0 \in \mathbb{Z}$. *Hence,*

$$aa_0 + bb_0 = (nb + mf)(nb_0 + mk_0) + bb_0 = n^2bb_0 + nb_0mf + nbmk_0 + m^2fk_0 + bb_0 =$$

$$bb_0(n^2 + 1) + m(nb_0f + nbk_0 + mfk_0) = bb_0md + m(nb_0f + nbk_0 + mfk_0),$$

so $m \mid (aa_0 + bb_0) \leq m$. *Therefore,* $aa_0 + bb_0 = m$ *from which it follows that* $ab_0 - ba_0 = 0$. *Since* $\gcd(a, b) = \gcd(a_0, b_0) = 1$, *then by employing Euclid's Lemma 1.1 on page 18 we get that* $a = a_0$ *and* $b = b_0$, *which completes the task.*

Remark 1.6 *In Equation (1.21) we used the fact that there is a way to write a product of a sum of two squares as a sum of two squares itself. There is in fact a more general result that we write now for the reader to verify as a straightforward calculation — (see Exercise 1.63 on page 48). If $x, y, u, v, D \in \mathbb{Z}$, then*

$$(x^2 + Dy^2)(u^2 + Dv^2) = (xu + Dyv)^2 + D(xv - yu)^2 =$$

$$(xu - Dyv)^2 + D(xv + yu)^2. \tag{1.22}$$

We will use Equation (1.22) in our next illustration.

We return to the above in Chapter 6 and add to it in our quest for representations of integers as sums of squares. At this juncture, it provides an insight into the study and shows the power of Thue's Theorem, as does the following.

Example 1.22 *Suppose that p is an odd prime such that $p \mid (n^2 + 2)$ for some $n \in \mathbb{N}$ relatively prime to p. We will demonstrate that there exist unique $a, b \in \mathbb{N}$ such that $p = a^2 + 2b^2$.*

Since $p \mid (n^2 + 2)$, then by Thue's Theorem, there exist $x \in \mathbb{N}$ and $y \in \mathbb{Z}$ with $x, |y| < \sqrt{p}$ such that $p \mid (nx - y)$. Thus, we have $u, v \in \mathbb{Z}$ such that $n^2 + 2 = pu$ and $nx - y = pv$. Hence,

$$pu = n^2 + 2 = \left(\frac{pv + y}{x}\right)^2 + 2,$$

and rewriting, we get

$$p(x^2 u - pv^2 - 2vy) = y^2 + 2x^2,$$

so $p \mid (y^2 + 2x^2)$. However, since $x, |y| < \sqrt{p}$, then $y^2 + 2x^2 < 3p$. Therefore, there are only two possibilities. Either $y^2 + 2x^2 = p$, in which case we have our representation with $a = |y|$ and $b = x$, or $y^2 + 2x^2 = 2p$. In the latter case, y must be even so $p = x^2 + 2(y/2)^2$, which is our representation with $a = x$ and $b = |y/2|$. This proves the existence of such representations. We now establish uniqueness.

Suppose there exist $a_0, b_0 \in \mathbb{N}$ such that $p = a_0^2 + 2b_0^2$, with $p \mid (nb_0 - a_0)$, and $p = a^2 + 2b^2$ with $p \mid (nb - a)$. Then by Equation (1.22),

$$p^2 = (a_0^2 + 2b_0^2)(a^2 + 2b^2) = (aa_0 + 2bb_0)^2 + 2(a_0 b - ab_0)^2. \tag{1.23}$$

Therefore, $0 \le aa_0 + 2bb_0 \le p$. Also, since $p \mid (nb_0 - a_0)$ and $p \mid (nb - a)$, then there exist $w, z \in \mathbb{Z}$ such that $nb - a = pw$ and $nb_0 - a_0 = pz$. Hence,

$$aa_0 + 2bb_0 = (nb - pw)(nb_0 - pz) + 2bb_0 = n^2 bb_0 - nbpz - pwnb_0 + p^2 wz + 2bb_0 =$$

$$bb_0(n^2 + 2) + p(pwz - nbz - wnb_0) = bb_0 pu + p(pwz - nbz - wnb_0),$$

so $p \mid (aa_0 + 2bb_0)$, which implies that $aa_0 + 2bb_0 = p$. It follows from (1.23) that $a_0 b - ab_0 = 0$. Since we clearly have that $\gcd(a, b) = 1 = \gcd(a_0, b_0)$, then as above $a = a_0$ and $b = b_0$, thereby securing our initial assertion.

Example 1.22 is another illustration of representation problems that we will study in detail in Chapter 6. This will give the reader further illustrations of the above techniques and the applicability of Thue's Theorem.

Example 1.23 *Suppose that $p > 2$ is a prime such that $p \mid (n^2 - 2)$. We will prove p can be written in the form $a^2 - 2b^2$ in infinitely many ways.*

We leave it to the reader to use a similar argument to that used in Example 1.22, employing Thue's Theorem to show that

$$p = x^2 - 2y^2 \text{ for some } x, y \in \mathbb{N}.$$

To show that there are infinitely many such representations, we employ two sequences of numbers known to Theon of Smyrna — see Biography 1.13. They are given by

$$s_1 = 1 = d_1, \text{ and for any } n \in \mathbb{N}, s_{n+1} = s_n + d_n, \text{ and } d_{n+1} = 2s_n + d_n. \quad (1.24)$$

It is a straightforward induction for the reader to verify that

$$d_n^2 - 2s_n^2 = (-1)^n. \qquad (1.25)$$

Using Equations (1.22) and (1.25), we get

$$p = (d_{2n}^2 - 2s_{2n}^2)(x^2 - 2y^2) = (d_{2n}x - 2s_{2n}y)^2 - 2(d_{2n}y + s_{2n}x)^2,$$

so since there are infinitely many s_{2n}, d_{2n} for all $n \in \mathbb{N}$, we have our infinitely many representations.

Biography 1.13 Theon of Smyrna (ca. 70–135 A.D.) *was a Greek philosopher and mathematician, who was purported to have had, as a student, Ptolemy, the great Greek mathematician, astronomer, geographer, and astrologer. His work was deeply influenced by the Pythagoreans. We do not know with certainty the details of his life, but we do know a fair amount about his works. He is known to have written* On Mathematics Useful for the Understanding of Plato *which covered not only mathematics, primarily number theory, but also music, including the* music of numbers. *The latter was an exploration of harmony using ratio and proportion. In his work was also a covering of the* music of the cosmos, *as well as astronomy. Indeed our own moon has a crater named after him,* Theon Senior. *As well, we also have a lunar crater named after Ptolemy, the* Ptolemaeus crater.

The sequences in Equation (1.24) are called the side *and* diagonal *sequences, respectively, and the result in Equation (1.25) was proved by Theon in about 130 A.D. Some solutions to the latter equation were known to Pythagorus — see Biography 1.14 on the following page.*

Exercises

1.61. Use the techniques illustrated in this section to prove that when $p > 2$ is a prime such that $p \mid (n^2 + 3)$ for some $n \in \mathbb{N}$ with $\gcd(p, n) = 1$, then there exist unique $a, b \in \mathbb{N}$ such that $p = a^2 + 3b^2$.

Biography 1.14 Pythagorus *lived from roughly 580 to 500 B.C., although little is known about his life with any degree of accuracy. He is not known to have written any books, but his followers carried on his legacy. The most famous result bearing his name, although known to the Babylonians, is the theorem that says that the square of the hypotenuse of a right-angled triangle is equal to the sum of the squares of the other two sides. Nevertheless, Pythagorus is undoubtedly the first to* prove *this. He is thought to have traveled to Egypt and Babylonia and settled in Crotona on the southeastern coast of* Magna Graecia, *now Italy, where he founded a secret society that became known as the* Pythagoreans. *Their motto,* number rules the universe, *reflected the mysticism embraced by Pythagorus, who was more of a mystic and a prophet than a scholar. The Pythagoreans' belief that everything was based on the natural numbers was deeply rooted. The degree of their commitment to this belief is displayed by an anecdote about $\sqrt{2}$. Hippasus was a Pythagorean who revealed to outsiders the* secret that $\sqrt{2}$ *is irrational. For this indiscretion, he was drowned by his comrades.*

1.62. Let $m, n \in \mathbb{N}$ such that $m > 1$ and $m \mid (n^2 + 2)$ with $\gcd(m, n) = 1$. Prove that n is prime if and only if there are unique $a, b \in \mathbb{N}$ such that $n = a^2 + 2b^2$ with $\gcd(a, b) = 1$.

1.63. Verify Equation (1.22) on page 46. (Note that the case where $D = 1$ goes back to Diophantus.)

Biography 1.15 Diophantus of Alexandria *was a Greek mathematician, born around 200 A.D. in Alexandria, and died there around 284. However, details of his life are scarce and largely circumstantial. He is known for his renowned work* Arithmetica, *which was a collection of problems on number theory, especially on solutions of equations that bear his name,* Diophantine equations. *(See, for instance, Example 1.15 on page 28. We will study such equations in depth in Chapter 7.) The method for solving such equations is known as* Diophantine analysis. Arithmetica *was divided into 13 books. Of these, 6 were communicated in Greek by Byzantine scholars to Europe in the late 15th century. Four other books were discovered in 1968. These were Arabic translations by Qusta ibn Luqa. The importance of* Arithmetica *is that it is the first known work to use algebra in what we consider to be a modern style. Moreover, it inspired others to apply his methods to number-theoretic problems. For instance, al-Karaji (ca. 980–1030), an Arabian mathematician, used Diophantus' methods. The most famous application was by Fermat, who wrote in the margins of his copy of* Arithmetica *that he had solved what we now know as Fermat's Last Theorem — see Biography 1.10 on page 38.*

1.6 Combinatorial Number Theory

> *I have laboured to refine our language to grammatical purity, and to clear it from colloquial barbarisms, licentious idioms, and irregular combinations.*
> From no. 208 of **The Rambler (1752)**
> **Samuel Johnson (1709–1784), English poet, critic, and lexographer**

Combinatorial number theory, or simply *combinatorics* is that branch devoted to the study of arrangements of items according to specified patterns, to determine the total number of such patterns, and to establish techniques for the creation of patterns satisfying specified rules.

Some combinatorial problems can be solved using Dirichlet's Pigeonhole Principle introduced on page 35. Another basic combinatorial method is to compare elements in two finite sets forming a one-to-one correspondence to compare elements in an effort to find one of a desired type, for instance. The combination of these ideas was used in Claim 1.6 on page 44. We now employ that result in what follows to illustrate the method here applied to a new problem.

Example 1.24 *Let $z_1, z_2, \ldots, z_n$ for $n \in \mathbb{N}$ be distinct integers. We will demonstrate that there exist $k, \ell \in \mathbb{N}$ such that $1 \le k < \ell \le n$ with*

$$n \mid (z_k + z_{k+1} + \cdots + z_{k+\ell}).$$

For each $m \in \mathbb{N}$ with $1 \le m \le n$, let $s_m = z_1 + z_2 + \cdots + z_m$, let $s_0 = 0$, and set

$$\mathcal{S} = \{s_0, s_2, \ldots, s_n\}.$$

Since the cardinality of $\mathcal{S}$ is $n + 1$, then by Claim 1.6 there exist k, ℓ with $0 \le k < \ell \le n$ such that $n \mid (s_\ell - s_k)$, so $n \mid (z_{k+1} + z_{k+2} + \cdots + z_n)$, as required.

This illustrates that for a set of n integers, there is always a set of consecutive integers in the set whose sum is a multiple of n.

The binomial coefficient — (see Definition 1.2 on page 9) — is the vehicle for looking at combinatorial principles. We now define basic notions that we will show are based on it.

Definition 1.12 | **Permutations and Combinations**

A permutation *of $r \in \mathbb{N}$ elements in an $n \ge r$ element set $\mathcal{S}$ is an ordered selection of r elements from $\mathcal{S}$, and the total number of ways of so doing is*

$$P(n, r) = \frac{n!}{(n-r)!} = r! \binom{n}{r}.$$

A combination *of r objects in an n-element set $\mathcal{S}$ is a subset of $\mathcal{S}$ containing r elements, and the total number of ways of so doing is*

$$C(n, r) = \frac{n!}{r!(n-r)!} = \binom{n}{r}.$$

In Exercise 1.38 on page 38 it is stated that the product of any $k \in \mathbb{N}$ consecutive natural numbers is divisible by $k!$ and the hint given is that the binomial coefficient is the mechanism for proving this combinatorial statement. This may be demonstrated here in light of the above.

If M is the largest of k consecutive integers, then that product is given by

$$M(M-1)\cdots(M-k+1) = P(M,k).$$

Moreover, since to each of the $C(M,k)$ different combinations of k elements from an M-element set, we have $P(k,k)$ different orderings, then

$$P(M,k) = P(k,k)\binom{M}{k} = k!\binom{M}{k},$$

so $k! \mid M(M-1)\cdots(M-k+1)$ given that $\binom{M}{k} \in \mathbb{Z}$.

The above combinatorial technique is a valuable one for divisibility arguments. The following is an illustration of the combination notation.

Example 1.25 *The combination notion can be used for instance, to find the number of ways of choosing two objects from a set of five objects*, without regard for order, *namely* $\binom{5}{2} = 5!/(2!3!) = 10$ *distinct ways.*

We may also produce a proof of a famous result due to Fermat that we will revisit when we introduce congruences in Chapter 2. (See Biography 1.10 on page 38.) Here we use the binomial coefficient, namely the notion of a permutation.

Theorem 1.24 | **Fermat's Little Theorem** |

If p is a prime and $n \in \mathbb{N}$, then $p \mid (n^p - n)$.

Proof. By the division algorithm on page 16, $n = pq + r$ for integers q, r with $0 \le r < p$, then by the Binomial Theorem 1.6 on page 9,

$$n^p - n = (pq + r)^p - pq - r = \sum_{j=0}^{p}\binom{p}{j}(pq)^{p-j}r^j - pq - r =$$

$$p\left(\sum_{j=0}^{p-1}\binom{p}{j}p^{p-1-j}q^{p-j}r^j - q\right) + r^p - r.$$

Therefore, we need only show that $p \mid (r^p - r)$. If $r = 0$, then clearly the result holds. Now we use an induction argument by assuming the result holds for r and prove it holds for $r + 1$. Since neither j nor $p - j$ divides p for any $j = 1, 2, \ldots, p - 1$, then

$$p \;\left|\; \binom{p}{j} = \frac{p(p-1)\cdots(p-j+1)}{j!} \right. \quad \text{for } j = 1, 2, \ldots, p-1. \qquad (1.26)$$

For each such j set $\binom{p}{j} = pa_j$ for some $a_j \in \mathbb{Z}$.

$$(r+1)^p = \sum_{j=0}^{p} \binom{p}{j} r^j = r^p + 1 + p \left(\sum_{j=1}^{p-1} a_j r^j \right).$$

Therefore, by the induction hypothesis, $(r+1)^p - r - 1$ is divisible by p, thereby securing the result. □

The following illustration demonstrates the use of Fermat's result.

Example 1.26 *Suppose that $n = 4m + 3$ where $m \in \mathbb{Z}$. We will prove that there does not exist an integer x such that $n \mid (x^2 + 1)$.*

Suppose, to the contrary, that there is such an x. First we show that this implies there is a prime $p = 4y + 3$, $y \geq 0$, dividing $x^2 + 1$. Assume no such prime divides n. Then since the only other prime divisors of n are of the form $4s+1$, and all products of primes of the form $4t+1$ are clearly also of that form, then $n = 4t + 1$ for some integer t. Hence, $4t + 1 = 4m + 3$, so $4(t - m) = 2$, forcing $4 \mid 2$, which is absurd. We have shown the existence of a prime $p = 4y+3$ dividing n, which must therefore divide $x^2 + 1$. Set $x^2 + 1 = p\ell$ for some $\ell \in \mathbb{Z}$.

By Theorem 1.24, $p \mid (x^{p-1} - 1)$, so there exists a $z \in \mathbb{Z}$ such that $pz = x^{p-1} - 1 = (x^2)^{(p-1)/2} - 1 = (p\ell - 1)^{(p-1)/2} - 1$, and by the Binomial Theorem, this equals,

$$\sum_{j=0}^{2y+1} \binom{2y+1}{j} (p\ell)^{2y+1-j}(-1)^j - 1 =$$

$$p \left(\sum_{j=0}^{2y} \binom{2y+1}{j} p^{2y-j} \ell^{2y+1-j}(-1)^j \right) + (-1)^{2y+1} - 1 =$$

$$p \left(\sum_{j=0}^{2y} \binom{2y+1}{j} p^{2y-j} \ell^{2y+1-j}(-1)^j \right) - 2.$$

Hence $p \mid 2$, a contradiction, which secures the result.

Now we introduce one more famous result that we will revisit in Chapter 2 — see Biography 1.16 on the following page.

Theorem 1.25 │ **Wilson's Theorem** │

If $n \in \mathbb{N}$ with $n > 1$, then n is prime if and only if

$$n \mid ((n-1)! + 1).$$

Proof. If $n = 2$ or 3, the result is clear so we assume that $n > 3$. If $n = p$ is prime, then we first require the following.

Claim 1.7 *For each natural number $j < p$, there exists a unique natural number $r_j < p$ such that $jr_j = 1 + px_j$, and $j = r_j$ if and only if $j = p$ or $j = p - 1$.*

Since $\gcd(p, j) = 1$, then by Theorem 1.10 on page 22, there exist integers x_j, y_j such that $jx_j + py_j = 1$, where we may assume without loss of generality that $x_j > 0$. If $x_j > p$, then by the division algorithm, there exist integers q_j, r_j such that $x_j = q_j p + r_j$ with $0 < r_j < p$, so $jr_j + p(y_j + jq_j) = 1$. This proves existence of such a representation. If $r_j = r_k = r$ for some natural numbers $j, k < p$, then $jr - px_j = 1 = kr - px_k$, so $r(j-k) = p(x_j - x_k)$. Since p is prime to k, then $p \mid (j - k) < p$. Hence, $j = k$. This is uniqueness of representation.

Now if $j = r_j$ for any such j, then $j^2 - px_j = 1$, whence $p \mid (j^2 - 1) = (j - 1)(j + 1)$, so $p \mid (j - 1)$ or $p \mid (j + 1)$. In other words, either $j = 1$ or $j = p - 1$, which secures the claim.

By Claim 1.7, each of the r_j for $j = 2, 3, \ldots, p - 2$ is one of the elements in the same set. Thus, we may pair the (even number) of the $p - 3 \geq 2$ integers so,

$$(p - 2)(p - 3) \cdots 3 \cdot 2 = 1 + pz$$

for some integer z. Now multiplying by $p-1$ we get $(p-1)! = p-1+(p-1)pz = -1 + p[1 + (p - 1)z]$, so $p \mid [(p - 1)! + 1]$, as required.

Conversely, if $n \mid ((n - 1)! + 1)$, and n is composite, then there is a prime $p < n$ dividing n, so $p \mid (n - 1)!$. Hence, $p \mid [(n - 1)! + 1 - (n - 1)!] = 1$, a contradiction, so n is prime. $\qquad\square$

Remark 1.7 *Wilson's Theorem is typically stated in one direction only, namely if p is prime, then $n \mid ((n - 1)! + 1)$. However, we prove both directions since it is indeed both a necessary and a sufficient condition. One could, theoretically view this as a test for primality. However, it is highly inefficient. In fact, it can be shown that the number of bit operations to calculate $n!$ is $O(n^2 \log_e^2(n))$, whereas for more efficient algorithms that we will study later, this complexity is much reduced. Indeed, there now exists an unconditional polynomial-time algorithm for primality testing which we present in Appendix F.*

Biography 1.16 *John Wilson was born on August 6, 1741, in Applethwaite, Westmoreland, England. On July 7, 1764, he was elected as a Fellow of Peterhouse, Cambridge, where he studied. On March 13, 1782, he was elected Fellow of the Royal Society, and was appointed king's counsel on April 24 of that year. The latter was part of his legal career, which he began on January 22, 1763. On November 15, 1786, he was knighted for his numerous accomplishments. Wilson married Mary Ann Adair on April 7, 1788, and the marriage produced a son and two daughters. However, he died only five years later, on October 18, 1793, in Kendal, Westmoreland, where he was raised. Although Theorem 1.25 bears his name, it was actually first proved by Lagrange in the early 1770s, (see Biography 2.7 on page 114). Another concept also bears his name — see Exercise 1.65 on page 54.*

We now apply Wilson's Theorem to a result related to Example 1.26 on page 51.

Example 1.27 *In Example 1.26, we demonstrated that no positive integer of the form $4n+3$ satisfies that $n \mid (x^2+1)$ for any integer x. Now we demonstrate which primes do satisfy this divisibility, using Theorem 1.25 as the vehicle for so doing.*

If p is prime, we will prove that $p \mid (x^2+1)$ for some integer x if and only if either $p = 2$ or $p = 4m+1$ for some positive integer m.

Using Wilson's Theorem, we may write, for any odd prime p,

$$\left(\prod_{j=1}^{(p-1)/2} j \right) \left(\prod_{j=(p+1)/2}^{p-1} j \right) = -1 + pz$$

for some integer z. Now we pair the elements j in the first product with the elements $p - j$ in the second product and rewrite it as,

$$\prod_{j=1}^{(p-1)/2} j(p - j) = -1 + pz.$$

However, $j(p - j) = pj - j^2$, so for some integer w,

$$\prod_{j=1}^{(p-1)/2} j(p - j) = pw + \prod_{j=1}^{(p-1)/2} (-j^2) = (-1)^{(p-1)/2} \left(\prod_{j=1}^{(p-1)/2} j \right)^2 + pw.$$

Now if $p = 4m+1$, then $(p-1)/2$ is even so the above equals, $\left(\prod_{j=1}^{(p-1)/2} j \right)^2 + pw$. Therefore, we have shown that

$$-1 + pz = \left(\prod_{j=1}^{(p-1)/2} j \right)^2 + pw,$$

so by letting $x = \prod_{j=1}^{(p-1)/2} j$ we have that $p \mid (x^2+1)$.

Conversely, if $p \mid (x^2+1)$, we know from Example 1.26 that p cannot be of the form $4m+3$, so p must be either 2 or of the form $4m+1$ for some integer m.

In the above we have seen several instances of the combinatorial principle in action, which we mentioned at the outset of this section, namely a one-to-one correspondence setup where we pair off elements in two finite sets to determine, perhaps, the number of elements, or to establish the existence of an element with given properties that we are seeking. Throughout the text we will have opportunities to use the principles displayed in this section as tools in our voyage to understand the fundamentals of number theory.

Exercises

1.64. Let $n \in \mathbb{N}$ and let N be the greatest integer less than or equal to $(n-1)/2$. Prove the following Fibonacci number identity (see page 3).

$$F_n = \sum_{j=0}^{N} \binom{n-j-1}{j}.$$

(*Hint: Use Pascal's Identity established in Exercise 1.14 on page 14.*)

1.65. A prime p is known as a *Wilson prime* if $p^2 \mid [(p-1)!+1]$. Find all Wilson primes less than 564. (See Biography 1.16 on page 52.)

1.66. Prove that if p is an odd prime, then

$$p \, \left| \, \left[\left(\frac{p-1}{2}! \right)^2 + (-1)^{(p+3)/2} \right] \right..$$

(*Hint: Use Wilson's Theorem.*)

1.67. Let $a_j \in \mathbb{Z}$ for $j = 1, 2, \ldots, n \in \mathbb{N}$ and assume that p is prime. Prove that

$$p \, \left| \, \left[\left(\sum_{j=1}^{n} a_j \right)^p - \sum_{j=1}^{n} a_j^p \right] \right..$$

(*Hint: Use the Binomial Theorem and the fact (1.26), established in the proof of Theorem 1.24 on page 50, that $p \mid \binom{p}{j}$ for all natural numbers $j < p$.*)

☆ 1.68. Let $\mathcal{F}_n$ denote the nth Fermat number introduced on page 37. Prove that the smallest natural number m such that $\mathcal{F}_n \mid (2^m - 1)$ is $m = 2^{n+1}$.

(*Hint: Use the division algorithm on 2^{n+1} and m.*)

☆ 1.69. Let $\mathcal{F}_n$ be the nth Fermat number. Prove that if p is a prime divisor of $\mathcal{F}_n$, then $p = 2^{n+1}m + 1$ for some $m \in \mathbb{N}$.

(*Hint: Use Fermat's Little Theorem and Exercise 1.68.*)

1.70. Let $M_n = 2^n - 1$ be a Mersenne number introduced on page 36. Given a prime p, prove that any prime divisor of M_p is of the form $2mp + 1$ for some $m \in \mathbb{N}$.

(*Hint: Use Fermat's Little Theorem and Example 1.14 on page 27.*)

1.71. Let m, n be natural numbers both larger than 1. Prove that if $m^n - 1$ is prime, then it is a Mersenne prime. Conclude that n is prime.

(*Hint: See Theorem 1.20 on page 36.*)

1.7 Partitions and Generating Functions

> *Great wits are sure to madness near allied, And thin partitions do their bounds divide.*
>
> From part I, line 163 of **Absalom and Achitophel (1681)**
> **John Dryden (1631–1700), English poet, critic, and playwright**

To complement the coverage in §1.6, this section is devoted to some aspects of what is known as partition theory. This is an area of additive number theory, which is that branch dealing with the representation of integers as sums of other integers. For instance, the Base Representation Theorem on page 8 is a simple example of such an additive result.

We begin with some elementary definitions.

Definition 1.13 | **Partitions**

A partition *of a nonnegative integer n is a representation of n as a sum of natural numbers, called* parts *or* summands *of the partitions. Two sums which only differ in the order of their summands are considered to be the same partition. Thus, we write $n = z_1 + z_2 + \cdots + z_\ell$, where $z_1 \geq z_2 \geq \cdots \geq z_\ell$. We denote the number of such partitions by $p(n)$.*

Example 1.28 *Since $4 = 1 + 1 + 1 + 1 = 1 + 2 + 1 = 2 + 1 + 1 = 1 + 1 + 2 = 2 + 2 = 3 + 1 = 1 + 3$, but order is irrelevant, we have only 5 partitions of 4, namely $4 = 1 + 1 + 1 + 1 = 2 + 1 + 1 = 2 + 2 = 3 + 1$. Thus, $p(4) = 5$.*

Euler first proved the fundamental properties of $p(n)$ in 1748 in his book *Introductio in Analysin Infinitorum* — see Biography 1.17 on the following page. One of the chief goals of this section is to prove Euler's result on $p(n)$. First, we need to define the other name in the header of this section.

Definition 1.14 | **Generating Functions**

Suppose that $f : \mathbb{N} \cup \{0\} \to \mathbb{N} \cup \{0\}$ is a function. Then

$$G_f(x) = \sum_{n=0}^{\infty} f(n)x^n$$

is called the generating function *of f.*

The generating function is defined for all values of x for which the sequence converges. (See Definition A.23 on page 307 and the discussion surrounding it.)

Theorem 1.26 | Generating Function for p(n) |

The generating function for $p(n)$ *is*

$$G_p(x) = \prod_{j=1}^{\infty} (1 - x^j)^{-1}.$$

Proof. From Example A.10 on page 308, we have that

$$(1 - x^j)^{-1} = \sum_{k=0}^{\infty} x^{jk}.$$

Thus,

$$\prod_{j=1}^{\infty} \left(\sum_{k=0}^{\infty} x^{jk} \right) = \sum_{n=0}^{\infty} b_n x^n,$$

where b_n is the number of times we may write n as a sum of terms of the form jk for $j = 1, 2, 3, \ldots$ and $k \geq 0$. In other words, b_n is the number of solutions of the Diophantine equation $k_1 + 2k_2 + \cdots + tk_t = n$ where the k_i for $i = 1, 2, \ldots, t$ are nonnegative integers, but this is just $p(n)$, which completes the proof. $\square$

Remark 1.8 *What Theorem 1.26 shows is that*

$$G_p(x) = \sum_{n=0}^{\infty} p(n)x^n = \prod_{j=1}^{\infty} (1 - x^j)^{-1}.$$

Moreover, it can be shown that this generating function converges for $|x| < 1$. *In fact, it can be shown that*

$$\lim_{n \to \infty} \frac{p(n+1)}{p(n)} = 1 = \lim_{n \to \infty} p(n)^{1/n},$$

and the latter coupled with Cauchy's test is enough to establish convergence within the unit circle. (See Theorem A.15 on page 307 for a description of the latter test.)

Now we aim at proving a result by Euler that employs the above. First we need to develop some tools.

Definition 1.15 | Distinct Parts and Odd Parts |

If $n \in \mathbb{N}$, then $d(n)$ denotes the number of partitions of n into distinct parts (summands), and $o(n)$ denotes the number of partitions of n into only odd parts. Also, let $d_r(n)$ denote the number of partitions of n into distinct parts where none of the parts is larger than r.

Example 1.29 *Since $4 = 3 + 1$ are the only partitions of 4 into distinct parts, then $d(4) = 2$, and since $1 + 1 + 1 + 1 = 1 + 3$ are the only partitions of 4 into odd parts, then $o(4) = 2$, so $d(4) = o(4)$. Also, $d_2(4) = 0$ since no partitions into distinct parts have all the parts less than 2, but $d_3(4) = 1$, and $d_4(4) = 2 = d(4)$. Indeed it can be seen that $d_r(n) = d(n)$ for any $r \geq n$, and in general, $0 \leq d_r(n) \leq d(n) \leq p(n)$.*

Lemma 1.3 | The Generating Functions for $d_r(n)$ and $o(n)$ |

For $r, n \in \mathbb{N}$

$$\sum_{n=0}^{\infty} d_r(n) x^n = \prod_{j=1}^{r} (1 + x^j),$$

and

$$\sum_{n=0}^{\infty} o(n) x^n = \prod_{j=1}^{\infty} \frac{1}{1 - x^{2j-1}}.$$

Proof. We have,

$$\prod_{j=1}^{r}(1+x^j) = (1+x)(1+x^2)(1+x^3)\cdots(1+x^r) = 1+x+x^2+(x^{2+1}+x^3)+(x^4+x^{3+1})+$$

$$(x^5 + x^{4+1} + x^{3+2}) + (x^6 + x^{5+1} + x^{4+2}) + \cdots,$$

so we see that x^n is a summand the same number of times that n can be partitioned into distinct summands no bigger than r, namely $d_r(n)$ times. Hence,

$$\prod_{j=1}^{r}(1 + x^j) = \sum_{n=0}^{\infty} d_r(n) x^n,$$

as required.

The same argument works for $o(n)$ so we leave that as an exercise for the reader.   □

Now we are able to prove Euler's result that shows the illustration in Example 1.29 is no accident. In fact, we have the following.

Theorem 1.27 | **Euler's Theorem on Parts** |

For a given natural number n, $d(n) = o(n)$, namely the number of partitions of $n \in \mathbb{N}$ in which all parts are odd equals the number of partitions of n in which all parts are distinct.

Proof. First we recall the remark made in Example 1.29 on the preceding page that $d_r(n) = d(n)$ for all $r \geq n$, and in general, $0 \leq d_r(n) \leq d(n) \leq p(n)$. We use these facts in the following.

Claim 1.8 $\sum_{n=0}^{\infty} d(n)x^n = \prod_{j=1}^{\infty} (1 + x^j)$ *where* $|x| < 1$.

By Lemma 1.3 on the page before, we have,

$$\left| \sum_{n=0}^{\infty} d(n)x^n - \prod_{j=1}^{r} (1 + x^j) \right| = \left| \sum_{n=0}^{\infty} d(n)x^n - \sum_{n=0}^{r} d_r(n)x^n \right| =$$

$$\left| \sum_{n=r+1}^{\infty} (d(n) - d_r(n))x^n \right| \leq \sum_{n=r+1}^{\infty} d(n)|x|^n \leq \sum_{n=r+1}^{\infty} p(n)|x|^n.$$

However, for $|x| < 1$, we have $\lim_{r \to \infty} \sum_{n=r+1}^{\infty} p(n)|x|^n = 0$, so

$$\sum_{n=0}^{\infty} d(n)x^n = \lim_{r \to \infty} \prod_{j=1}^{r} (1 + x^j) = \prod_{j=1}^{\infty} (1 + x^j) \prod_{j=1}^{\infty} (1 + x^j),$$

which establishes the claim.

Now we use Claim 1.8 to finish the proof. We have,

$$\sum_{n=0}^{\infty} d(n)x^n = \prod_{j=1}^{\infty} (1 + x^j) = \prod_{j=1}^{\infty} \frac{(1 + x^j)(1 - x^j)}{(1 - x^j)} = \prod_{j=1}^{\infty} \frac{(1 + x^{2j})}{(1 - x^j)} =$$

$$\prod_{j=1}^{\infty} (1 - x^{2j}) \prod_{j=1}^{\infty} \frac{1}{1 - x^j} = \prod_{j=1}^{\infty} (1 - x^{2j}) \prod_{j=1}^{\infty} \frac{1}{(1 - x^{2j-1})(1 - x^{2j})} =$$

$$\prod_{j=1}^{\infty} \frac{1}{(1 - x^{2j-1})} = \sum_{n=0}^{\infty} o(n)x^n,$$

where the last equality comes from Lemma 1.3. By the definition of Maclaurin series (see Definition A.24 on page 308), no function can have more than one such representation. Thus, we must have that $d(n) = o(n)$ for all $n \in \mathbb{N}$.   □

Exercises

1.72. Prove that the generating function for the Fibonacci sequence, F_n (see page 3) is given by,

$$\frac{x}{1 - x - x^2},$$

namely show that

$$\sum_{n=1}^{\infty} F_n x^n = \frac{x}{1 - x - x^2}.$$

1.73. Let $m, n \in \mathbb{N}$ with $n \geq m$, and let $p_m(n)$ denote the number of partitions of n into no more than m parts. Prove that if ℓ is the greatest integer less than or equal to n/m, then

$$p_m(n) = \sum_{j=0}^{\ell} p_{m-1}(n - mj).$$

(*Hint: First establish that $p_m(n) = p_{m-1}(n) + p_m(n - m)$. Then use an induction argument.*)

1.74. Euler proved the following known as the *Pentagonal Numbers Theorem*,

$$\prod_{j=1}^{\infty}(1 - x^j) = \sum_{n=-\infty}^{\infty} (-1)^n x^{n(3n+1)/2}.$$

Use Euler's result to prove the following fact. For any $n \in \mathbb{N}$,

$$p(n) = p(n-1) + p(n-2) - p(n-5) - p(n-7) + \cdots + (-1)^{j+1} p(n-n_j) + \cdots,$$

where $n_j = j(3j \pm 1)/2$ are called *Pentagonal numbers*.

(*A proof of Euler's aforementioned result may be found in [20], where this and deeper results on partition theory are explored. Pentagonal numbers get their name from the fact that if we border a regular pentagon, marked by 5 dots, then successively form pentagons outward with $3, 4, \ldots, j, \ldots$ dots on each side, then the total number of dots is $j(3j - 1)/2$.*)

(*Hint: Use Theorem 1.26 on page 56 in conjunction with Euler's result above.*)

1.75. Let $E(n)$ denote the number of partitions of n into an even number of distinct parts and let $U(n)$ denote the number of partitions of n into an odd number of distinct parts. Prove that $E(n) = U(n)$ except for the cases $n = j(3j \pm 1)/2$, in which case

$$E(n) - U(n) = (-1)^j.$$

(*Hint: Use Exercise 1.74.*)

(**§1.7 Closing Remark:** For the reader interested in a recreational application of generating functions, see the entertaining [11, pp. 217–231], where *Conway's Napkin Problem* is discussed.)

1.8   True Primality Tests

> *Beauty is the first test; there is no permanent place in the world for ugly*
> *mathematics.*
>
>   From **A Mathematician's Apology (1940)** — see [19]
>   **Godfrey Harold Hardy (1877–1947), English mathematician**

We have already encountered some instances of what are known as primality tests. For instance, Example 1.16 on page 31 illustrated the sieve of Eratosthenes, and in Theorem 1.25 on page 51. However both of these are highly inefficient. Moreover, we have not agreed upon a formal definition of what constitutes a primality test. There are two different types of such tests, the first of which we study in this section, and formalize now.

In what follows, we may view an *algorithm* as any methodology following a set of rules to achieve a goal. We may think of a *deterministic* algorithm as an algorithm which always terminates with a "yes" or "no" answer.

Definition 1.16 | **Primality Proofs**

A Primality Proving Algorithm, *also known as a* True Primality Test, *is a deterministic algorithm that, given an input n, verifies the hypothesis of a theorem whose conclusion is that n is prime. A* Primality Proof *is the computational verification of such a theorem. In this case, we call n a* provable prime — *a prime that is verified by a Primality Proving Algorithm.*

The classical example of a True Primality Test is the following — see page 36, where we introduced Mersenne numbers. The following is a true primality test of the numbers M_n. The following proof is a simplification of that given in [9], that uses some elementary group theory which is, at this juncture, considered to be "advanced" material since we will not develop the full force of these tools until Chapter 2. The reader with a knowledge of some elementary group theory can first review Remark A.1 on page 304, although for the novice reader, this will not be necessary.

Remark 1.9 *Let q be prime and set*

$$G = \{a + b\sqrt{3} : a, b \in \mathbb{Z}_q\},$$

where, $\mathbb{Z}_q$ is the set of those remainders upon division by q of any integer a. In other words,

$$\mathbb{Z}_q = \{\overline{0}, \overline{1}, \overline{2}, \ldots, \overline{q-1}\},$$

where $\overline{r}$ is defined as follows. By the division algorithm, for any integer a there exist $q_a, r_a \in \mathbb{Z}$ such that $a = qq_a + r_a$ with $0 \le r_a < q$. Clearly, r_a has only q possible values for any $a \in \mathbb{Z}$, so r_a does not depend on a. Rather it is one of the q elements of $\mathbb{Z}_q$. Therefore, for all those infinitely many values of $a \in \mathbb{Z}$

with remainder r upon division by q, we write $\bar{r}$. Therefore, with this fact in hand, we easily see that the cardinality of G is q^2. Also, we see that G is closed under multiplication and addition since if $x + y\sqrt{3}, z + w\sqrt{3} \in G$, then

$$(x + y\sqrt{3})(z + w\sqrt{3}) = (xz + 3yw) + (z + x)\sqrt{3},$$

where $xz + 3yw, z + x \in \mathbb{Z}_q$. Furthermore, it is easily checked that $\mathbb{Z}_q$ is closed under multiplication and addition, namely for any integers r, s, with $\bar{r}, \bar{s} \in \mathbb{Z}_q$, then $\bar{r} \cdot \bar{s} \in \mathbb{Z}_q$ and $\bar{r} \pm \bar{s} \in \mathbb{Z}_q$. We need this development in the following proof.

Theorem 1.28 | Lucas-Lehmer Test

If $p > 2$ is prime, then M_p is prime whenever M_p divides the $(p-1)$-st term of the sequence s_j, which is defined by the recurrence $s_j = s_{j-1}^2 - 2$, for any $j > 1$, where $s_1 = 4$.

Proof. Recall first Theorem 1.20 on page 36, which told us that if M_n is prime, then n is prime. Also, we recall the recurrence sequences introduced in Exercise 1.12 on page 12, and let $\alpha = 2 + \sqrt{3}$ and $\beta = 2 - \sqrt{3}$ be the roots of the polynomial $f(x) = x^2 - 4x + 1$. In the notation of that exercise, $u = 4$ and $v = -1$. Also, $\alpha + \beta = 4$, $\alpha\beta = 1$, and $V_n = \alpha^n + \beta^n$. We first prove the following.

Claim 1.9 *For any $n \in \mathbb{N}$, $V_{2^{n-1}} = s_n$.*

We use induction on n. If $n = 1$, then

$$V_{2^{n-1}} = V_1 = \alpha + \beta = 4 = s_1 = s_n.$$

Now assume that $V_{2^{n-1}} = s_n$. Thus,

$$s_{n+1} = s_n^2 - 2 = V_{2^{n-1}}^2 - 2 = (\alpha^{2^{n-1}} + \beta^{2^{n-1}})^2 - 2 =$$

$$\alpha^{2^n} + \beta^{2^n} + 2\alpha^{2n}\beta^{2n} - 2 = \alpha^{2^n} + \beta^{2^n} = V_{2^n},$$

and we have the claim.

If $M_p \mid s_{p-1}$, then there is an integer x such that $s_{p-1} = xM_p$, so by Claim 1.9,

$$\alpha^{2^{p-2}} + \beta^{2^{p-2}} = xM_p.$$

Multiplying through by $\alpha^{2^{p-2}}$ we get that

$$\alpha^{2^{p-1}} + (\alpha\beta)^{2^{p-2}} = xM_p\alpha^{2^{p-2}},$$

which implies

$$\alpha^{2^{p-1}} = xM_p\alpha^{2^{p-2}} - 1. \tag{1.27}$$

Now by squaring both sides,

$$\alpha^{2^p} = (xM_p\alpha^{2^{p-2}} - 1)^2. \tag{1.28}$$

Suppose that q is a prime dividing M_p with $q \leq \sqrt{M_p}$. In the notation of Remark 1.9, $\overline{\alpha} \in G$. Thus, by Equation (1.27), $\overline{\alpha}^{2^{p-1}} = \overline{q-1}$, since this is the value once we throw away all multiples of q in $xM_p\alpha^{2^{p-2}}$. Similarly, by Equation (1.28), $\overline{\alpha}^{2^p} = \overline{1}$. Now we need one more result to finish the proof.

Claim 1.10 *The values,*

$$\overline{1}, \overline{\alpha}, \overline{\alpha}^2, \overline{\alpha}^3, \overline{\alpha}^4, \ldots, \overline{\alpha}^j, \ldots, \overline{\alpha}^{2^p-1}$$

are all in G and are distinct.

Since $\overline{\alpha} \in G$, then $\overline{\alpha}^j \in G$ for all nonnegative integers $j \leq q^2$ since G is closed under multiplication, by the development in Remark 1.9. Let $t \in \mathbb{N}$ be the smallest value such that $\overline{\alpha}^t = \overline{1}$, where $t \leq 2^p$, given that $\overline{\alpha}^{2^p} = \overline{1}$. By the division algorithm, there exist $s, r \in \mathbb{Z}$ such that $2^p = st + r$ where $0 \leq r < t$. Hence,

$$\left(\overline{\alpha}^{2^{p-1}}\right)^2 = \overline{\alpha}^{2^p} = \overline{\alpha}^{st+r} = (\overline{\alpha}^t)^s \overline{\alpha}^r = \overline{1}\,\overline{\alpha}^r = \overline{\alpha}^r.$$

However,

$$\left(\overline{\alpha}^{2^{p-1}}\right)^2 = (\overline{q-1})^2 = \overline{q^2 - 2q + 1} = \overline{1},$$

so by the minimality of t, we must have $r = 0$. Hence, $2^p = ts$. If $s > 1$, then s is even, so

$$\overline{q-1} = \overline{\alpha}^{2^{p-1}} = (\overline{\alpha}^t)^{s/2} = \overline{1},$$

which is a contradiction since $\overline{q-1} = \overline{1}$, implies that $q = 2$, and we know that q is odd. Thus, $s = 1$, which secures the claim.

By Claim 1.10, $2^p \leq q^2 \leq M_p = 2^p - 1$, a contradiction that secures the result. $\square$

Remark 1.10 *The converse of Theorem 1.28 also holds. However, it requires a bit more development so we leave this for later in the text. The above sufficiency was proved by Lucas in 1878, but only for primes of the form $p = 4m + 1$. The necessity and sufficiency for all exponents was proved by Lehmer in the early 1930s. See Biographies 1.18 on the next page and 1.19 on page 64.*

Concerning the complexity of the Lucas-Lehmer result, it can be shown that using Theorem 1.28, we can determine whether M_p is prime or composite in deterministic polynomial time. In fact, the test can be accomplished in $O(p^3)$ bit operations.

Example 1.30 Let $M_{13} = 8191$, and let $\overline{s_j}$ denote the remainder after division by M_{13} of s_j. Then we compute $\overline{s_j}$, the least nonnegative residue of s_j modulo M_{13}, as follows: $\overline{s_2} = 14$, $\overline{s_3} = 194$, $\overline{s_4} = 4870$, $\overline{s_5} = 3953$, $\overline{s_6} = 5970$, $\overline{s_7} = 1857$, $\overline{s_8} = 36$, $\overline{s_9} = 1294$, $\overline{s_{10}} = 3470$, $\overline{s_{11}} = 128$, and $\overline{s_{12}} = 0$. Thus, M_{13} is prime by the Lucas-Lehmer test.

The above is the most we can do on primality testing with the tools we have available now. We will leave further primality tests, including the *probabilistic* kind for the next chapter, when we have developed modular arithmetic to make our task much easier. Indeed, we will revisit some of what is covered in this chapter, with an eye to demonstrating the power of the modular method.

Remark 1.11 *Recently a very sophisticated, yet relatively elementary method was found that is the first deterministic polynomial-time algorithm for primality testing. We have presented this in its entirety in Appendix F, for the more adventurous reader who may be considering a second course in number theory, or merely looking for some more challenging exposure to the concept of primality testing.*

Biography 1.18 François Édouard Anatole Lucas (1842–1891) *was born on April 4, 1842, in Amiens, France. In 1864, he graduated from École Normale as Agrégé des sciences mathématiques, meaning that he had passed the state agrégation examination required for a teaching position at French lycées (high schools). However, his first position was assistant astronomer at the Observatory of Paris. He remained there until the Franco-Prussian war in 1870 in which he served as an auxiliary artillery officer. After the war, he became a mathematics teacher at various high schools in Paris. He had interests in recreational mathematics, but his serious interest was in number theory, especially Diophantine analysis. Although he spent only the years 1875–1878 on the problems of factoring and primality testing, his contribution was impressive. Some of the ideas developed by Lucas may be interpreted today as the beginnings of computer design. His death was untimely and unfortunate. While attending a social function, a plate fell and a chip from it cut his face. Later he died from an infection that developed from that cut. See [53] for a book devoted to the life of Lucas and primality testing methods.*

Exercises

1.76. With reference to Exercise 1.12 on page 12, let $\alpha = 2 + \sqrt{3}$ and $\beta = 2 - \sqrt{3}$ be the roots of the polynomial $f(x) = x^2 - 4x + 1$. Set

$$U_n = (\alpha^n - \beta^n)/(\alpha - \beta) \text{ and } V_n = \alpha^n + \beta^n.$$

Prove that If $n = q > 3$ is prime, then both

$$q \mid (U_q - 3^{(q-1)/2}) \text{ and } q \mid (V_q - 4).$$

1.77. Prove that all Mersenne numbers M_p are relatively prime for distinct primes p.

(*Hint: See Example 1.14 on page 27.*)

1.78. Given $n \in \mathbb{N}$, set $(n-1)! = q(n)n(n-1)/2 + r(n)$, where $q(n), r(n) \in \mathbb{N}$ with $0 \le r(n) < n(n-1)/2$. In other words, $r(n)$ is the remainder after dividing $(n-1)!$ by $n(n-1)/2$. Prove that

$$\{r(n) + 1 : r(n) > 0\} = \{p : p > 2 \text{ is prime}\}.$$

(*Hint: Use Wilson's Theorem on page 51.*) This result was first proved by J. de Barinaga in 1912 (see [14, p. 428]).

1.79. Define a recurrence sequence by $r_1 = 2$ and $r_j = 2r_{j-1}^2 - 1$ for all $j \ge 2$. Prove that if $M_p \mid r_{p-1}$ where p is prime then M_p is prime.

In Exercises 1.80–1.84, prove the results for each $m, n \in \mathbb{N}$, where the notation is that of Exercise 1.76.

1.80. $2U_{m+m} = U_n V_m + V_n U_m$.

1.81. $2V_{m+n} = V_m V_n + 12 U_m U_n$.

1.82. For $n \ge m$, $2U_{n-m} = U_n V_m - V_n U_m$.

1.83. Both $U_{n+2} = 4U_{n+1} - U_n$ and $V_{n+2} = 4V_{n+1} - V_n$.

1.84. For $n \ge m$, $2V_{n-m} = V_m V_n - 12 U_m U_n$.

Biography 1.19 Derrick Henry Lehmer (1905–1991) *was born in Berkeley, California, on February 23, 1905. After graduating with his bachelor's degree from Berkeley in 1927, he went to the University of Chicago. There he studied under L. E. Dickson, but he left after only a few months. Neither the Chicago weather nor the working environment suited him. Brown University offered him a better situation with an instructorship, and he completed both his master's degree and his Ph.D., the latter in 1930. During the period 1930–1940, he had brief stints at the California Institute of Technology; Stanford University; Lehigh University; and Cambridge, England, the latter on a Guggenheim Fellowship. In 1940, he accepted a position at the University of California at Berkeley where he remained until his retirement in 1972. He was a pioneering giant in the world of computational number theory and was widely respected in the mathematical community. The reader is advised to look into his contributions given in his selected works [25]. He was also known for his valued sense of humour, as attested by John Selfridge in the foreword to the aforementioned selected works, as well as by one of Lehmer's students, Ron Graham. In particular, Selfridge concludes with an apt description of Lehmer's contributions, saying that he "has shown us this beauty with the sure hand of a master."*

1.9 Distribution of Primes

> *Therefore it is necessary to arrive at a prime mover, put in motion by no other; and this everyone understands to be God.*
> From part 1 in article 3 of **Summa Theologicae (ca. 1265)**
> **St Thomas Aquinas (ca. 1225–1274), Italian Dominican Friar**

Gauss (see Biography 1.7 on page 33) first studied the number of primes less than x, denoted by $\pi(x)$. It is known (see [23, p. 37]., for instance) that Gauss wrote a letter to the astronomer Encke, on Christmas eve of 1849, where he reveals that when he was in his mid teens, in around 1793, he observed that as x gets large, $\pi(x)$ behaves akin to $x/\log_e(x)$, and he conjectured that

$$\lim_{x \to \infty} \frac{\pi(x)}{x/\log_e(x)} = 1, \qquad (1.28)$$

denoted by

$$\pi(x) \sim x/\log_e(x).$$

Note that, in general, if f and g are functions of a real variable x, then

$$f(x) \sim g(x) \text{ means } \lim_{x \to \infty} f(x)/g(x) = 1.$$

Such functions are said to be *asymptotic*. The Prime Number Theorem says that the proportion of prime natural numbers below x is about $1/\log_e(x)$.

Equation (1.28) is called the *Prime Number Theorem*, a proof of which is beyond the scope of this book. This celebrated theorem was first proved, independently, by Hadamard, and Vallée-Poussin — see Biographies 1.20 on the following page and 1.22 on page 68. Their proofs depended heavily upon complex analysis. However, proofs without the use of complex variables were given independently in 1949 by Erdös and Selberg — see [46] as well as [48] for a modern perspective. (Also, see Biographies 1.21 on page 67 and 1.23 on page 70.) Nevertheless, even these so-called *elementary* proofs are *more* difficult in some ways than those involving complex analysis, but they do provide insights into *why* the Prime Number Theorem is true. (However, a rather palatable and quite elementary sketch of the proof appears in the recent article [35], where very little is used beyond elementary calculus. However, the proof relies on an as-yet-unproved hypothesis that $x/\pi(x)$ is asymptotic to an increasing function. Albeit, a candidate for one such function is provided in the paper. Moreover, D.J. Newman devised a simple proof based upon an analytic argument. See [54] for a description of this proof (in under four pages).)

The proof by Hadamard and Poussin made use of Hadamard's theory of integral functions applied to the *Riemann zeta function*, given by

$$\zeta(s) = \sum_{n=1}^{\infty} n^{-s}, \text{ for } Re(s) > 1,$$

(see Biography 1.26 on page 72). Selberg's proof used what is now known as *Selberg sieve methods*, and for this he received the Field's medal in 1950. For his work, Erdös received the Cole Prize in 1952. Selberg's sieve results are now a foundational aspect of elementary number theory. Indeed, Chen [10], used Selberg's methods to prove the important result that every even natural number is the sum of a prime and another natural number having at most two prime factors. This is the closest we have come to proving the Goldbach conjecture, which says that every even natural number bigger than 2 is a sum of two primes. Selberg is best known for his classification of all arithmetic zeta functions. Erdös' contributions were also far reaching given that he provided the basics for graph theory. Moreover his probabilistic methods have seen applications in combinatorics, and other aspects of number theory.

There are better approximations to $\pi(x)$ such as the *logarithmic integral*

$$\mathrm{li}(x) = \int_2^x dt / \log_e(t),$$

which Gauss also conjectured, after postulating the validity of Equation (1.28). If fact, J.E. Littlewood showed that $\pi(x) - \mathrm{li}(x)$ takes on infinitely many positive and infinitely many negative values as x ranges over $\mathbb{N}$. However, this is an existence result since no actual value of x has ever been found for which $\pi(x) - \mathrm{li}(x)$ is positive. Yet, in 1933, Samuel Skewes, who was a student of Littlewood, demonstrated that $\mathrm{li}(x) - \pi(x)$ changes sign for at least one x with $x < 10^{10^{10^{34}}}$, a monster of a number now known as *Skewes' constant*.

We now look at some applications of $\pi(x)$.

Theorem 1.29 | **Infinitude of Primes Via $\pi(\mathbf{x})$**

$$\lim_{x \to \infty} \pi(x) = \infty,$$

namely, there exist infinitely many primes.

Biography 1.20 Jacques Hadamard (1865–1963) *was born in Versailles, France. He obtained his doctorate in 1892 for a thesis on functions defined by Taylor series — see Biography A.4 on page 310. This ground-breaking thesis was one of the pioneering efforts in the general theory of analytic functions. Arguably his greatest contribution was a proof of the Prime Number Theorem, published in 1896. However, his name is attached to an important aspect of the theory of integral equations and coding theory, to mention a few, namely* Hadamard matrices. *He is also known for his work on geodesics for surfaces of negative curvature that began what we know as* symbolic dynamics. *In 1912, he was elected to the Academy of Sciences where he succeeded Poincaré. He continued to work until his ninety-seventh year when he died, leaving approximately 300 publications including books and research papers.*

The following essentially says that almost all positive integers are composite. We use an idea of Tchebychev in the proof in order to present an elementary verification, without directly using the Prime Number Theorem — see Biography 1.25 on page 71.

Biography 1.21 Atle Selberg (1917–2007) *was born on June 14, 1917 in Langesund, Norway. At an early age he had already read Ramanujan's collected works. By 1942, he was appointed as a research fellow at the University of Oslo, where he earned his doctorate in 1943. He emigrated to the United States with his new bride in 1947, and he spent the last years of the 1940s at the Institute for Advanced Study at Princeton. By 1951, the year after he won the Field's medal, he was promoted to full professor at Princeton. His work on the proof of the Prime Number Theorem using "elementary" techniques, earning him the Field's medal, was not his most important achievement. He later developed what has come to be known as* Selberg's Trace Formula *for* $SL_2(\mathbb{R})$. *He used this formula to prove that the* Selberg zeta function *of a Riemann surface satisfies an analogue of the Riemann hypothesis — see page 72. His other distinctions include being elected to the* American Academy of Arts and Sciences, *as well as both the* Norwegian Academy of Sciences *and the* Royal Danish Academy of Sciences. *He is considered to be one of the best analytic number theorists to have ever lived. Indeed, after reviewing Selberg's collected works in 1989–1991, Matti Jutila a Finnish mathematician at the University of Turku, said "...the author is a living classic who has profoundly influenced mathematics, especially analytic number theory in a broad sense, for about fifty years..." He died from a heart ailment at the age of 90 on August, 6, 2007, in his home at Princeton. In a statement released from the Institute at Princeton, Peter Sarnak said that Selberg was "a mathematician's mathematician".*

Theorem 1.30 | **Almost All Integers are Composite**

$$\lim_{x \to \infty} \frac{\pi(x)}{x} = 0.$$

Proof. For any $n \in \mathbb{N}$,

$$4^n = (1+1)^{2n} > \binom{2n}{n} \geq \prod_{n < p \leq 2n} p > \prod_{n < p \leq 2n} n = n^{\pi(2n) - \pi(n)},$$

where p denotes the primes in the specified range and the first inequality follows from the Binomial Theorem. Now take the $\log_n$ of both sides and we get,

$$\pi(2n) - \pi(n) \leq \frac{\log_e(4)n}{\log_e(n)} < \frac{7n}{5\log_e(n)}, \qquad (1.29)$$

using the result from elementary calculus that

$$\log_n(y) = \log_e(y)/\log_e(n)$$

for the first inequality and the easily verifiable fact that $\log_e 4 < 7/5$ for the last inequality. Therefore, a straightforward induction argument, left to the reader, shows that for any $m \in \mathbb{N}$, we have

$$\pi(2^m) < \frac{2^{m+1}}{m \log_e(2)}.$$

Therefore, if m is that integer such that $2^m \le n < 2^{m+1}$, then

$$(m+1) \log_e(2) > \log_e(n),$$

so

$$\pi(n) \le \pi(2^{m+1}) < \frac{2^{m+2}}{(m+1) \log_e(2)} < \frac{4n}{\log_e(n)}.$$

Hence, for any $n \in \mathbb{N}$,

$$\frac{\pi(n)}{n} < \frac{4}{\log_e(n)}, \tag{1.30}$$

from which the result immediately follows. $\qquad\qquad\qquad\qquad\qquad\square$

Remark 1.12 *A task for the reader, using the above proof, is Exercise 1.86 on page 70 which says that if p_n is the n-th prime, then*

$$\lim_{n \to \infty} \log_e(p_n) / \log_e(n) = 1. \tag{1.31}$$

Now we apply the Prime Number Theorem to $x = p_n$ and we get that

$$\lim_{n \to \infty} \frac{n \log_e(p_n)}{p_n} = 1,$$

so by Equation (1.31),

$$\lim_{n \to \infty} \frac{p_n}{n \log_e(p_n)} = 1. \tag{1.32}$$

What the latter says is that for sufficiently large n, p_n is approximated by $n \log_e(n)$. Moreoever, from Equation (1.32), we see that

$$\lim_{n \to \infty} \frac{p_{n+1}}{p_n} = 1.$$

Below we look at more consequences of the Prime Number Theorem.

Biography 1.22 Charles-Jean-Gustave-Nicholas De La Vallée-Poussin (1866–1962) *was born in Louvain, Belgium. He is perhaps best known for his proof of the Prime Number Theorem, and his important, fundamental textbook* Cours d'analyse, *which went through several editions culminating in a seventh edition in 1938. He also worked on approximations to functions by algebraic and trigonometric functions in the decade from 1908 to 1918. He held the chair of mathematics at the University of Louvain for half a century. He died in his mid 90s on March 2, 1962 in Louvain.*

Tchebychev was the first to establish, in 1849, that for $x \geq 2$,

$$0.92x/\log_e(x) < \pi(x) < 1.7x/\log_e(x), \tag{1.33}$$

— see [47] for a proof of this result. However, a proof of the Prime Number Theorem eluded him. Yet this result may be used to verify the following renowned result on prime distribution, first a conjectured by Joseph Bertrand in 1845 — see Biography 1.24 on page 71. It was proved by Tchebychev in 1850.

◆ **Bertrand's Postulate**

For any $n \in \mathbb{N}$, there is at least one prime p such that $n < p \leq 2n$.

Related to the above is the following result, which is more far reaching than Bertrand's Postulate, and shows the power of the Prime Number Theorem.

Theorem 1.31 | **Primes Between ax and bx**

If a, b are positive real numbers with $a < b$, then for sufficiently large $x \in \mathbb{R}$, there is at least one prime between ax and bx.

Proof. Clearly,

$$\lim_{x \to \infty} \frac{\log_e(ax)}{\log_e(bx)} = 1,$$

so by Equation (1.28),

$$\lim_{x \to \infty} \frac{\pi(bx)}{\pi(ax)} = \frac{b}{a}.$$

Hence, for $0 < a < b$, $\pi(bx) > \pi(ax)$, for sufficiently large x. □

Remark 1.13 *Theorem 1.31 says that if $a = 1$, and $b = 1 + \varepsilon$, where ε is any positive real number, there is at least one prime between n and $n(1+\varepsilon)$ for $n \in \mathbb{N}$ sufficiently large. Hence, we may conclude that for sufficiently large $n \in \mathbb{N}$,*

$$\pi(an) < \pi((a+1)n).$$

This may be viewed as a significant generalization of Bertrand's Postulate.

Exercises

1.85. Prove that if $n > 1$ is an integer, then

$$n\pi(n-1) < (n-1)\pi(n)$$

if and only if n is prime.

Biography 1.23 Pál Erdös (1913–1996) *was born in Budapest, Hungary o* *March 26, 1913. His early years were made more difficult by virtue of his bein* *of Jewish descent in an anti-semitic environment. Indeed, in 1920, anti-Jewis* *laws were enacted in Hungary, and these laws were not very different from thos* *enacted by Hitler over a decade later. Nevertheless, even with restrictions o* *Jewish entry into universities, he was allowed to enter in 1930. In 1934, h* *earned his doctorate from the University of Pázmány Péter in Budapest. H* *was eventually forced out of Hungary because of his Jewish heritage, and mad* *his way to the United States. In 1948, he returned to Hungary where he agai* *saw family and friends after the long absence. In the early 1950s he travelle* *between England and the United States, then spent a decade in Israel. In 196* *he met Ron Graham at a conference, which began a mathematical collaboratio* *that lasted the rest of his life. In fact, Graham provided a room in his house fc* *Erdös to stay whenever he chose to do so. He was incredibly prolific, havin* *over 700 papers to his credit by the time he was 60, and he was well know* *for his eccentricities. Furthermore, he had hundreds of collaborators, eventual* *leading to the famous* Erdös number, *which is defined as follows. Erdös himse* *has Erdös number zero. In order to be assigned an Erdös number, an authc* *must cowrite a mathematical paper with an author having an Erdös number.* *the lowest Erdös number of a coauthor is n, then the author's Erdös number* *$n + 1$. For instance, this author's Erdös number is 2 since I have coauthore* *papers with Andrew Granville, who coauthored with Erdös. For an amusin* *confusion of this author's name, involving Erdös, see page 4 of the memoir o* *Erdös by Cameron at* http://www.maths.qmw.ac.uk/~pjc/preprints/erd.pdf

1.86. In [15], it was proved that

$$\pi(2n) - \pi(n) > \frac{n}{3\log_e(2n)}, \tag{1.34}$$

for any $n \in \mathbb{N}$. Use this and the inequalities (1.29)–(1.30) in the proof of Theorem 1.30 on page 67 to prove that if p_n is the n-th prime, then

$$12n\log_e(p_n) > p_n > \frac{n\log_e(p_n)}{4},$$

for any integer $n > 1$. Conclude that $\lim_{n\to\infty}\log_e(p_n)/\log_e(n) = 1$.

1.87. It can be shown that $\lim_{x\to\infty}(\pi(2x) - \pi(x)) = \infty$. Use this fact to show that for any $n \in \mathbb{N}$, there exists an $m_n \in \mathbb{N}$ such that there are at least n primes between t and $2t$ when $t \geq m_n$.

Biography 1.24 Joseph Louis François Bertrand (1822–1900) *was a French mathematician. His areas of interest were not only number theory, but also differential geometry, probability, and thermodynamics. He was a professor at the École Polytechnique from 1856 until he was appointed professor of analysis at the Collège de France in 1862. He also was appointed a member of the Paris Academy of Sciences in 1856 and served as its permanent secretary from 1874 until he died in Paris on April 3, 1900.*
Other than the conjecture/postulate that is cited above, there is a paradox in probability theory that bears his name as well. Bertrand's Paradox *asks for the probability that a chord, chosen at random, in a circle is longer than the side of an equilateral triangle inscribed in the circle. He gave three answers, all ostensibly valid, but with inconsistent conclusions. Yet, this is not a true paradox since the problem has a well-defined solution, although not unique, once the notion of* chosen at random *is made precise and clearly specified.*

Biography 1.25 Pafnuty Lvovich Tchebychev (1821–1894) *was born in Oka- tovo, Russia. He began his post-secondary education at Moscow University in 1837, and graduated in 1841. His first position was at St. Petersburg Univer- sity in 1843, where he stayed until his retirement in 1882. His contributions were not only to number theory, but also to probability theory, numerical anal- ysis, and real analysis. He had a profound influence on Russian mathematics including the fact that his doctoral thesis was used as a textbook in Russian universities. He had many honours bestowed upon him in his lifetime. Of the many was being elected as a member of the Berlin Academy of Sciences in 1871, the Royal Society of London in 1877, and the Swedish Academy of Sciences in 1893. He was even awarded the French Légion d'Honneur.*

1.88. Prove that the Prime Number Theorem follows from the inequality:

$$\frac{n}{\log_e(n) - 1/2} < \pi(n) < \frac{n}{\log_e(n) - 3/2}.$$

(*This above inequality was shown to hold for all natural numbers $n \geq 67$ in* [45].)

1.89. Let p_n denote the nth prime. Assuming Bertrand's postulate, prove that $p_n \leq 2^n$.

1.90. Let $m, n \in \mathbb{N}$, with $m \geq n$, and set $Q = \sum_{j=0}^{m} \frac{1}{n+j}$. Assuming Bertrand's postulate, prove that $Q \notin \mathbb{N}$.

Biography 1.26 Georg Friedrich Bernhard Riemann (1826–1866) *was born i Breselenz, Germany. His post-secondary education began at Göttingen Univer sity, where he intended to study theology. Eventually, his interests turned t mathematics and he transfered to Berlin University in the spring of 1847, wher he had advisors such as Dirichlet, Eisenstein, Jacobi, and Steiner. However he returned to Göttingen in 1849 to obtain his doctorate, supervised by Gauss and submitted in 1851. His thesis was on the study of complex variables, an had the creation of what we now call* Riemann surfaces, *via the introductio of topological methods into complex function theory. Eventually, he was ap pointed to the chair of mathematics at Göttingen on July 30, 1859, after th death of Dirichlet. He died on July 20, 1866 in Selasca, Italy, where he finall succumbed to tuberculosis before he turned 40.*

The Riemann Hypothesis

On page 65, we defined the Riemann zeta function as $\zeta(s) = \sum_{n=1}^{\infty} n^{-s}$ fo $re(s) > 1$. This function has zeros at all $-2m$ for any $m \in \mathbb{N}$, which as know as the *trivial zeros*, sometimes called the *real zeros* since they are real and n other zeros are real. The *nontrivial zeros* of the zeta function, those lying i the *critical strip* $0 < Re(s) < 1$, and there are infinitely many such nontrivia zeros, none of which are real, so they are often called the *complex zeros* c the zeta function. Also, these zeros are symmetric about the real axis and th *critical line* $\mathrm{Re}(s) = 1/2$. For example, if $s = 3/4 + ix$ is a zero, then so $s = 1/4 + ix$. Zeros lying on the critical line are of the form $s = 1/2 + ix$ ar nontrivial, complex, and there are infinitely many of them. Indeed Selberg contribution is his verification that a positive proportion of the zeros of $\zeta(s)$ i the critical strip actually lie on the critical line – see Biography 1.21 on page 6 In 1859, Riemann published a now famous result in [41] wherein he provided formula for $\pi(x)$ in terms of the zeros of $\zeta(s)$, thereby linking the zeta functio with the distribution of primes. The *Riemann hypothesis* says that *all* th *nontrivial zeros* of the zeta function lie on the critical line $Re(s) = 1/2$. Note that there is an equivalent formulation using $\mathrm{li}(x)$ introduced on page 6 namely that

$$\pi(x) = \mathrm{li}(x) + O(\sqrt{x}\log_e(x)),$$

which holds if and only if the zeta function does not vanish on the hal plane $Re(s) > 1/2$. In other words, the Riemann hypothesis is equivaler to the statement that the error that occurs when $\pi(x)$ is estimated by $\mathrm{li}(x)$ $O(\sqrt{x}\log_e(x))$.

The Riemann hypothesis is widely believed to hold given the preponderance c evidence in its favour. Much heuristic evidence has been amassed in terms c computing the zeros up to large bounds and all have shown to have their re part equal to 1/2. The Riemann hypothesis is arguably the most importar unsolved problem in mathematics today.

Chapter 2

Modular Arithmetic

> *Population, when unchecked, increases in a geometric ratio. Subsistence only increases in an arithmetical ratio.*
> From Chapter 1 of **Essay on the Principle of Population (1798)**
> **Thomas Robert Malthus (1766–1834), English political economist**

In this chapter, we learn about the important topic of congruences and the modular arithmetic underlying the theory.

2.1   Basic Properties

We now turn to a concept called *congruences*, invented by Gauss (see Biography 1.7 on page 33). The stage is set by the discussion of divisibility given in §1.2.

Gauss sought a convenient tool for abbreviating the family of expressions $a = b + nk$, called an *arithmetic progression* with modulus n, wherein k varies over all natural numbers, $n \in \mathbb{N}$ is fixed, as are $a, b \in \mathbb{Z}$. He did this as follows.

Definition 2.1 | Congruences |

If $n \in \mathbb{N}$, then we say that a is congruent *to b modulo n if $n|(a-b)$, denoted by*

$$a \equiv b \pmod{n}.$$

On the other hand, if $n \nmid (a-b)$, then we write

$$a \not\equiv b \pmod{n}$$

and say that a and b are incongruent *modulo n, or that a is* not congruent *to b modulo n. The integer n is the* modulus *of the congruence. The set of all integers that are congruent to a given integer r modulo n, denoted by $\bar{r}$, is* *called the* congruence class *or* residue class *of r modulo n. (Note that since the notation $\overline{r}$ does not specify the modulus n, then the bar notation will always be taken in context.)*

Remark 2.1 *Note that if $n = 1$, then for any integers a, b, $a \equiv b \pmod{n}$. This is the trivial case, which is uninteresting, so usually we assume that $n > 1$. Also, see Remark 2.5 on page 79.*

Remark 2.2 *We have already seen some modular arithmetic in action without having explicitly said so. In Remark 1.9 on page 60, we looked at the set*

$$\mathbb{Z}_q = \{\overline{0}, \overline{1}, \overline{2}, \ldots, \overline{q-1}\},$$

where q is prime and defined $\overline{r}$ to essentially be that given more generally in Definition 2.1. We will have more to say about this set later in this section.

Also in Remark 1.9, it is stated that there is closure under addition and multiplication which we now prove for all moduli.

Proposition 2.1 | **Closure Under Addition and Multiplication** |

Let $n \in \mathbb{N}$ and $a, b, c, d \in \mathbb{Z}$. If $a \equiv b \pmod{n}$ and $c \equiv d \pmod{n}$, then $a + c \equiv b + d \pmod{n}$, $a - c \equiv b - d \pmod{n}$, and $ac \equiv bd \pmod{n}$.

Proof. Since there exist integers $k, \ell \in \mathbb{Z}$ such that $a = b + kn$ and $c = d + \ell n$, then (grouping the $\pm$ into a single proof)

$$a \pm c = b + kn \pm (d + \ell n) = b \pm d + (k \pm \ell)n,$$

so

$$a \pm c \equiv b \pm d \pmod{n}.$$

Similarly,

$$ac \equiv (b + kn)(d + \ell n) \equiv bd \pmod{n}.$$

$\square$

Remark 2.3 *Essentially what Proposition 2.1 says is that the stipulations for multiplication and addition given by $\overline{ab} = \overline{a} \cdot \overline{b}$ and $\overline{a \pm b} = \overline{a} \pm \overline{b}$ are indeed well defined. (For the reader needing a reminder of the meaning of* well defined *see the discussion on page 285.)*

What the following verifies is that congruences satisfy the three defining properties of an *equivalence relation*.

Proposition 2.2 | Congruences are Equivalence Relations

Let $n \in \mathbb{N}$. Then each of the following holds.

(a) For each $a \in \mathbb{Z}$, $a \equiv a \,(\mathrm{mod}\ n)$, called the *reflexive property*.

(b) For any $a, b \in \mathbb{Z}$, if $a \equiv b \,(\mathrm{mod}\ n)$, then $b \equiv a \,(\mathrm{mod}\ n)$, called the *symmetric property*.

(c) For any $a, b, c \in \mathbb{Z}$, if $a \equiv b \,(\mathrm{mod}\ n)$, and $b \equiv c \,(\mathrm{mod}\ n)$, then $a \equiv c \,(\mathrm{mod}\ n)$, called the *transitive property*.

Proof. (a) If $n \in \mathbb{N}$, then $n \mid 0 = a - a$, so $a \equiv a \,(\mathrm{mod}\ n)$, which establishes the reflexive property.

(b) Let $n \in \mathbb{N}$, $a, b, c \in \mathbb{Z}$, $a \equiv b \,(\mathrm{mod}\ n)$, so $a - b = kn$ for some $k \in \mathbb{Z}$. By rewriting, $b - a = (-k)n$, implying $b \equiv a \,(\mathrm{mod}\ n)$, which establishes the symmetric property.

To prove part (c), we use Definition 2.1. Given that $a \equiv b \,(\mathrm{mod}\ n)$, as well as $b \equiv c \,(\mathrm{mod}\ n)$, then $n \mid (a - b)$ and $n \mid (b - c)$. Therefore,

$$n \mid (a - b) + (b - c) = (a - c),$$

which is to say

$$a \equiv c \pmod{n}.$$

$\square$

The next result tells us how to divide using congruences.

Proposition 2.3 | Congruence Cancellation Law

If $\gcd(c, n) = g$, *then*

$$ac \equiv bc \pmod{n}$$

if and only if

$$a \equiv b \pmod{n/g}.$$

Proof. If $ac - bc = kn$ for some $k \in \mathbb{Z}$, then $(a - b)c/g = kn/g$. By Claim 1.3 on page 26, $\gcd(c/g, n/g) = 1$. Therefore, (n/g) divides $(a - b)$, namely

$$a \equiv b \pmod{n/g}.$$

Conversely, if $a \equiv b \,(\mathrm{mod}\ n/g)$, then there exists an integer $d \in \mathbb{Z}$ such that $a = b + dn/g$, so $ac = bc + d(c/g)n$. Hence, $ac \equiv bc \,(\mathrm{mod}\ n)$. $\square$

Notice that Proposition 2.3 tells us that we cannot simply cancel the value of c from both sides of the congruence, if $\gcd(c, n) = g > 1$, since the modulus must be taken into consideration. Only when $g = 1$ may we cancel and leave the modulus unchanged.

Some more properties of congruences are given in the next result.

Proposition 2.4 | Additional Properties of Congruences |

Let $a, b, c \in \mathbb{Z}$, $m, n \in \mathbb{N}$, and $a \equiv b \,(\text{mod } n)$. Then each of the following holds.

(a) $am \equiv bm \,(\text{mod } mn)$.

(b) $a^m \equiv b^m \,(\text{mod } n)$.

(c) If m divides n, then $a \equiv b \,(\text{mod } m)$.

Proof. (a) Given that $a \equiv b \,(\text{mod } n)$, $a - b = kn$ for some integer k. Multiplying by m, we get $(a - b)m = knm$, so $am - bm = (km)n$, namely $am \equiv bm$ $(\text{mod } n)$.

(b) Since $n \,|\, (a - b)$, then

$$n \,|\, (a - b)(a^{m-1} + a^{m-2}b + \cdots + b^{m-1}) = a^m - b^m.$$

In other words,

$$a^m \equiv b^m \quad (\text{mod } n).$$

(c) Since $a = b + kn$ for some $k \in \mathbb{Z}$ and $n = \ell m$ for some $\ell \in \mathbb{N}$, then $a = b + k\ell m$, so $a - b = (k\ell)m$, whence $a \equiv b \,(\text{mod } m)$.   □

Propositions 2.2–2.4 can be employed to establish a modular arithmetic. First we need another couple of notions.

Definition 2.2 | Residues |

If $n \subset \mathbb{N}$ and $\ell \equiv r \,(\text{mod } n)$, then r is called a residue of ℓ modulo n. Any set of integers $\{r_1, r_2, \ldots, r_n\}$ is called a complete residue system modulo n if, for any $z \in \mathbb{Z}$, there is exactly one r_j such that $z \equiv r_j \,(\text{mod } n)$.

Remark 2.4 Note that if $\mathcal{R} = \{r_1, r_2, \ldots, r_n\}$ is a complete residue system modulo n, and $r_i \equiv r_j \,(\text{mod } n)$, then $i = j$ since otherwise $z = r_i$ is congruent to two distinct elements of $\mathcal{R}$, itself and r_j, contradicting Definition 2.2.

Proposition 2.5 | Complete Residue Systems and Congruences |

(a) *For all $n \in \mathbb{N}$, congruence modulo n partitions the integers $\mathbb{Z}$ into disjoint subsets. In other words, $\mathbb{Z} = \overline{m_1} \cup \overline{m_2} \cup \cdots \cup \overline{m_n}$, where $\overline{m_j}$ for $j = 1, 2, \ldots, n$ are congruence classes modulo n with $\overline{m_i} \cap \overline{m_j} = \varnothing$ if $i \neq j$.*

(b) *For any $n \in \mathbb{N}$, the set $\{0, 1, 2, \ldots, n - 1\}$ is a complete residue system modulo n.*

Proof. For part (a), we need to show that every $m \in \mathbb{Z}$ is in *exactly* one residue class modulo n. Since $m \in \overline{m}$, then m is in *some* congruence class. We must prove that it is in *no more than one* such class.

If $m \in \overline{m}_1$ and $m \in \overline{m}_2$, both $m \equiv m_1 \pmod{n}$ and $m \equiv m_2 \pmod{n}$. Thus, $m_1 \equiv m_2 \pmod{n}$ by Proposition 2.2 (c), so $\overline{m}_1 = \overline{m}_2$, and we are done.

For part (b), we let $\mathcal{S} = \{s_0, s_1, \ldots, s_{n-1}\}$, where $s_j = j$ for $j = 0, 1, \ldots, n-1$. Since the division algorithm tells us that, for any $z \in \mathbb{Z}$, there exist unique $q_z, s_z \in \mathbb{Z}$ such that $z = q_z n + s_z$ where $0 \leq s_z < n$, then $z \equiv s_z \pmod{n}$, and $s_z \in \mathcal{S}$. Moreover, $s_j \not\equiv s_i \pmod{n}$ for any $s_i, s_j \in \mathcal{S}$ with $i \neq j$ since $n \mid (s_i - s_j)$ for $n > s_j \geq s_i \geq 1$ implies $s_i = s_j$ since $s_j - s_i < n$. Hence, $\mathcal{S}$ is a complete residue system modulo n. $\qquad\square$

Proposition 2.5 motivates the following.

Definition 2.3 | Least Residue Systems

For any $n \in \mathbb{N}$, the complete residue system $\{0, 1, 2, \ldots, n-1\}$ is called the least residue system modulo n.

When dealing with problems involving congruences, it is most often best to deal with the least residues since this can simplify the work to be done.

Example 2.1 *There are four congruence classes modulo 4, namely*

$$\overline{0} = \{\ldots, -4, 0, 4, \ldots\},$$
$$\overline{1} = \{\ldots, -3, 1, 5, \ldots\},$$
$$\overline{2} = \{\ldots, -2, 2, 6, \ldots\},$$

and

$$\overline{3} = \{\ldots, -1, 3, 7, \ldots\},$$

since each element of $\mathbb{Z}$ is in exactly one of these disjoint sets, namely

$$\mathbb{Z} = \overline{0} \cup \overline{1} \cup \overline{2} \cup \overline{3}.$$

Moreover, the least residue system modulo 4 is $\{0, 1, 2, 3\}$.

The following gathers together the totality of the laws underlying modular arithmetic that will give us the structure to dive deeper into number-theoretic concepts and their applications. The reader is encouraged to review the fundamental laws for arithmetic beginning on page 289, so that we will see that these seemingly trivial laws have a generalization to the following important scenario.

Theorem 2.1 | Modular Arithmetic

Let $n \in \mathbb{N}$ and suppose that for any $x \in \mathbb{Z}$, $\overline{x}$ denotes the congruence class of x modulo n. Then for any $a, b, c \in \mathbb{Z}$ the following hold.

(a) $\overline{a} \pm \overline{b} = \overline{a \pm b}$.   (Modular additive closure)

(b) $\overline{a}\overline{b} = \overline{ab}$.   (Modular multiplicative closure)

(c) $\overline{a} + \overline{b} = \overline{b} + \overline{a}$.   (Commutativity of modular addition)

(d) $(\overline{a} + \overline{b}) + \overline{c} = \overline{a} + (\overline{b} + \overline{c})$.   (Associativity of modular addition)

(e) $\overline{0} + \overline{a} = \overline{a} + \overline{0} = \overline{a}$.   (Additive modular identity)

(f) $\overline{a} + \overline{-a} = \overline{-a} + \overline{a} = -\overline{a} + \overline{a} = \overline{0}$.   (Additive modular inverse)

(g) $\overline{a}\overline{b} = \overline{b}\overline{a}$.   (Commutativity of modular multiplication)

(h) $(\overline{a}\overline{b})\overline{c} = \overline{a}(\overline{b}\overline{c})$.   (Associativity of modular multiplication)

(i) $\overline{1}\overline{a} = \overline{a}\overline{1} = \overline{a}$.   (Multiplicative modular identity)

(j) $\overline{a}(\overline{b} + \overline{c}) = \overline{a}\overline{b} + \overline{a}\,\overline{c}$.   (Modular distributivity)

Proof. Parts (a)–(b) are a consequence of Proposition 2.1. Also see Remark 2.3 on page 74. Part (c) can be established using part (a) since

$$\overline{a} + \overline{b} = \overline{a + b} = \overline{b + a} = \overline{b} + \overline{a}.$$

In other words, the commutativity property is inherited from the integers $\mathbb{Z}$. Part (d) also follows from part (a) since

$$(\overline{a} + \overline{b}) + \overline{c} = \overline{a + b} + \overline{c} = \overline{a + b + c} = \overline{a} + \overline{(b + c)} = \overline{a} + (\overline{b} + \overline{c}).$$

Part (e) is a consequence of parts (c) and (a) since $\overline{0} + \overline{a} = \overline{a} + \overline{0} = \overline{a + 0} = \overline{a}$, where the first equality holds by part (c) and the second equality holds by part (a). The first equality of part (f) follows from parts (a) and (c) in exactly the same fashion, whereas the second part follows from part (b). Part (g) follows from the ordinary commutativity of multiplication of integers and part (b), since $\overline{a}\overline{b} = \overline{ab} = \overline{ba} = \overline{b}\overline{a}$. Part (h) may now be deduced from part (b) and ordinary associativity of the integers since

$$(\overline{a}\overline{b})\overline{c} = \overline{(ab)}\overline{c} = \overline{(ab)c} = \overline{a(bc)} = \overline{a}\overline{(bc)} = \overline{a}(\overline{b}\overline{c}).$$

Part (i) is a simple consequence of parts (b), (g), and the multiplicative identity of the integers since $\overline{1}\overline{a} = \overline{1\,a} = \overline{a\,1} = \overline{a}$. Lastly, part (j) is a consequence of parts (a), (b), and the ordinary distributivity of multiplication over addition given that $\overline{a}(\overline{b} + \overline{c}) = \overline{a}\overline{(b + c)} = \overline{a(b + c)} = \overline{ab + ac} = \overline{ab} + \overline{ac} = \overline{a}\overline{b} + \overline{a}\,\overline{c}$. $\square$

Any set that satisfies the (named) properties (a)–(j) of Theorem 2.1 is called a *commutative ring with identity*. Now we look at a specific such ring that has important consequences.

Definition 2.4 | The Ring $\mathbb{Z}/n\mathbb{Z}$

For $n \in \mathbb{N}$, the set

$$\mathbb{Z}/n\mathbb{Z} = \{\overline{0}, \overline{1}, \overline{2}, \ldots, \overline{n-1}\}$$

is called the Ring of Integers Modulo n, *where $\overline{m}$ denotes the congruence class of m modulo n.*

Remark 2.5 *Notice that since $\{0, \ldots, n-1\}$ is the least residue system modulo n, then every $z \in \mathbb{Z}$ has a representative in the ring of integers modulo n, namely an element $j \in \{0, \ldots, n-1\}$ such that $z \equiv j \,(\mathrm{mod}\ n)$. The ring $\mathbb{Z}/n\mathbb{Z}$ will play an important role in the applications that we study later in the text. There are other structures hidden within the properties listed in Theorem 2.1 that are worth mentioning, since we will also encounter them in our number-theoretic travels. Any set satisfying the properties* (a), (d)–(f) *is called an* additive group, *and if additionally it satisfies* (c), *then it is called an* additive abelian group. *A fortiori, $\mathbb{Z}/n\mathbb{Z}$ is an additive abelian group as is $\mathbb{Z}$. Any set satisfying* (a)–(f), (h), *and* (j) *is called a* ring, *and if in addition it satisfies* (g), *then it is a* commutative ring. *As we have seen, any set satisfying all of the conditions* (a)–(j) *is a commutative ring with identity. The case $n = 1$ leads to the trivial ring $R = \{\overline{0}\}$, where $\overline{0} = \overline{1}$. Although trivial rings have two binary operations, multiplication provides no new content to the additive group. Typically, we avoid this trivial case by stipulating the condition $\overline{0} \neq \overline{1}$. See Remark 2.1 on page 74.*

In general, we would use symbols other than the bar operation and possibly binary symbols other than the multiplication and addition symbols, but the listed properties in Theorem 2.1 would remain essentially the same for the algebraic structures defined above.

There is a multiplicative property of $\mathbb{Z}$ that $\mathbb{Z}/n\mathbb{Z}$ does not have. On page 290, the Cancellation Law for $\mathbb{Z}$ is listed. This is not the case for $\mathbb{Z}/n\mathbb{Z}$ in general. For instance, $2 \cdot 3 \equiv 2 \cdot 8 \,(\mathrm{mod}\ 10)$, but $3 \not\equiv 8 \,(\mathrm{mod}\ 10)$. In other words, $2 \cdot 3 = 2 \cdot 8$ in $\mathbb{Z}/10\mathbb{Z}$, but $3 \neq 8$ in $\mathbb{Z}/10\mathbb{Z}$, and we know by the Cancellation Law for Congruences, Proposition 2.3 on page 75, why this is the case. Indeed Proposition 2.3 answers the question: For which $n \in \mathbb{N}$ does it hold that:

$$\text{for any } a, b, c \in \mathbb{Z}/n\mathbb{Z} \text{ with } a \neq 0, \ ab = ac \text{ if and only if } b = c? \qquad (2.1)$$

By Proposition 2.3, (2.1) cannot hold if $\gcd(a, n) > 1$. Thus, since any composite n will have elements $a \in \mathbb{Z}/n\mathbb{Z}$ with $\gcd(a, n) > 1$, then (2.1) will hold if and only if n is prime. We now explore in more detail what this means.

When $\gcd(a, n) = 1$, there is a solution $x \in \mathbb{Z}$ to $ax \equiv 1 \,(\mathrm{mod}\ n)$ which follows immediately from Example 1.15 on page 28. Therein it is proved that the linear Diophantine equation $ax + by = c$ has solutions $x, y \in \mathbb{Z}$ if and only if $\gcd(a, b) \mid c$. With $c = 1$ and $b = n$, we have the case $ax \equiv 1 \,(\mathrm{mod}\ n)$. This motivates the following.

Definition 2.5 | Modular Multiplicative Inverses |

*Suppose that $a \in \mathbb{Z}$ and $n \in \mathbb{N}$. A multiplicative inverse of the integer a
modulo n is an integer x such that $ax \equiv 1 \pmod{n}$. If x is the least positive
such inverse, then we call it the least multiplicative inverse of the integer a
modulo n, denoted by $x = a^{-1}$.*

Example 2.2 *Consider $n = 11$ and $a = -3$, and suppose that we want to find
the least multiplicative inverse of a modulo n. Since $-3 \cdot 7 \equiv 1 \pmod{11}$ and no
smaller natural number than 7 satisfies this congruence, then $a^{-1} = 7$ modulo
11.*

Example 2.3 *If $n = 22$ and $a = 6$, then no multiplicative inverse of a modulo
n exists since $\gcd(a, n) = 2$. Asking for a multiplicative inverse of such a
value a modulo n is similar to asking for division by 0 with ordinary division
of integers. In other words, multiplicative inverses of an element a modulo n,
when $\gcd(a, n) > 1$, is undefined.*

Example 2.4 *This example is designed to classify those integers a which are
their own multiplicative inverse modulo a prime p.*

If a is its own inverse modulo p, then $a^2 \equiv 1 \pmod{p}$. In other words,

$$p \mid (a^2 - 1) = (a-1)(a+1),$$

*so either $p \mid (a-1)$ or $p \mid (a+1)$, namely $a \equiv \pm 1 \pmod{p}$. Conversely, if $a \equiv \pm 1$
$\pmod{p}$, then $a^2 \equiv 1 \pmod{p}$ by part (b) of Proposition 2.4 on page 76. Hence,
we have shown:*

$$a^2 \equiv 1 \pmod{p} \text{ if and only if } a \equiv 1 \pmod{p} \text{ or } a \equiv -1 \pmod{p}.$$

In the above case a is said to be a self-multiplicative inverse.

Earlier in this discussion, we saw that (2.1) holds for all $a \in \mathbb{Z}/n\mathbb{Z}$, $a \neq 0$, if
and only if n is prime. Another way of stating this is as follows. Every nonzero
$z \in \mathbb{Z}/n\mathbb{Z}$ has a multiplicative inverse if and only if n is prime.

If the existence of multiplicative inverses is satisfied for any given element
along with (b), (h)–(i) of Theorem 2.1 for a given set, then that set is called a
multiplicative group. In addition, if the set satisfies (g) of Theorem 2.1, then it
is called an *abelian multiplicative group*. Notice that $\mathbb{Z}$ is *not* a multiplicative
group since any nonzero $a \in \mathbb{Z}$ with $a \neq \pm 1$ has no multiplicative inverse.

There is one property that is held by $\mathbb{Z}$ that is of particular importance
to the ring $\mathbb{Z}/n\mathbb{Z}$. There are mathematical structures $\mathcal{S}$ that have what are
called *zero divisors*. These are elements $s, t \in \mathcal{S}$ such that both s and t are
nonzero, yet $st = 0$. For instance, in the ring $\mathbb{Z}/6\mathbb{Z}$, $2 \cdot 3 = 0$, so this ring has

zero divisors. The integers $\mathbb{Z}$ have no zero divisors. What is the situation for $\mathbb{Z}/n\mathbb{Z}$ with respect to zero divisors? If n is composite, then there are natural numbers $n > n_1 > 1$ and $n > n_2 > 1$ such that $n = n_1 n_2$. Hence, $n_1 n_2 = 0$ in $\mathbb{Z}/n\mathbb{Z}$. Therefore, $\mathbb{Z}/n\mathbb{Z}$ has *no zero divisors* if and only if n is prime. Any set that satisfies all the conditions (a)–(j) of Theorem 2.1 together with having no zero divisors and having multiplicative inverses for all of its nonzero elements is called a *field*. Hence, we have established the following.

Theorem 2.2 $\boxed{\text{The Field } \mathbb{Z}/p\mathbb{Z}}$

If $n \in \mathbb{N}$, then $\mathbb{Z}/n\mathbb{Z}$ is a field if and only if $n = p$ is prime.

In Theorem A.6 on page 300, we employed the notation F^* to denote the multiplicative group of nonzero elements of a given field F. In particular, when we have a finite field $\mathbb{Z}/p\mathbb{Z} = \mathbb{F}_p$ of p elements for a given prime p, then

$(\mathbb{Z}/p\mathbb{Z})^*$ denotes the multiplicative group of nonzero elements of $\mathbb{F}_p$.

This is tantamount to saying that $(\mathbb{Z}/p\mathbb{Z})^*$ is the group of units in $\mathbb{F}_p$, and $(\mathbb{Z}/p\mathbb{Z})^*$ is cyclic by Theorem A.6. Thus, this notation and notion may be generalized as follows. Let $n \in \mathbb{N}$ and let the group of units of $\mathbb{Z}/n\mathbb{Z}$ be denoted by $(\mathbb{Z}/n\mathbb{Z})^*$. Then

$$(\mathbb{Z}/n\mathbb{Z})^* = \{\bar{a} \in \mathbb{Z}/n\mathbb{Z} : 0 < a < n \text{ and } \gcd(a, n) = 1\}. \qquad (2.2)$$

The structure of $(\mathbb{Z}/n\mathbb{Z})^*$ is going to be of vital importance as we move through the text.

Now we go on to look at some of the consequences of this notion of modular division, which is implicit in the above. Definition 2.5 gives us the means to do modular division since multiplication by a^{-1} is equivalent to division in $\mathbb{Z}/n\mathbb{Z}$. As an application, we note that if, in Theorem 1.24 on page 50, we have that $\gcd(n, p) = 1$, then the translation into modular arithmetic is that

$$n^{p-1} \equiv 1 \pmod{p},$$

since, by Theorem 2.3 on page 75, we may multiply both sides of the congruence $n^p \equiv n \pmod{p}$ by n^{-1}.

Remark 2.6 *Remark 1.9 on page 60 may now be put into a modular perspective, since we see that $\mathbb{Z}_q$ therein is the field $\mathbb{Z}/q\mathbb{Z}$ by Theorem 2.2. Hence, the statements in that remark about closure of addition, subtraction, and multiplication are clear from Theorem 2.1 on page 77. Moreover, any commutative ring with identity having no zero divisors is called an* Integral Domain. *The set G in Remark 2.1, usually denoted by $\mathbb{Z}[\sqrt{3}]$ is an integral domain. Thus, we see that the proof of Theorem 1.28 on page 61 actually was doing computations inside the integral domain $\mathbb{Z}[\sqrt{3}]$. The modular version of Theorem 1.28 is that M_p is prime whenever $s_{p-1} \equiv 0 \pmod{M_p}$. In Remark 1.26 on page 50, we commented that the converse of Theorem 1.28 is also true but we will wait until Chapter 4, when we have the power of* quadratic reciprocity *at our disposal.*

The next aspect of modular arithmetic that we will need later in the text is called *modular exponentiation*. For $b, r \in \mathbb{N}$, this involves the finding of a least nonnegative residue of b^r modulo a given $n \in \mathbb{N}$, especially when the given natural numbers r and n are large. There is an algorithm for doing this that is far more efficient than repeated multiplication of b by itself. The algorithm begins with a tacit use of the Base Representation Theorem in the binary case — see Theorem 1.5 on page 8.

$$\boxed{\blacklozenge \ \textbf{The Repeated Squaring Method}}$$

Given $d, n \in \mathbb{N}$, $d > 1$, $x \in \mathbb{Z}$, and

$$d = \sum_{j=0}^{k} d_j 2^j, \ d_j \in \{0, 1\},$$

the goal is to find $x^d \pmod{n}$.

First, we initialize by setting $c_0 = x$ if $d_0 = 1$ and set $c_0 = 1$ if $d_0 = 0$. Also, set $x_0 = x$, $j = 1$, and execute the following steps:

(1) Compute $x_j \equiv x_{j-1}^2 \pmod{n}$.

(2) If $d_j = 1$, set $c_j = x_j c_{j-1} \pmod{n}$.

(3) If $d_j = 0$, then set $c_j \equiv c_{j-1} \pmod{n}$.

(4) Reset j to $j + 1$. If $j = k + 1$, output $c_k \equiv x^d \pmod{n}$ and terminate the algorithm. Otherwise, go to step (1).

The above algorithm will be valuable later in the text we look at applications to cryptography. For now we illustrate the algorithm with a worked example involving modular arithmetic.

Example 2.5 *Let $n = 101$, and compute $3^{61} \pmod{101}$ as follows. Since $61 = 1 + 2^2 + 2^3 + 2^4 + 2^5$, then $k = 5$, $d_j = 1$ for $j = 0, 2, 3, 4, 5$ and $d_1 = 0$. Also, since $d_0 = 1$, we set $c_0 = 3$, $x_0 = 3$, and $j = 1$. The following are the jth steps for $j = 1, 2, 3, 4, 5$.*

(1) $x_1 \equiv x_0^2 \equiv 9 \pmod{101}$. *Since $d_1 = 0$, $c_1 = c_0 = 3$.*

(2) $x_2 \equiv 9^2 \equiv 81 \pmod{101}$. *Since $d_2 = 1$, $c_2 \equiv 3 \cdot 81 \equiv 41 \pmod{101}$.*

(3) $x_3 \equiv 81^2 \equiv 97 \pmod{101}$. *Since $d_3 = 1$, $c_3 \equiv 97 \cdot 41 \equiv 38 \pmod{101}$.*

(4) $x_4 \equiv 97^2 \equiv 16 \pmod{101}$. *Since $d_4 = 1$, $c_4 \equiv 16 \cdot 38 \equiv 2 \pmod{101}$.*

(5) $x_5 \equiv 16^2 \equiv 54 \,(\text{mod } 101)$. *Since* $d_5 = d_k = 1$, $c_5 \equiv 54 \cdot 2 \equiv 7 \,(\text{mod } 101)$. *Hence,* $3^{61} \equiv 7 \,(\text{mod } 101)$.

Exercises

2.1. Suppose that we altered the definition of a complete residue system given on page 76 to read: *Any set of integers* $\mathcal{R} = \{r_1, r_2, \ldots, r_m, m \in \mathbb{N}\}$, *is called a* complete residue system modulo n *if, for any* $z \in \mathbb{Z}$, *there is exactly* one r_j *such that* $z \equiv r_j \,(\text{mod } n)$.

Prove that $m = n$.

2.2. Prove that if $a \in \mathbb{Z}$ is odd, then $a^2 \equiv 1 \,(\text{mod } 8)$.

2.3. Prove that if $a \in \mathbb{Z}$ is even, then $a^2 \equiv 0 \,(\text{mod } 4)$.

2.4. Let $a \in \mathbb{Z}$ and $n \in \mathbb{N}$. Provide a *counter*example to show that the following assertion is false.

If $a \equiv \pm 1 \,(\text{mod } p)$ for all primes p dividing n, then $a^2 \equiv 1 \,(\text{mod } n)$.

2.5. Prove that if $a^2 \equiv 1 \,(\text{mod } n)$, then $a \equiv \pm 1 \,(\text{mod } p)$ for all primes $p \mid n$.

2.6. Prove that any set of n consecutive integers determines a complete residue system modulo n.

2.7. Prove that if $\mathcal{R} = \{\overline{r_1}, \overline{r_2}, \ldots, \overline{r_n}\}$ is a set of congruence classes modulo $n \in \mathbb{N}$ such that $\overline{r_i} = \overline{r_j}$ if and only if $i = j$, then $\{r_1, r_2, \ldots, r_n\}$ is a complete residue system modulo n.

2.8. Let $\mathcal{R} = \{r_1, r_2, \ldots, r_n\}$ be a complete residue system modulo $n \in \mathbb{N}$, and $a \in \mathbb{N}$ with $\gcd(a, n) = 1$. Prove that for any integer b, $\{ar_1 + b, \ldots, ar_n + b\}$ is a complete residue system modulo n.

2.9. Suppose that $\mathcal{S} = \{s_1, s_2, \ldots, s_m\}$ is a complete residue system modulo $m \in \mathbb{N}$, and $\mathcal{R} = \{r_1, r_2, \ldots, r_n\}$ is a complete residue system modulo $n \in \mathbb{N}$. Prove that if $\gcd(m, n) = 1$, then

$$\{mr_i + ns_j : r_i \in \mathcal{R}, s_j \in \mathcal{S}, \text{ for } 1 \leq i \leq n, \text{ and } 1 \leq j \leq m\}$$

forms a complete residue system modulo mn.

☆ 2.10. If p is prime, prove that all the coefficients in $(1 - x)^{-p}$ are divisible by p except those coefficeints of x^j for $j = 0, 1, 2, \ldots$, which are congruent to 1 modulo p. (See Appendix A for a discussion of the results on series needed. In particular, see Examples A.10 on page 308 and A.12 on page 309 which will be needed for the solution. Also, you will need the fact (1.26), established in the proof of Fermat's Little Theorem on page 50. See the related Remark 2.7 on page 88.)

2.2 Modular Perspective

> *There is nothing ugly;* I never saw an ugly thing in my life: *for let the form of an object be what it may, — light, shade, and perspective will always make it beautiful.*
>   From C.R. Leslie **Memoirs of the Life of John Constable (1843) Chapter 17**
>   **John Constable (1776–1837), English painter**

The principal task of this section is to look back to some of the results that we obtained in Chapter 1 from our new perspective of modular arithmetic. First of all, we already reinterpreted one result in terms of congruences to verify the existence of solutions to

$$ax \equiv 1 \pmod{n}$$

on page 79. Now we restate it in its entirety.

Theorem 2.3 $\boxed{\text{Solutions of Linear Congruences}}$

Let $a, b \in \mathbb{Z}$ and $n \in \mathbb{N}$. Then

$$ax \equiv b \pmod{n} \tag{2.3}$$

has a solution $x \in \mathbb{Z}$ if and only if $g = \gcd(a, n) \mid b$. Furthermore, if such a solution exists, then there are exactly g incongruent solutions modulo n, and exactly one of these is in the least residue system modulo n/g, this being the unique solution modulo n/g of *(2.3).*

Proof. See Example 1.15 on page 28 for a proof of the first assertion. We also proved that if x and y are integers such that $ax + ny = g$, and we let $x_0 = bx/g$, then all solutions are of the form $x_0 + nz/g$ as z ranges over the integers. Now we need only verify that x_0 is the unique solution sitting in the least residue system modulo n/g and that there are exactly g incongruent solutions modulo n. Clearly from what was established in Example 1.15, we know that (2.3) has a solution if and only if $x \equiv x_0 \pmod{n/g}$, so we may select x_0 to be in the least residue system modulo n/g. Moreover, from the form of the solutions given above, we know that the solutions modulo n are precisely the values,

$$x = x_0 + mn/g \text{ for } m = 0, 1, 2, \ldots, g - 1,$$

where $m = 0$ gives the unique solution x_0.   □

Theorem 2.3, in conjunction with the notion of modular multiplicative inverses introduced in Definition 2.5 on page 80 is an excellent vehicle for solving linear congruences.

Example 2.6 *Suppose that we wish to solve $7x \equiv 1 \,(\text{mod } 9)$. Then we have that $x \equiv 7^{-1} \equiv 4 \,(\text{mod } 9)$ as the solution.*

The next result that lends itself well to modular interpretation is the Chinese Remainder Theorem introduced on page 40.

Theorem 2.4 | **Chinese Remainder Theorem** |

Let $n_i \in \mathbb{N}$ for natural numbers $j \leq k \in \mathbb{N}$ be pairwise relatively prime, set

$$n = \prod_{j=1}^{k} n_j$$

and let $r_i \in \mathbb{Z}$ for $i \leq k$. Then the system of k simultaneous linear congruences given by

$$x \equiv r_1 \pmod{n_1},$$
$$x \equiv r_2 \pmod{n_2},$$
$$\vdots$$
$$x \equiv r_k \pmod{n_k},$$

has a unique solution modulo n.

Proof. See Theorem 1.22 on page 40.   □

An application of the above was another problem that we introduced in Chapter 1, which we may now reinterpret and use the tools of this section to solve.

Example 2.7 *With reference to the statement of the Coconut problem on page 41, we begin by observing that the first sailor began with a pile of $n \equiv 1 \,(\text{mod } 3)$ coconuts. The second sailor began with a pile of*

$$m_1 = \frac{2(n-1)}{3} \equiv 1 \pmod{3}$$

coconuts, and the third sailor began with a pile of

$$m_2 = \frac{2(m_1 - 1)}{3} \equiv 1 \pmod{3}$$

coconuts, after which the three of them divided up the remaining pile of

$$m_3 = \frac{2(m_2 - 1)}{3} \equiv 1 \pmod{3}$$

coconuts. We calculate m_3 and get

$$m_3 = \frac{8}{27}n - \frac{38}{27} \equiv 1 \pmod{3}.$$

We now solve for n by multiplying through both sides and the modulus by 27, then simplifying to get $8n \equiv 65 \,(\text{mod } 81)$. (Note that each of n, m_1, m_2, and m_3 must be natural numbers.) Since the multiplicative inverse of 8 modulo 81 is 71, namely $8^{-1} \equiv 71 \,(\text{mod } 81)$, then $n \equiv 8^{-1} \cdot 65 \equiv 71 \cdot 65 \equiv 79 \,(\text{mod } 81)$, and the smallest solution is 79.

We generalized the Chinese Remainder Theorem in Example 1.20 on page 42 and we can now reinterpret that in terms of congruences.

Theorem 2.5 $\boxed{\textbf{Generalized Chinese Remainder Theorem}}$

Let $n_j \in \mathbb{N}$, set $\ell = \text{lcm}(n_1, n_2, \ldots, n_k)$, and let $r_j \in \mathbb{Z}$ be any integers for $j = 1, 2, \ldots, k$. Then the system of k simultaneous linear congruences given by

$$x \equiv r_1 \ (\text{mod } n_1),$$

$$x \equiv r_2 \ (\text{mod } n_2),$$

$$\vdots$$

$$x \equiv r_k \ (\text{mod } n_k),$$

has a solution if and only if

$$\gcd(n_i, n_j) \mid (r_i - r_j) \text{ for each pair of natural numbers } i, j \leq k.$$

Moreover, if a solution exists, then it is unique modulo ℓ. Additionally, if there exist integer divisors $m_j \geq 1$ of n_j with $\ell = m_1 \cdot m_2 \cdots m_k$ such that the m_j are pairwise relatively prime, and there exist integers

$$s_j \equiv 0 \ (\text{mod } \ell/m_j) \text{ and } s_j \equiv 1 \ (\text{mod } m_j) \text{ for } 1 \leq j \leq k,$$

then

$$x = \sum_{j=1}^{k} s_j r_j$$

is a solution of the above congruence system.

Proof. See Example 1.20 for a proof of the first assertion. We now establish uniqueness of solutions modulo ℓ.

Suppose that $x \equiv r_j \,(\text{mod } n_j)$ and $y \equiv r_j \,(\text{mod } n_j)$ for $1 \leq j \leq k$. Then $x - y \equiv 0 \,(\text{mod } n_j)$ for each such j. This means that $\ell \mid (x - y)$. Hence, any solution x is unique modulo ℓ.

The last statement of the theorem is clear since if such m_j and s_j exist, then

$$x = \sum_{j=1}^{k} s_j r_j \equiv r_j \ (\text{mod } m_j) \text{ for } 1 \leq j \leq k$$

has a unique solution modulo ℓ by the Chinese Remainder Theorem 1.22, and the proof is secured. □

In Example 1.20, we noted that Theorem 2.5 was a generalization of Theorem 1.22 by Yih-hing in 717 AD. We now present the problem for which he designed that generalization.

◆ The Units of Work Problem

Determine the number of completed units of work when the same number x of units to be performed by each of four sets of $2, 3, 6,$ and 12 workers performing their duties for certain numbers of whole days such that there remain $1, 2, 5,$ and 5 units of work not completed by the respective sets. We assume further that no set of workers is lazy, namely each completes a nonzero number of units of work.

Here we are looking to solve

$$x \equiv 1 \pmod{2}, \ x \equiv 2 \pmod{3}, \ x \equiv 5 \pmod{6}, \text{ and } x \equiv 5 \pmod{12}.$$

Since $\ell = \operatorname{lcm}(2,3,6,12) = 12$, then we let $m_1 = m_2 = 1$, $m_3 = 3$, and $m_4 = 4$. Thus, $s_1 = s_2 = 0$ since $m_1 = m_2 = 1$. Also, $s_3 = 4$ since $s_3 \equiv 0 \pmod{4}$ and $s_3 \equiv 1 \pmod{3}$; and $s_4 = 9$, since $s_4 \equiv 0 \pmod{3}$ and $s_4 \equiv 1 \pmod{4}$. Since $(r_1, r_2, r_3, r_4) = (1, 2, 5, 5)$, then $x = \sum_{j=1}^{4} r_j s_j = 5 \cdot 4 + 5 \cdot 9 = 65 \equiv 17 \pmod{12}$.

Note that we cannot choose $x = 5$ since this would mean that no units of work had been completed by the last two sets of workers. For $x = 17$, the completed units of work must be $8 \cdot 2 = 16$ for the first set since they do not complete one unit, $5 \cdot 3 = 15$ for the second set since they do not complete two units, $2 \cdot 6 = 12$ for the third set since they do not complete five units, and $1 \cdot 12 = 12$ for the fourth set for the same reason. Hence, the total completed units of work is 55, and Yih-hing's problem is solved.

Another result most amenable to restatement in terms of modular arithmetic is the following. In Section 1.5, we proved Thue's Theorem that we may now state in congruential terms.

Theorem 2.6 | Thue's Theorem

Suppose that $m \in \mathbb{N}$, $n \in \mathbb{Z}$ with $\gcd(m, n) = 1$, $m > 1$. Then there exist $x, y \in \mathbb{Z}$ such that $1 \leq x < \sqrt{m}$, and $1 \leq |y| < \sqrt{m}$ such that $nx \equiv y \pmod{m}$.

Proof. See Theorem 1.23 on page 44. □

The illustrations of Thue's Theorem in Section 1.5 are actually better stated in terms of congruences.

Example 2.8 *Example 1.21 on page 45, restated in terms of congruences, is given as follows. Let $m, n \in \mathbb{N}$, $m > 1$, such that $n^2 \equiv -1 \pmod{m}$. Then there exist unique $a, b \in \mathbb{N}$ such that $m = a^2 + b^2$ with $\gcd(a, b) = 1$, and $nb \equiv a \pmod{m}$.*

Example 2.9 *This is a reinterpretation of the result proved in Example 1.22 on page 46. Suppose that p is an odd prime such that $x^2 \equiv -2 \,(\text{mod } p)$ for some $x \in \mathbb{Z}$. Then there exist unique $a, b \in \mathbb{N}$ such that $p = a^2 + 2b^2$.*

Example 2.10 *Example 1.23 on page 47 has an interpretation in terms of modular arithmetic in the following fashion. Suppose that $p > 2$ is a prime such that $n^2 \equiv 2 \,(\text{mod } p)$. Then p can be written in the form $a^2 - 2b^2$ in infinitely many ways.*

Now we look at some famous results from §1.6 in congruential form.

Theorem 2.7 | Fermat's Little Theorem |

  If p is a prime and $n \in \mathbb{N}$ relatively prime to p, then $n^{p-1} \equiv 1 \,(\text{mod } p)$.

Proof. See Theorem 1.24 on page 50.   □

Remark 2.7 *We also observe that in the proof of Theorem 1.24, we proved the fact, which we express in terms of congruences here, that*

$$\binom{p}{j} \equiv 0 \pmod{p}$$

for any prime p and any integer $j = 1, 2, \ldots, p-1$.

Example 2.11 *If $n \equiv 3 \,(\text{mod } 4)$, then $x^2 \equiv -1 \,(\text{mod } n)$ has no solution $x \in \mathbb{Z}$. This is the modular version of Example 1.26 on page 51.*

We also have the following renowned result.

Theorem 2.8 | Wilson's Theorem |

  If $n \in \mathbb{N}$ with $n > 1$, then n is prime if and only if $(n-1)! \equiv -1 \,(\text{mod } n)$.

Proof. See Theorem 1.25 on page 51.   □

Example 2.12 *The modular version of Example 1.27 on page 53, as an application of Theorem 2.8 is as follows. If p is prime, then $x^2 \equiv -1 \,(\text{mod } p)$ for some integer x if and only if either $p = 2$ or $p = 4m+1$ for some positive integer m.*

Remark 2.8 *Examples 2.8–2.12 are all instances of the solvability, or lack thereof, of quadratic congruences, to get what are called quadratic residues, or nonresidues, respectively. This is a principal topic of Chapter 4 where we study quadratic reciprocity in depth. See also Remark 2.6 on page 81.*

◆ **Residue Computers**

We close this section with a brief discussion of applications of modular arithmetic to computer design. *Residue computers* are specific, high-speed computers based upon the CRT, and so they employ modular arithmetic. These computers are far more efficient than ordinary binary computers since they can perform faster additions, subtractions, and multiplications. The reason is that if $n = \prod_{j=1}^{k} p_j^{a_j}$ is the canonical prime factorization of a modulus, n, then residue computers perform computations in each $\mathbb{Z}/p_j^{a_j}\mathbb{Z}$ (or indeed in any $\mathbb{Z}/n_i\mathbb{Z}$ for $n_i | n$), which is easier than performing computations in $\mathbb{Z}/n\mathbb{Z}$. Moreover, additions, subtractions, and multiplications in $\mathbb{Z}/p_j^{a_j}\mathbb{Z}$ are *carry free*, which makes them faster than binary computers that are hampered by *carry propagation* and time delays. Indeed, the CRT allows the decomposition of big computations in $\mathbb{Z}/n\mathbb{Z}$ into small computations in each $\mathbb{Z}/p_j^{a_j}\mathbb{Z}$ that is an example of what computer scientists call *divide-and-conquer* used in their design of algorithms. In fact, these residue computers have important applications in digital signal processing, and *very large-scale integration* (VLSI), which is the current level of computer microchip miniaturization. For instance, see [28].

Exercises

2.11. Prove that $n > 1$ is prime if and only if $(n-2)! \equiv 1 \pmod{n}$.

In Exercises 2.12–2.15, find the solution in the least residue class for the given modulus.

2.12. $2x \equiv 1 \pmod 9$.

2.13. $7x \equiv 5 \pmod{27}$.

2.14. $4x \equiv 15 \pmod{101}$.

2.15. $101x \equiv 5 \pmod{103}$.

In Exercises 2.16–2.19, find the unique solution for each modulus in the given systems of linear congruences.

2.16. $x \equiv 3 \pmod 4$, $x \equiv 4 \pmod 5$, $x \equiv 5 \pmod 7$.

2.17. $x \equiv 2 \pmod 3$, $x \equiv 6 \pmod 7$, $x \equiv 7 \pmod{11}$.

2.18. $x \equiv 1 \pmod{13}$, $x \equiv 2 \pmod{19}$, $x \equiv 3 \pmod{23}$.

2.19. $x \equiv 1 \pmod 3$, $x \equiv 2 \pmod 7$, $x \equiv 3 \pmod{11}$, $x \equiv 4 \pmod{13}$.

☆ 2.20. Prove that if p is prime then $p + 2$ is prime if and only if

$$4((p-1)! + 1) \equiv -p \pmod{p^2 + 2p}.$$

2.21. Prove that Theorems 2.7–2.8 on page 88 both hold if and only if

$$(p-1)! n^p \equiv -n \pmod p.$$

2.3   Arithmetic Functions: Euler, Carmichael, and Möbius

Still glides the Stream, and shall for ever glide; The Form remains, the Function never dies.

From **The River Duddon (1820) no. 34 "After-Thought"**
William Wordsworth (1770–1850), English poet

First we formalize what the title of this section means.

Definition 2.6 | **Arithmetic/Number-Theoretic Functions**

Arithmetic *or* number-theoretic *functions are any functions whose domain is* $\mathbb{N}$ *and whose range is a subset of* $\mathbb{C}$. *An arithmetic function* f *is said to be* multiplicative *if*

$$f(mn) = f(m)f(n) \text{ for all } m, n \in \mathbb{N} \text{ such that } \gcd(m,n) = 1.$$

If

$$f(mn) = f(m)f(n) \text{ for all } m, n \in \mathbb{N},$$

f *is said to be* completely multiplicative.

Note that multiplicative functions are completely determined by their values on prime powers, so once it is known that a function is multiplicative, the next step is to formulate its values on prime powers. Most often the arithmetic functions we study are real-valued functions. The first that we will study is a renowned entity in mathematics with a multitude of applications.

Definition 2.7 | **Euler's ϕ-Function**

For any $n \in \mathbb{N}$ *the* Euler ϕ-function, *also known as* Euler's Totient (*see Biographies 1.17 on page 56 and 2.1 on the facing page*), $\phi(n)$ *is defined to be the number of* $m \in \mathbb{N}$ *such that* $m < n$ *and* $\gcd(m,n) = 1$.

Note that Gauss introduced the symbol $\phi(n)$ (see [16, Articles 38–39, pp. 20–21]) to denote the totient.

Example 2.13 If p is prime, then any $j \in \mathbb{N}$ with $j < p$ is relatively prime to p, so $\phi(p) = p - 1$.

Example 2.14 *Let* $n \in \mathbb{N}$. *Then the cardinality of* $(\mathbb{Z}/n\mathbb{Z})^*$ *is* $\phi(n)$. *See* (2.2) *on page 81.*

Biography 2.1 James Joseph Sylvester (1814–1897) *gave the name* totient *to the function* $\phi(n)$. *He defined the* totatives *of n to be the natural numbers* $m < n$ *relatively prime to n. Sylvester was born in London, England, on September 3, 1814. He taught at University of London from 1838 to 1841 with his former teacher Augustus De Morgan (1806–1871). Later he left mathematics to work as an actuary and a lawyer. This brought him into contact with Arthur Cayley (1821–1895) who also worked the courts of Lincoln's Inn in London, and thereafter they remained friends. Sylvester returned to mathematics, being appointed professor of mathematics at the Military Academy at Woolrich in 1854. In 1876 he accepted a position at the newly established Johns Hopkins University, where he founded the first mathematical journal in the U.S.A., the* American Journal of Mathematics. *In 1883, he was offered a professorship at Oxford University. This position was to fill the chair left vacant by the death of the Irish number theorist* Henry John Stephen Smith (1826–1883). *When his eyesight began to deteriorate in 1893, he retired to live in London. Nevertheless, his enthusiasm for mathematics remained until the end as evidenced by the fact that in 1896 he began work on Goldbach's Conjecture (which says that every even integer $n > 2$ is a sum of two primes — see also page 66). He died in London on March 15, 1897, from complications involving a stroke.*

Euler's totient allows us to extend our theory of residues introduced in §2.1 as follows.

Definition 2.8 | Reduced Residue Systems |

If $n \in \mathbb{N}$, then any set of $\phi(n)$ integers, incongruent modulo n, and relatively prime to n, is called a reduced residue system modulo n.

Example 2.15 *The set $\{1, 3, 7, 9\}$ is a reduced residue system modulo 10 since $\phi(10) = 4$, each element of the set is relatively prime to 10, and they are incongruent modulo 10.*

Definition 2.8 allows us to prove the multiplicativity of the totient.

Theorem 2.9 | The Totient is Multiplicative |

If $m, n \in \mathbb{N}$ are relatively prime, then

$$\phi(mn) = \phi(m)\phi(n).$$

Therefore, if $n = \prod_{j=1}^{k} p_j^{a_j}$ where the p_j are distinct primes, then

$$\phi(n) = \prod_{j=1}^{k} (p_j^{a_j} - p_j^{a_j - 1}) = \prod_{j=1}^{k} \phi(p_j^{a_j}).$$

Proof. We begin by proving the following assertion.

Claim 2.1 *Let* $S = \{s_1, s_2, \ldots, s_{\phi(m)}\}$ *be a reduced residue system modulo* $m \in \mathbb{N}$, *and* $\mathcal{R} = \{r_1, r_2, \ldots, r_{\phi(n)}\}$ *be a reduced residue system modulo* $n \in \mathbb{N}$. *Then a reduced residue system modulo* mn *is obtained from all solutions of the pair of congruences*

$$x \equiv r_i \ (\text{mod } n) \ \text{and } x \equiv s_j \ (\text{mod } m) \ \text{for some } i, j \in \mathbb{N}. \qquad (2.4)$$

If x is in a reduced residue system modulo mn, then $\gcd(x, m) = \gcd(x, n) = 1$, so there exist $i, j \in \mathbb{N}$ such that $x \equiv r_i \,(\text{mod } n)$ and $x \equiv s_j \,(\text{mod } m)$. Conversely, if there exist $i, j \in \mathbb{N}$ such that $x \equiv r_i \,(\text{mod } n)$ and $x \equiv s_j \,(\text{mod } m)$, then $\gcd(x, mn) = 1$. By Theorem 2.4 on page 85, the Chinese Remainder Theorem expressed in congruential form, there is a unique solution $x \in \mathbb{Z}$ to (2.4). This is Claim 2.1.

Since each distinct pair (i, j) yields a distinct solution x of (2.4), and since there are $\phi(m)\phi(n)$ such pairs, then by Claim 2.1,

$$\phi(mn) = \phi(m)\phi(n),$$

and we have the multiplicativity of the totient.

It follows immediately from the mutiplicative property that $\phi(n) = \prod_{j=1}^{k} \phi(p_j^{a_j})$. Thus, we need only prove that for each $j = 1, 2, \ldots, k$,

$$\phi(p_j^{a_j}) = p_j^{a_j} - p_j^{a_j - 1}. \qquad (2.5)$$

However, this is true since those natural numbers less than or equal to $p_j^{a_j}$ and divisible by p_j are precisely those $k = ip$ for $i = 1, 2, \ldots, p_j^{a_j - 1}$, so there are $p_j^{a_j - 1}$ of them. Hence, Equation (2.5) holds and the theorem is proved. □

Remark 2.9 *What is implicit in Claim 2.1 is that if* $\gcd(m, n) = 1$, *then when* $S = \{s_1, s_2, \ldots, s_{\phi(m)}\}$ *is a reduced residue system modulo* $m \in \mathbb{N}$, *and* $\mathcal{R} = \{r_1, r_2, \ldots, r_{\phi(n)}\}$ *is a reduced residue system modulo* $n \in \mathbb{N}$, *we must have that*

$$\{mr_i + ns_j : r_i \in \mathcal{R}, s_j \in S, \text{ for } 1 \leq i \leq \phi(n), \text{ and } 1 \leq j \leq \phi(m)\}$$

forms a reduced residue system modulo mn.

Corollary 2.1 *For any* $n = \prod_{j=1}^{k} p_j^{a_j}$,

$$\phi(n) = n \prod_{j=1}^{k} \left(1 - \frac{1}{p_j}\right).$$

Proof. By Theorem 2.9

$$\phi(n) = \prod_{j=1}^{k} \left(p_j^{a_j} - p_j^{a_j-1}\right) = \prod_{j=1}^{k} p_j^{a_j}\left(1 - \frac{1}{p_j}\right) = n\prod_{j=1}^{k}\left(1 - \frac{1}{p_j}\right),$$

as required. □

Remark 2.10 *If the set*

$$\mathcal{R} = \{r_1, \ldots, r_{\phi(n)}\}$$

is a reduced residue system modulo n, then so is

$$\mathfrak{R} = \{mr_1, \ldots, mr_{\phi(n)}\}$$

for $m \in \mathbb{N}$ with $\gcd(m, n) = 1$. To see this, note that since

$$\gcd(m, n) = \gcd(r_j, n) = 1,$$

then

$$\gcd(mr_j, n) = 1 \text{ for all natural numbers } j \leq \phi(n).$$

If

$$mr_j \equiv mr_k \pmod{n}$$

for some $j \neq k$ with $1 \leq j, k \leq \phi(n)$, then

$$r_j \equiv r_k \pmod{n},$$

by Proposition 2.3 on page 75, a contradiction.

Remark 2.10 sets the stage for one of the features of this section, namely a generalization of Fermat's Little Theorem.

Theorem 2.10 | **Euler's Generalization of Fermat's Little Theorem**

If $n \in \mathbb{N}$ and $m \in \mathbb{Z}$ such that $\gcd(m, n) = 1$, then

$$m^{\phi(n)} \equiv 1 \pmod{n}.$$

Proof. By the discussion in Remark 2.10, each element in $\mathcal{R}$ is congruent to a unique element in $\mathfrak{R}$ modulo n. Hence,

$$\prod_{j=1}^{\phi(n)} r_j \equiv \prod_{j=1}^{\phi(n)} mr_j \equiv m^{\phi(n)} \prod_{j=1}^{\phi(n)} r_j \pmod{n},$$

and $\gcd(\prod_{j=1}^{\phi(n)} r_j, n) = 1$, so

$$m^{\phi(n)} \equiv 1 \pmod{n},$$

by Proposition 2.3 on page 75. □

Remark 2.11 *By Euler's Theorem, any integer m relatively prime to $n \in \mathbb{N}$ satisfies that $m^{\phi(n)} \equiv 1 \pmod{n}$. In other words, $m \cdot m^{\phi(n)-1} \equiv 1 \pmod{n}$. This says that $m^{\phi(n)-1}$ is a multiplicative inverse of m modulo n. This makes the finding of such inverses a simpler task, especially in view of the ability to use the repeated squaring method introduced on page 82. For instance, we have $3^{\phi(35)-1} \equiv 3^{24-1} = 3^{23} \equiv 12 \pmod{35}$, and 12 is a (least) multiplicative inverse of 3 modulo 35.*

Another elegant result on the totient was proved by Gauss as follows. In this result, we introduce the symbols $\sum_{d|n}$ and $\bigcup_{d|n}$. These denote the sum and set-theoretic union, respectively, expressed over all positive divisors d of n.

Theorem 2.11 | **Sum of the Totients at Divisors** |
   If $n \in \mathbb{N}$, then $\sum_{d|n} \phi(d) = n$.

Proof. By Claim 1.3 on page 26, $\gcd(m, n) = d$ if and only if $\gcd(m/d, n/d) = 1$. Therefore, $\phi(n/d)$ is the *cardinality* of $\mathcal{T}_d = \{m \in \mathbb{N} : m \leq n, \gcd(m, n) = d\}$, namely the number of elements in $\mathcal{T}_d$, denoted by $|\mathcal{T}_d|$. Since

$$\{1, 2, \ldots, n\} = \bigcup_{d|n} \mathcal{T}_d,$$

the set-theoretic union of all $\mathcal{T}_d$ as d ranges over all divisors of n, then

$$n = \left| \bigcup_{d|n} \mathcal{T}_d \right| = \sum_{d|n} |\mathcal{T}_d| = \sum_{d|n} \phi(d)$$

where the second equality follows from the fact that the $\mathcal{T}_d$ are disjoint. This is the desired result. $\square$

Another important arithmetic function related to the totient is given as follows — see Biography 2.2 on the next page.

Definition 2.9 | **Carmichael's Lambda Function** |
   Let $n = 2^a \prod_{j=1}^{k} p_j^{a_j}$ be the canonical prime factorization of n. If $\lambda(1) = 1$, and

$$\lambda(n) = \begin{cases} \phi(n) & \text{if } n = 2^a, \text{ and, } 1 \leq a \leq 2, \\ 2^{a-2} = \phi(n)/2 & \text{if } n = 2^a, a > 2, \\ \mathrm{lcm}(\lambda(2^a), \phi(p_1^{a_1}), \ldots, \phi(p_k^{a_k})) & \text{if } k \geq 1, \end{cases}$$

   then λ is called Carmichael's function.

Later we will see the role that the Carmichael function plays in cryptography, in particular for factoring — see §4.3, especially Exercise 4.27 on page 208.

> **Biography 2.2** Robert Daniel Carmichael (1879–1967) *was born in Goodwa-ter, Alabama. He earned his bachelor's degree from Lineville College in 1898. In 1911, he received his doctorate from Princeton under the direction of G.D. Birkhoff. His thesis on differential equations was praised as an extraordinary contribution to the area. In 1912, he conjectured that there are infinitely many of the numbers that now bear his name. In 1992, W. Alford, A. Granville, and C. Pomerance proved his conjecture, (see [3] and [18, p.30].) Other than Carmichael numbers (see the parenthetical comment at the end of Exercise 2.23 on page 101), and Carmichael's function, there is also Carmichael's Theorem, the latter stating that whenever* $\gcd(m,n) = 1$*, we have* $m^{\lambda(n)} \equiv 1 \pmod{n}$*. However, he worked, not only in number theory, but also in differential equa-tions, group theory, and physics. Yet he is known, as well, for his very acces-sible books* Theory of Numbers, *published in 1914, and* Diophantine analysis, *published in 1915.*

There is a basic property of arithmetic functions that we now establish as a segue to the introduction of another important such function. First, we need a new concept.

Definition 2.10 | **Summatory Functions**

If f is an arithmetic function, then

$$F(n) = \sum_{d|n} f(d),$$

is called the summatory function *of f.*

Theorem 2.12 | **Summatory of Multiplicative Functions**

If f is a multiplicative arithmetic function, then $F(n) = \sum_{d|n} f(d)$ is multi-plicative.

Proof. Suppose that f is a multiplicative function. If $\gcd(m,n) = 1$, then any divisor d of mn can be written uniquely as the product of $d_1|m$, $d_2|n$, where $\gcd(d_1, d_2) = 1$. Therefore,

$$F(mn) = \sum_{d|mn} f(d) = \sum_{d_1|m, d_2|n} f(d_1 d_2) = \sum_{d_1|m, d_2|n} f(d_1)f(d_2) =$$

$$\sum_{d_1|m} f(d_1) \sum_{d_2|n} f(d_2) = F(m)F(n),$$

as required. □

Now we introduce the important arithmetic function that will allow us to characterize other arithmetic functions in terms of it — see Biography 2.3 on page 99.

Definition 2.11 | The Möbius Function |

If $n \in \mathbb{N}$, then the Möbius function is given by

$$\mu(n) = \begin{cases} 1 & \text{if } n = 1, \\ 0 & \text{if } n \text{ is not squarefree}, \\ (-1)^k & \text{if } n = \prod_{j=1}^{k} p_j \text{ where the } p_j \text{ are distinct primes}. \end{cases}$$

Theorem 2.13 | Möbius is Multiplicative |

The Möbius function μ is multiplicative.

Proof. Let $m = \prod_{i=1}^{r} p_i^{a_i}$ and $n = \prod_{j=1}^{s} q_j^{b_j}$ be the canonical prime factorizations of the relatively prime natural numbers m and n. Then $mn = \prod_{i=1}^{r} p_i^{a_i} \prod_{j=1}^{s} q_j^{b_j}$, where $p_i \neq q_j$ for any i, j since $\gcd(m, n) = 1$. If both $m > 1$ and $n > 1$, then whenever any $a_i > 1$ or $b_j > 1$, say $a_i > 1$ without loss of generality, then we have $\mu(mn) = 0 = 0 \cdot \mu(n) = \mu(m)\mu(n)$. If all $a_i = b_j = 1$, then $\mu(m) = (-1)^r$ and $\mu(n) = (-1)^s$, so $\mu(m)\mu(n) = (-1)^r(-1)^s = (-1)^{r+s} = \mu(mn)$. If $m = 1$ or $n = 1$, say, $m = 1$ without loss of generality, then $\mu(mn) = \mu(n) = 1 \cdot \mu(n) = \mu(m)\mu(n)$. Hence, μ is multiplicative. $\square$

Based upon the above, we may now prove the following.

Theorem 2.14 | Möbius Summatory |

For any $n \in \mathbb{N}$,

$$\sum_{d \mid n} \mu(d) = \begin{cases} 0 & \text{if } n > 1, \\ 1 & \text{if } n = 1. \end{cases}$$

Proof. By Theorems 2.12–2.13, the Möbius summatory $M(n) = \sum_{d \mid n} \mu(d)$ is multiplicative, and since for any prime p such that $p \mid n$,

$$M(p^a) = \begin{cases} 1 & \text{if } a = 0, \\ 1 - 1 + 0 + 0 + \cdots + 0 = 0 & \text{if } a > 1, \end{cases}$$

then $M(n) = 0$ if $p \mid n$ for any prime p. $\square$

Remark 2.12 *There is another way of viewing the Möbius sum in Theorem 2.14 that will shed more light on the issue. Let $n = \prod_{j=1}^{k} p_j^{a_j}$ be the canonical prime factorization of the natural number $n > 1$, and for a nonnegative integer $m \in \mathbb{Z}$, set $N = \prod_{j=1}^{k} (1 - p_j^m)$. When written as a sum, N consists of the terms 1 and $\pm d^m$ where d is a divisor of n consisting of distinct prime divisors. The coefficient of d^m is $+1$ when d has an even number of prime factors and is -1 otherwise. Hence, by the definition of the Möbius function, that coefficient* is $\mu(d)$. *Furthermore, we have from the definition of μ, that $a_j = 1$ for all $j = 1, 2, \ldots, k$ if and only if $d = 1$ or d is a product of distinct primes. Hence,*

$$N = (1 - p_1^m)(1 - p_2^m) \cdots (1 - p_k^m) = \sum_{d|n} \mu(d) d^m. \tag{2.6}$$

Thus, for $m = 0$ we get that $\sum_{d|n} \mu(d) = 0$ whenever $n > 1$.

Given the above, we may ask what happens when we sum over only the absolute values of the Möbius function. The answer is given as follows.

Theorem 2.15 | Möbius on Absolute Values |

Let $n \in \mathbb{N}$ be divisible by $k \geq 0$ distinct prime divisors. Then

$$\sum_{d|n} |\mu(d)| = 2^k.$$

Proof. If $n = 1$, then the result is clear from the definition of μ, so we assume that $n > 1$. Set $n_1 = p_1 p_2 \cdots p_k$, where the p_j are the distinct prime divisors of n. Then $\sum_{d|n} |\mu(d)| = \sum_{d|n_1} |\mu(d)|$. Since there are $\binom{k}{\ell}$ distinct divisors $d_j^{(\ell)}$ of n_1 containing exactly $\ell \leq k$ prime factors, then $\mu(d_j^{(\ell)}) = (-1)^\ell$. Therefore, $\sum_{j=1}^{\binom{k}{\ell}} |\mu(d_j^{(\ell)})| = \binom{k}{\ell}$, so $\sum_{d|n} |\mu(d)| = \sum_{j=0}^{k} \binom{k}{\ell} = 2^k$, where the last equality comes from the full summation property of the binomial coeffiicient — see Exercise 1.17 on page 14. □

Remark 2.13 *We may even look at a slightly more general version of the Möbius summatory by setting*

$$\mathcal{M}(x) = \sum_{n \leq x} \mu(n),$$

where x is any real value and the n ranges over all positive integers no bigger than x. (Thus, $\mathcal{M}(x) = 0$ for $x < 1$.) In this case, we get, for $x \geq 2$,

$$|\mathcal{M}(x)| \leq \sum_{n \leq x} |\mu(n)| \leq \sum_{n \leq x} 1 \leq x,$$

namely that $\mathcal{M}(x) = O(x)$, sometimes referenced as $\mathcal{M}(x)$ belongs to $O(x)$. In fact, more can be proved, namely,

$$\mathcal{M}(x) = \sum_{n \leq x} \mu(n) = o(x), \tag{2.7}$$

where $o(x)$ means $\lim_{x \to \infty} \mathcal{M}(x)/x = 0$, sometimes referenced as $\mathcal{M}(x)$ belongs to $o(x)$. (See Appendix B for a detailed discussion of the little "oh" and big "oh" notations.) *Equation (2.7) follows from the Prime Number Theorem and the verifiable fact that if $\varepsilon > 0$ is arbitrarily small, then for all sufficiently large x, $|\mathcal{M}(x)| < \varepsilon x$. If we write $\mathcal{M}(n) = \sum_{j=1}^n \mu(j)$, then this is known as the Mertens function — see Biography 2.4 on page 100. Merten's conjectured that $|\mathcal{M}(n)| < \sqrt{n}$ for any $n > 1$. Indeed, in 1983, a computer search verified the conjecture for all $n \leq 10^9$. However, it fell in 1984, when Andrew Odlyzko and Herman te Riele provided an indirect proof that the conjecture is false. We say indirect since no actual counterexample was found. Instead they proved that a value of n must exist for which $|\mathcal{M}(n)| \geq \sqrt{n}$, and later this was refined to show that the value must sit in the range of $n \leq 3.21 \cdot 10^{64}$. For another interpretation of Merten's function, see Exercise 2.26 on page 101.*

Theorem 2.16 | **The Möbius Inversion Formula**

If f and g are arithmetic functions, then

$$f(n) = \sum_{d|n} g(d) \text{ for every } n \in \mathbb{N},$$

if and only if

$$g(n) = \sum_{d|n} \mu(d) f\left(\frac{n}{d}\right) \text{ for every } n \in \mathbb{N}.$$

Proof. If $f(n) = \sum_{d|n} g(d)$ for every $n \in \mathbb{N}$, then

$$\sum_{d|n} \mu(d) f\left(\frac{n}{d}\right) = \sum_{d_1 d_2 = n} \mu(d_1) f(d_2) = \sum_{d_1 d_2 = n} \mu(d_1) \sum_{d'|d_2} g(d') =$$

$$\sum_{d_1 d' e = n} \mu(d_1) g(d') = \sum_{d' e' = n} g(d') \sum_{d_1 | e'} \mu(d_1).$$

However, by Theorem 2.14, $\sum_{d_1 | e'} \mu(d_1) = 0$ if $e' > 1$, and is 1 if $e' = 1$. Hence, $\sum_{d|n} \mu(d) f\left(\frac{n}{d}\right) = g(n)$.

Conversely, if $g(n) = \sum_{d|n} \mu(d) f\left(\frac{n}{d}\right)$, then

$$\sum_{d|n} g(d) = \sum_{d|n} \sum_{d_1|d} \mu(d_1) f\left(\frac{d}{d_1}\right) = \sum_{d_1 d_2 e = n} \mu(d_1) f(d_2) = \sum_{d_2 d' = n} f(d_2) \sum_{d_1 | d'} \mu(d_1).$$

Again, by Theorem 2.14, $\sum_{d_1 | d'} \mu(d_1) = 0$ if $d' > 1$, and is 1 if $d' = 1$. Hence, $\sum_{d|n} g(d) = f(n)$, and we have the formula. $\square$

Biography 2.3 August Ferdinand Möbius (1790–1868) *was born in Schulp-forta in Prussia, close to Naumburg, Germany. His mother was a descendant of Martin Luther, and his father was a dancing teacher. He had some informal mathematical training at home until his early teen years. In 1809, he began his formal mathematical training when he entered Leipzig University to study law. However, his interests turned to mathematics and related areas such as astronomy. His interests eventually led him to Göttingen, where he studied astronomy with Gauss. He also studied mathematics at Halle with Pfaff. In 1815, he was appointed lecturer at Leipzig, and was promoted to professor of astronomy in 1844, a position he held for the rest of his life. He is best known for his work in topology, especially for the one-sided surface that bears his name — the Möbius strip obtained by joining the ends of a strip of paper together after giving it a half twist. He is also well-known for his map-colouring ideas that led to the four-colour problem. His contributions were not only to number theory and astronomy, but also to mechanics, projective geometry, optics, and statistics.*

We may now apply the Möbius formula to the totient.

Theorem 2.17 | **Möbius and the Totient**

For any $n \in \mathbb{N}$, and distinct primes p dividing n,

$$\phi(n) = n \sum_{d|n} \frac{\mu(d)}{d} = n \prod_{p|n} \left(1 - \frac{1}{p}\right).$$

Proof. First, set $f(n) = n = \sum_{d|n} \phi(d)$, where the last equality comes from Theorem 2.11 on page 94. Then we invoke the Möbius Inversion Formula, Theorem 2.16 on the facing page to get

$$\phi(n) = \sum_{d|n} \mu(d)f(n/d) = \sum_{d|n} \frac{\mu(d)n}{d} = n \sum_{d|n} \frac{\mu(d)}{d}.$$

The second equality follows from Theorem 2.11 on page 94. □

We close this section with an illustration of how to use the technique of interchanging an order of summation and the, rarely mentioned, fact that the Möbius function is a detector of relatively prime integers due to the summatory property given in Theorem 2.14.

Example 2.16 *We wish to evaluate the following sum via the above-mentioned technique. We have, for $n > 1$ that*

$$\sum_{\substack{1 \le k \le n \\ \gcd(k,n)=1}} k = \sum_{1 \le k \le n} k \sum_{d | \gcd(k,n)} \mu(d) = \sum_{1 \le k \le n} \sum_{\substack{d|n \\ d|k}} k\mu(d) = \sum_{d|n} \sum_{1 \le q \le n/d} dq\mu(d) =$$

$$\sum_{d|n} d\mu(d) \sum_{1 \le q \le n/d} q = \sum_{d|n} d\mu(d) \frac{\frac{n}{d}(\frac{n}{d}+1)}{2} = \frac{1}{2}\left(\sum_{d|n} \frac{\mu(d)n^2}{d} + \sum_{d|n} n\mu(d)\right),$$

where the penultimate equality comes from Theorem 1.1 on page 2 acting on the previous second sum. We now invoke Theorem 2.17 on the page before and Theorem 2.14 on page 96 to get that the above equals $\frac{1}{2}n^2 \frac{\phi(n)}{n} = \frac{n\phi(n)}{2}$. Hence, we have demonstrated the important fact that for $n > 1$,

$$\sum_{\substack{1 \le k \le n \\ \gcd(k,n)=1}} k = \frac{n\phi(n)}{2}.$$

Biography 2.4 Franz Carl Joseph Mertens (1840–1927) *was born on March 20, 1840 in Schroda, Posen, Prussia (now Sroda, Poland). Mertens studied at the University of Berlin with Kronecker and Kummer as his advisors, obtaining his doctorate, on potential theory, in 1865. His first position was at the Jagiellonian University at Cracow, and he worked his way up to ordinary professor by 1870. He also held positions at the Polytechnic in Graz, Austria, and the University of Vienna from which he retired in 1911. Among his students at Vienna were Ernst Fischer (1875–1954), and Eduard Helly (1884–1943). Fischer is best known for the Riesz-Fischer theorem in the theory of Lebesgue integration, and Helly proved the Hahn-Banach theorem in 1912, some fifteen to twenty years before Hahn and Banach provided their versions. Mertens' areas of interest included not only number theory and potential theory, but also geometric applications to algebra and matrix theory. Other than the conjecture that bears his name, he is known for his elementary proof of Dirichlet's Theorem — see Biography 1.8 on page 35 and Theorem 1.19 on page 35. He also has his name attached to three number-theoretic results on density of primes, one of which is an asymptotic formula for the fraction of natural numbers not divisible by the primes less than a given x. Although his conjecture was proved to be false, as noted above, it stood for almost a century before it fell. It is unfortunate since a proof of his conjecture would have meant that the Riemann hypothesis is true — see page 72. Merten's died on March 5, 1927 in Vienna.*

Exercises

2.22. Prove that if $d \mid n \in \mathbb{N}$, then $\phi(d) \mid \phi(n)$. Use this fact to show that $2|\phi(n)$ for any $n > 2$.

2.23. Prove that

$$\sum_{j=1}^{p-1} j^{p-1} \equiv -1 \pmod{p}$$

for any prime p. (*It is an open question as to whether* $\sum_{j=1}^{n-1} j^{n-1} \equiv -1$ (mod n) *for a given* $n \in \mathbb{N}$ *implies that* n *is prime. However, it has been verified up to* 10^{1700}. *See* [18, p. 37]).

(It has been observed that if the converse to Exercise 2.23 *fails* to hold for some n, then that number would be a *Carmichael number*, which are defined to be those composite integers $n \in \mathbb{N}$ such that $b^{n-1} \equiv 1 \pmod{n}$ for all $b \in \mathbb{N}$ such that $\gcd(b, n) = 1$. These are also called *absolute pseudoprimes*. We will study these values later in §2.7, when we look at applications to primality testing.)

☆ 2.24. Let $n \in \mathbb{N}$. Prove that for all $a \in \mathbb{Z}$, $b^b \equiv a \pmod{n}$ for some $b \in \mathbb{N}$ if and only if $\gcd(n, \phi(n)) = 1$.

2.25. Prove that if n is composite and $\phi(n) \mid (n-1)$, then n is squarefree — see Definition 1.10 on page 30.

2.26. Given $n \in \mathbb{N}$, let $m_o \leq n$ be the number of squarefree natural numbers with an odd number of prime divisors, and let $m_e \leq n$ be the number with an even number of prime divisors. Prove that

$$\mathcal{M}(n) = m_e - m_o,$$

where $\mathcal{M}(n)$ is defined in Remark 2.13 on page 97.

2.27. Evaluate $\sum_{j=1}^{\infty} \mu(j!)$.

2.28. Prove that for any $n \in \mathbb{N}$,

$$\prod_{j=0}^{3} \mu(n+j) = 0.$$

2.29. Prove that if $k \geq 0$ is the number of distinct prime factors of n, then

$$\sum_{d \mid n} \mu^2(d) = 2^k.$$

In Exercises 2.30–2.31,

$$f * g = \sum_{d \mid n} f(d) g(n/d) = \sum_{d_1 d_2 = n} f(d_1) g(d_2)$$

is defined as the *Dirichlet product* for given arithmetic functions f and g. It is straightforward to see that $f*g$ is associative, commutative, and is an arithmetic function itself. (Indeed it can be shown that the set of all arithmetic functions f with $f(1) \neq 0$ forms a group under Dirichlet multiplication.) If $f * g = I = g * f$ where $I(n) = 1$ if $n = 1$ and $I(n) = 0$ otherwise, then f and g are *inverse functions* of one another, each called the Dirichlet inverse of the other. Also, let $u(n) = n$ for all $n \in \mathbb{N}$. Prove the following.

2.30. $f * u(n) = \sum_{d \mid n} f(d)$.

2.31. $\mu(n)$ is the Dirichlet inverse of $u(n)$.

2.4   Number and Sums of Divisors

> *That all things are changed, and that nothing really perishes, and that the sum of matter remains exactly the same, is sufficiently certain.*
>   From **The Works of Francis Bacon (1858)**, J. Spedding (ed.)
> **Francis Bacon (1561–1626), English lawyer, courtier, philosopher, and essayist**

We continue the study of arithmetic functions introduced in §2.3. The two functions in the header of this section are given as follows.

Definition 2.12 | Sum of Divisors |

For any $n \in \mathbb{N}$, the sum of the positive divisors of n is denoted by $\sigma(n)$, called the sum of divisors *function*.

Example 2.17 $\sigma(24) = 1 + 2 + 3 + 4 + 6 + 8 + 12 + 24 = 60$, and $\sigma(21) = 1 + 3 + 7 + 21 = 32$.

Definition 2.13 | Number of Divisors |

For any $n \in \mathbb{N}$, the number of positive divisors of n is denoted by $\tau(n)$, called the number of divisors *function*.

Example 2.18 $\tau(5) = 2$, since only 1 and 5 divide 5. Indeed, for any prime p, $\tau(p) = 2$. $\tau(8) = 4$ since only $1, 2, 4, 8$ divide 8.

Remark 2.14 *We do not need to prove that σ and τ are multiplicative, since we have proved a general result from which this follows, namely, Theorem 2.12 on page 95. To see this, merely let $f(n) = n$ and $g(n) = 1$, both of which are multiplicative, so $\sigma(n) = \sum_{d|n} f(d)$ and $\tau(n) = \sum_{d|n} g(n)$ are multiplicative by Theorem 2.12.*

Now we seek formulas for σ and τ. Given Remark 2.14, we need only work with prime powers to achieve this goal.

Theorem 2.18 | σ and τ on Prime Powers |

If p is a prime and $k \in \mathbb{N}$, then

$$\sigma(p^k) = \frac{p^{k+1} - 1}{p - 1},$$

and

$$\tau(p^k) = k + 1.$$

Proof. For $n \in \mathbb{N}$,

$$\sigma(p^k) = 1 + p + p^2 + \cdots + p^k = \frac{p^{k+1} - 1}{p - 1}, \tag{2.8}$$

by Theorem 1.2 on page 2, and since Equation (2.8) shows that p^k has exactly $k + 1$ divisors, then

$$\tau(p^k) = k + 1,$$

as required. $\qquad\square$

Corollary 2.2 *For any* $n \in \mathbb{N}$ *with prime factorization* $n = \prod_{j=1}^{m} p_j^{k_j}$, *for distinct primes* p_j *and* $k_j \in \mathbb{N}$ *for* $j = 1, 2, \ldots, m$,

$$\sigma(n) = \prod_{j=1}^{m} \frac{p_j^{k_j+1} - 1}{p_j - 1},$$

and

$$\tau(n) = \prod_{j=1}^{m} (k_j + 1).$$

Proof. This immediate from the multiplicativity of both functions and Theorem 2.18. $\qquad\square$

Example 2.19 $\sigma(2^k) = 2^{k+1} - 1$ *for any* $k \in \mathbb{N}$, *and* $\tau(2^k) = k + 1$. *Also,*

$$\sigma(1000) = \sigma(2^3 \cdot 5^3) = \left(\frac{2^4 - 1}{2 - 1} \right) \left(\frac{5^4 - 1}{5 - 1} \right) = 2340,$$

and

$$\tau(1000) = \tau(2^3 \cdot 5^3) = (3 + 1)(3 + 1) = 16.$$

The sum of divisors function allows us to introduce an important topic with a rich history and an abundance of famous postulates and open questions.

Definition 2.14 | Perfect Numbers |

An $n \in \mathbb{N}$ *is called* perfect *if* $\sigma(n) = 2n$. *In other words,* n *is perfect if it is equal to the sum of all its divisors less than itself.*

Example 2.20 *The smallest perfect number is* $6 = 1 + 2 + 3$, *and the next smallest is* $28 = 1 + 2 + 4 + 7 + 14$ — *see Biography 2.5 on the next page. Also,*

$$\sigma(6) = \sigma(2 \cdot 3) = (2^2 - 1) \cdot (3^2 - 1)/(3 - 1) = 12 = 2 \cdot 6,$$

and

$$\sigma(28) = \sigma(2^2 \cdot 7) = (2^3 - 1) \cdot (7^2 - 1)/(7 - 1) = 56 = 2 \cdot 28.$$

The sufficiency of the following condition is in Euclid's *Elements* (Book IX, Proposition 36) from some 2000 years ago, and links our discussion with the notion of Mersenne numbers introduced on page 36 — see Biography 1.4 on page 17. The necessity was proved by Euler in a work published posthumously — see Biography 1.17 on page 56.

Theorem 2.19 | **Even Perfect Numbers** |

If $n \in \mathbb{N}$, then n is an even perfect number if and only if

$$n = 2^{k-1}(2^k - 1), \text{ where } k \geq 2 \text{ is an integer and } 2^k - 1 \text{ is prime.} \qquad (2.9)$$

Proof. First we assume that condition (2.9) holds, where $p = 2^k - 1$ is prime. Then

$$\sigma(n) = \sigma(2^{k-1}p) = \sigma(2^{k-1})\sigma(p) = (2^k - 1)(p + 1) = (2^k - 1)2^k = 2n,$$

so n is perfect. (This was Euclid's contribution.)

Now assume that n is an even perfect number, and write $n = 2^{k-1}\ell$ where ℓ is odd and $k \geq 2$. Thus, since σ is multiplicative,

$$2^k \ell = 2n = \sigma(n) = \sigma(2^{k-1}\ell) = \sigma(2^{k-1})\sigma(\ell) = (2^k - 1)\sigma(\ell).$$

Hence, $(2^k - 1) \mid \ell$, so there exists $r \in \mathbb{N}$ such that $\ell = (2^k - 1)r$. Substituting the latter into the last displayed equation we get $2^k \ell = 2^k(2^k - 1)r = (2^k - 1)\sigma(\ell)$. Therefore, $\sigma(\ell) = 2^k r$. However, r and ℓ are both divisors of n, and $\ell + r = 2^k r$, so

$$2^k r = \sigma(\ell) \geq r + \ell = 2^k r,$$

forcing $\sigma(\ell) = r + \ell$. This means that ℓ has only two positive divisors, namely ℓ itself and r. It follows that ℓ is prime and $r = 1$. In other words, $\ell = 2^k - 1$ is a Mersenne prime, thereby securing the result. $\qquad \square$

Biography 2.5 Saint Augustine of Hippo (354–430 AD) *is purported to have said: "Six is a number perfect in itself, and not because God created the world in six days; rather the contrary is true. God created the world in six days because this number is perfect, and it would remain perfect, even if the work of the six days did not exist." Augustine, who was considered to be the greatest Christian philosopher of antiquity, merged the religion of the new testament with Platonic philosophy. Perfect numbers were known to the ancient Greeks in Euclid's time, although they only knew of the four smallest ones: $6, 28, 496, 8128$. They also attributed mystical properties to these numbers. (Note that the moon orbits the earth every 28 days.)*

Saint Augustine (Aurelius Augustinus) was certainly one of the most important figures in the foundations of Western Christianity. Even Protestants believe him to be one of the founders of the teachings of Reformation. His work, The Confessions, *considered to be one of the first Western autobiographies, is still in circulation. The name Hippo comes from the fact that in the late fourth century he was made bishop of Hippo Regius (now Annaba in Algeria), a position held until his death on August 28, 430, at the age of 75.*

Remark 2.15 | **Open Questions on Perfect Numbers**

Given Theorem 2.19 on the facing page, it is natural to ask about odd perfect numbers. This is an open question and a search for them up to considerably high bounds has been computed without finding any. (See Exercises 2.40–2.41 on the next page for properties of odd perfect numbers, should they exist.) Moreover, it is an open question as to whether there are infinitely many perfect numbers. Indeed by the above, Theorem 2.19, the problem of finding perfect numbers is reduced to finding Mersenne primes, so if there are infinitely many Mersenne primes, then there would be infinitely many perfect numbers, but this is an open problem.

Exercises

2.32. Calculate $\sigma(n)$ for each of the following n.

    (a) 56   (b) 105

    (c) 278   (d) 1001

    (e) 3^{10} (f) 2000

2.33. Calculate $\tau(n)$ for each of the following n.

    (a) 23   (b) 133

    (c) 276   (d) 1011

    (e) 5^{10} (f) 3001

2.34. The integers $n_j \in \mathbb{N}$ for $j = 1, 2, \ldots, k$ are called an *amicable k-tuple* if

$$\sigma(n_1) = \sigma(n_2) = \cdots = \sigma(n_k) = \sum_{j=1}^{k} n_j,$$

and if $k = 2$, it is called an *amicable pair*, if $k = 3$, an *amicable triple*, and so forth.

Prove that if

$$p = 3 \cdot 2^{n-1} - 1, \quad q = 3 \cdot 2^n - 1 \text{, and } r = 9 \cdot 2^{2n-1} - 1$$

are all primes (for a given $n \in \mathbb{N}$), then $2^n pq$ and $2^n r$ form an amicable pair.

(*This is called* Thabit's rule for amicable pairs — *see Biography 2.6 on page 107. The term* amicable pair *also known as* friendly numbers *comes from the fascinating property that each number is contained in the other in the sense that each number is equal to the sum of all the positive divisors of the other, except for the number itself.*)

2.35. Use Thabit's rule in Exercise 2.34 to find the smallest amicable pair.

2.36. Show that $(17296, 18416)$ is an amicable pair.

(*Use* $n = 4$ *in Thabit's rule in Exercise 2.34. In fact, Fermat wrote a letter to Mersenne in 1636, announcing this pair of amicable numbers.*)

2.37. Show that $(9363584, 9437056)$ is an amicable pair.

(*Use* $n = 7$ *in Thabit's rule in Exercise 2.34. This is the last amicable pair found by Thabit's rule, and it was announced by Descartes in a letter to Mersenne in 1638.*)

2.38. Let $n \in \mathbb{N}$, and select a nonnegative integer $m < n$ such that $g = 2^{n-m}+1$. Prove that if $p = 2^m g - 1$, $q = 2^n g - 1$, and $r = 2^{n+m} g^2 - 1$ are all primes, then $2^n pq$ and $2^n r$ form an amicable pair.

(*This is called* Euler's rule *for amicable pairs. Euler generalized Thabit's rule, which is the case where* $n - m = 1$. *As we saw in Exercise 2.37, Thabit's rule holds for* $n = 7$. *However it does not hold for any other* n *with* $7 < n \leq 20,000$.)

2.39. Find two amicable triples.

☆ 2.40. Prove that if n is an odd perfect number, then

$$n = p_1^{a_1} p_2^{2b_2} p_3^{2b_3} \cdots p_k^{2b_k},$$

where the p_j are distinct primes and $p_1 \equiv a_1 \equiv 1 \,(\text{mod } 4)$.

2.41. Prove that if n is an odd perfect number, then $n = p^a m^2$, where p is a prime not dividing m and $p \equiv a \equiv 1 \,(\text{mod } 4)$. Conclude that $n \equiv 1 \,(\text{mod } 4)$.

2.42. If $n \in \mathbb{N}$, then n is *deficient* if $\sigma(n) < 2n$. Prove that all prime powers are deficient.

2.43. If $n \in \mathbb{N}$, then n is *abundant* if $\sigma(n) > 2n$. Prove that if $n \in \mathbb{N}$ is abundant, then so is kn for all $k \in \mathbb{N}$.

(*Note that by Exercises 2.42–2.43 and Definition 2.14, all numbers are one of deficient, abundant, or perfect.*)

2.44. A number $n \in \mathbb{N}$ is *almost perfect* if $\sigma(n) = 2n - 1$. Prove that all powers of 2 are almost perfect. (*It is not known if there are any other almost perfect numbers.*)

2.45. A number $n \in \mathbb{N}$ is *triangular* if $n = \sum_{j=1}^{k} j$ for some $k \in \mathbb{N}$. Prove that every even perfect number is triangular.

2.46. A number $n \in \mathbb{N}$ is *polygonal* if $n = m(a^2 - a)/2 + a$ for some $a, m \in \mathbb{N}$. Prove that a polygonal number is triangular if $m = 1$ (see Exercise 2.45).

2.47. Prove that if n is an even perfect number, then $8n + 1$ is a square.

2.48. Prove that $n \in \mathbb{N}$ is triangular if and only if $8n+1$ is a square (see Exercise 2.45).

2.49. Prove that $\sum_{d|n} \mu(d)\sigma(n/d) = n$.

(*Hint: Use the Möbius inversion formula, Theorem 2.16 on page 98.*)

2.50. A number $n \in \mathbb{N}$ is *superperfect* if $\sigma(\sigma(n)) = 2n$. Prove that if $2^p - 1$ is prime, then 2^{p-1} is superperfect.

2.51. Prove that if 2^k is superperfect, then $2^{k+1} - 1$ is a Mersenne prime.

2.52. Prove that if $n \in \mathbb{N}$, then $\left(\sum_{d|n} \tau(d) \right)^2 = \sum_{d|n} \tau(d)^3$.

(*Hint: Use Exercise 1.6 on page 11.*)

2.53. If $n \in \mathbb{N}$,

$$s_1(n) = \sigma(n) - n, \text{ and } s_{j+1}(n) = \sigma(s_j(n)) - s_j(n) \text{ for all } j \in \mathbb{N},$$

then the numbers $s_j(n)$ form an *aliquot sequence*. Prove that if n is perfect, then

$$s_j(n) = s_1(n) = n \text{ for all } j \geq 1.$$

(*The term* aliquot *means a quantity that divides into another an integral number of times. Essentially then the aliquot parts are the divisors of an integer.*)

2.54. Let m and n be an amicable pair. Prove that the aliquot sequence defined in Exercise 2.53 has period 2. In other words, show that $s_{2j}(m) = m$ for all $j \in \mathbb{N}$.

2.55. Let $a \in \mathbb{Z}$, $n > 1$ a natural number with $\gcd(a,n) = 1$, and let r be the smallest positive integer such that $a^r \equiv 1 \,(\mathrm{mod}\ n)$. Prove that $r | \phi(n)$.

(The notion in this exercise is the main topic of Chapter 3.)

Biography 2.6 Thabit Ibn Qurra Ibn Marwan al-Sabi al-Harrani (836–901 A.D.) *was an Arab mathematician born in Harran, Mesopotamia, now Turkey. He lived in Baghdad where he studied a wide variety of topics including not only mathematics, but also astronomy, mechanics, medicine, and philosophy, to name a few. Thabit's* Book on the Determination of Amicable Numbers *contained the rule that bears his name — see Exercise 2.34 on page 105. He also wrote a book:* On the Verification of the Problems of Algebra by Geometrical Proofs, *where he solved quadratic Diophantine equations using ideas from Euclid's* Elements. *Yet nobody, to that point, had considered such methods, so it was a remarkable achievement. Moreover, he translated, from Greek to Arabic, the works of Euclid, Archimedes, Apollonius, Ptolemy, and a number of other great scholars. In the later part of his life, he became the friend and courtier of the reigning Caliph, Abbasid Caliph al-M'utadid. He died in Baghdad on February 18, 901.*

2.5   The Floor and the Ceiling

> *Look how the floor of heaven is thick inlaid with patines of bright gold.*
>   From Act 5, Scene 1, line 54 of **The Merchant of Venice (1596–1598),**
> Oxford Standard Author's Edition
>   **William Shakespeare (1564–1616), English dramatist**

Although the functions to be studied herein are not, strictly speaking, arithmetic functions, in a study of which we initiated in §2.3, they are related in very important ways that will become clear as we proceed. Those functions with the name in the title have already been tacitly referenced earlier. For instance, in the proof of Thue's Theorem 1.23 on page 44, the value of c is an instance of what is known as the ceiling, defined below. As well, the value N in Exercise 1.64 on page 54 is an example of the floor, also defined below.

Definition 2.15 | Floor and Ceiling Functions

If $x \in \mathbb{R}$, then the greatest integer less than or equal to x, also known as the floor function, is denoted by $\lfloor x \rfloor$. The least integer greater than or equal to x, also known as the celing *function, is denoted by $\lceil x \rceil$. The fractional part of x of x, which is the difference between x and the greatest integer less than or equal to x, denoted by*

$$\{x\} = x - \lfloor x \rfloor,$$

so

$$x = \lfloor x \rfloor + \{x\},$$

and consequently, $\lfloor x \rfloor$ is often called the integral part *of x.*

Example 2.21 *Some explicit values of the floor, ceiling, and fractional parts are given as follows.*

$$\lfloor -1/2 \rfloor = -1, \lceil -1/2 \rceil = 0, \{-1/2\} = -1/2 - (-1) = 1/2.$$

$$\lfloor \pi \rfloor = 3, \lceil \pi \rceil = 4, \{\pi\} = \pi - 3 = 0.141592653589793238462643383\cdots$$

$$\lfloor -\sqrt{2} \rfloor = -2, \lceil -\sqrt{2} \rceil = -1,$$

and

$$\{-\sqrt{2}\} = -\sqrt{2} - (-2) = 0.585786437626904951198311275\cdots$$

Also, in general, if $x \in \mathbb{R}$ and $n \in \mathbb{Z}$ are arbitrary, then by Exercsie 2.57 on page 112,

$$\lfloor x + n \rfloor = \lfloor x \rfloor + n.$$

We may now establish a result based upon some properties of the binomial coefficient that we visited in Section §1.1.

Theorem 2.20 | **Binomial Coefficient Sums**

If $n \in \mathbb{N}$, then

(a) $\sum_{j=0}^{\lfloor n/2 \rfloor} \binom{n}{2j} = 2^{n-1}$,

(b) $\sum_{j=0}^{\lfloor (n+1)/2 \rfloor} \binom{n}{2j-1} = 2^{n-1}$.

Proof. By the full summation property established in Exercise 1.17 on page 14, we have,

$$2^n = \sum_{j=0}^{n} \binom{n}{j} = \sum_{j=0}^{\lfloor n/2 \rfloor} \binom{n}{2j} + \sum_{j=1}^{\lfloor (n+1)/2 \rfloor} \binom{n}{2j-1}, \qquad (2.10)$$

and by the null summation property given in Exercise 1.16,

$$0 = \sum_{j=0}^{n} (-1)^j \binom{n}{j} = \sum_{j=0}^{\lfloor n/2 \rfloor} \binom{n}{2j} - \sum_{j=1}^{\lfloor (n+1)/2 \rfloor} \binom{n}{2j-1}. \qquad (2.11)$$

Adding Equations (2.10)–(2.11), and dividing both sides by 2, we get,

$$\sum_{j=0}^{\lfloor n/2 \rfloor} \binom{n}{2j} = 2^{n-1} \quad \text{and} \quad \sum_{j=1}^{\lfloor (n+1)/2 \rfloor} \binom{n}{2j-1} = 2^{n-1},$$

as required. $\qquad \square$

Theorem 2.20 shows us that the full and null summation properties have a more detailed breakdown as given in the proof.

The following links the floor function with the arithmetic functions we have studied earlier in this chapter.

Theorem 2.21 | **Arithmetic Functions and the Floor**

If f is an arithmetic function, such that

$$F(n) = \sum_{d|n} f(d),$$

then for any $N \in \mathbb{N}$,

$$\sum_{n=1}^{N} F(n) = \sum_{k=1}^{N} f(k) \left\lfloor \frac{N}{k} \right\rfloor.$$

Proof. First we observe that

$$\sum_{n=1}^{N} F(n) = \sum_{n=1}^{N} \sum_{d|n} f(d).$$

For any natural number $k \leq N$, $f(k)$ will appear in $\sum_{d|n} f(d)$ if and only if $k|n$, and $f(k)$ is in that sum at most once for a given k. The number of $k \leq N$ such that $f(k)$ appears in

$$\sum_{d|n} f(d)$$

is $\lfloor N/k \rfloor$ since they are exactly the values: $k, 2k, 3k, \ldots, \lfloor N/k \rfloor k$. Hence, the number of times $f(k)$ appears in $\sum_{n=1}^{N} \sum_{d|n} f(d)$ is $\lfloor N/k \rfloor$. This shows that

$$\sum_{n=1}^{N} F(n) = \sum_{k=1}^{N} f(k) \left\lfloor \frac{N}{k} \right\rfloor,$$

which is what we set out to accomplish.   □

Theorem 2.21 has many consequences for arithmetic functions that we have studied. For instance, the number τ and sum σ of divisors studied in §2.4 are virtually immediate consequences as follows.

Corollary 2.3 *If $N \in \mathbb{N}$, then*

$$\sum_{n=1}^{N} \tau(n) = \sum_{k=1}^{N} \left\lfloor \frac{N}{k} \right\rfloor.$$

Proof. Let $f(n) = 1$ in Theorem 2.21, and let $F = \tau$, then we get the result. □

Corollary 2.4 *If $N \in \mathbb{N}$, then*

$$\sum_{n=1}^{N} \sigma(n) = \sum_{k=1}^{N} k \left\lfloor \frac{N}{k} \right\rfloor.$$

Proof. Let $f(d) = d$ in Theorem 2.21, and let $\sigma = F$, then the result follows. □

Also, the Möbius function studied in §2.3 has a similar consequence from Theorem 2.21.

Corollary 2.5 *If $N \in \mathbb{N}$, then*

$$\sum_{k=1}^{N} \mu(k) \left\lfloor \frac{N}{k} \right\rfloor = 1.$$

Proof. Let $f(d) = \mu(d)$ and let $F(n) = \sum_{d|n} \mu(d)$, then the result follows from
Theorem 2.14 on page 96, since $\sum_{d|n} \mu(d) = 0$ if $n > 1$ and is 1 if $n = 1$.   □

Remark 2.16 *It is of interest to note a consequence of Corollary 2.5 to infinite
series, in particular the zeta function — see page 65. Since we may look at any
real number x and set $N = \lfloor x \rfloor$, then since $0 \le x - \lfloor x \rfloor < 1$,*

$$\left| \mu(k) \left\lfloor \frac{x}{k} \right\rfloor - \mu(k)\frac{x}{k} \right| < 1.$$

*Now if we consider the sum in Corollary 2.5, we see that the difference between
that sum and the one without the floor function has error less than 1. Also, the
error for $k = 1$ is equal to $x - \lfloor x \rfloor$, and since there are a total of $\lfloor x \rfloor$ summands
including the first, we deduce that*

$$\left| \sum_{k=1}^{N} \mu(k) \left\lfloor \frac{x}{k} \right\rfloor - x \sum_{k=1}^{N} \frac{\mu(k)}{k} \right| < x - \lfloor x \rfloor + \lfloor x \rfloor - 1 = x - 1.$$

Thus, Corollary 2.5 tells us that

$$\left| 1 - x \sum_{k=1}^{N} \frac{\mu(k)}{k} \right| < x - 1,$$

from which it is immediate that

$$\left| x \sum_{k=1}^{N} \frac{\mu(k)}{k} \right| \le x.$$

Therefore, via division by x,

$$\left| \sum_{k=1}^{N} \frac{\mu(k)}{k} \right| \le 1. \tag{2.12}$$

Now consider the infinite series

$$S = \sum_{k=1}^{\infty} \frac{\mu(k)}{k},$$

*for which Equation (2.12) tells us that all partial sums are bounded above by 1.
This was a conjecture made by Euler in 1748, and proved in the late nineteenth
century — see Dickson [14, Chapter XIX].*

*Now we turn to the Dirichlet product defined in Exercises 2.30–2.31 on page
101, which we now apply to the above, as follows. By Theorem 2.14 on page 96,*

$$\sum_{k=1}^{\infty} \frac{\mu(k)}{k^s} * \sum_{j=1}^{\infty} \frac{1}{j^s} = 1,$$

so by the definition of the zeta function on page 65,

$$\sum_{k=1}^{\infty} \frac{\mu(k)}{k^s} = \frac{1}{\zeta(s)}, \qquad s \in \mathbb{R}, s > 1,$$

which is the relationship with the zeta function we were seeking to establish. Note that via Definition 1.14 on page 55, this speaks about the generating function for the Möbius function. Indeed it can be similarly shown that $\zeta(s-1)/\zeta(s) = \sum_{k=1}^{\infty} \phi(k)/k^s$, $(s > 1)$, $\zeta(s)^2 = \sum_{k=1}^{\infty} \tau(k)/k^s$, and $\zeta(s)\zeta(s-1) = \sum_{k=1}^{\infty} \sigma(k)/k^s$, $(s > 2$ in the latter two cases).

Exercises

2.56. Prove that for any $x \in \mathbb{R}$, $x - 1 < \lfloor x \rfloor \le x$.

2.57. Prove that for any $x \in \mathbb{R}$ and $n \in \mathbb{Z}$, $\lfloor x + n \rfloor = \lfloor x \rfloor + n$.

2.58. Prove that for any $x, y \in \mathbb{R}$,

$$\lfloor x \rfloor + \lfloor y \rfloor \le \lfloor x + y \rfloor \le \lfloor x \rfloor + \lfloor y \rfloor + 1.$$

2.59. Prove that for any $x \in \mathbb{R}$, $\lfloor x \rfloor + \lfloor -x \rfloor = \begin{cases} 0 & \text{if } x \in \mathbb{Z}, \\ -1 & \text{otherwise.} \end{cases}$

2.60. Calculate the following values of the floor function.
(a) $\lfloor -6/5 \rfloor$ (b) $\lfloor 3/4 \rfloor$
(c) $\lfloor -5 + \lfloor -4/3 \rfloor \rfloor$ (d) $\lfloor 3 + \lfloor -2.3 \rfloor \rfloor$
(e) $\lfloor 222/777 \rfloor$ (f) $\lfloor -0.2 + \lfloor 0.2 \rfloor \rfloor$

2.61. Calculate the following values of the floor function.
(a) $\lfloor 6/5 \rfloor$ (b) $\lfloor 34/3 \rfloor$
(c) $\lfloor 4/3 + \lfloor -4/3 \rfloor \rfloor$ (d) $\lfloor -2.3 + \lfloor 3.2 \rfloor \rfloor$
(e) $\lfloor 77/22 \rfloor$ (f) $\lfloor 2.1 + \lfloor -0.2 \rfloor \rfloor$

2.62. Calculate the following values of the ceiling function.
(a) $\lceil -7/5 \rceil$ (b) $\lceil 13/12 \rceil$
(c) $\lceil -15 + \lfloor -5/3 \rfloor \rceil$ (d) $\lceil -3 + \lfloor 2.7 \rfloor \rceil$

2.63. Calculate the following values of the ceiling function.
(a) $\lceil -55/32 \rceil$ (b) $\lceil 22/7 \rceil$
(c) $\lceil 25 + \lfloor -35/3 \rfloor \rceil$ (d) $\lceil -7 + \lfloor 7.7 \rfloor \rceil$

2.64. Calculate the following fractional parts.
(a) $\{-555/23\}$ (b) $\{22/7\}$
(c) $\{12 + \{-36/5\}\}$ (d) $\{-7 + \{34/21\}\}$

2.65. Calculate the following fractional parts.
(a) $\{222/3\}$ (b) $\{333/21\}$
(c) $\{\{-13/7\} + \{63/5\}\}$ (d) $\{\{77/22\} - \{43/21\}\}$

2.6   Polynomial Congruences

> *Let every soul be subject to the higher powers...the powers that are ordained of God.*
> From Romans, Chapter 13, Verse 1, the authorized version of **The Bible** (**1611**)

In §2.2, we completely characterized solutions of linear congruences in Theorem 2.3 on page 84. We also interpreted several quadratic congruences from earlier discussions in Examples 2.8–2.12 on pages 87–88. We look to explore other higher order congruences in this section via polynomials.

Definition 2.16 | Integral Polynomial Congruences |

A polynomial $f(x) = \sum_{j=0}^{d} a_j x^j$ for $a_j, d \in \mathbb{Z}$, $d \geq 0$ is an integral polynomial in a single variable x. If $\gcd(a_0, a_1, \ldots, a_k) = g$, then g is the content *of f. If $g = 1$, then f is* primitive. *If $c \in \mathbb{Z}$ and $f(c) \equiv 0 \pmod{n}$ for some $n \in \mathbb{N}$, then c is a* root *of f or a* solution *of f modulo n. The a_j are called the* coefficients *of f. If $a_d \not\equiv 0 \pmod{n}$, then f is said to have* degree d *modulo n. If*

$$f(c_1) \equiv f(c_2) \equiv 0 \pmod{n} \text{ where } c_1 \equiv c_2 \pmod{n},$$

then c_1 and c_2 are said to be congruent solutions *of f modulo n, whereas if $c_1 \not\equiv c_2 \pmod{n}$, they are called* incongruent solutions *modulo n.*

Note that solving $f(x) \equiv 0 \pmod{n}$ for some integral polynomial f and some $n \in \mathbb{N}$ is equivalent to solving $f(x) = ny$ for integers x and y. For instance, a root of $f(x) = x^2 + 1$ modulo 5 is $x = 3$, so $f(x) = 10 = 5 \cdot 2 = n \cdot y$. This is an instance of the simplest case where the modulus is prime. Lagrange was the first to solve polynomial congruences involving a prime modulus — see Biography 2.7 on the following page.

Theorem 2.22 | Lagrange's Theorem |

Suppose that p is a prime and f is an integral polynomial of degree $d \geq 1$ modulo p. Then $f(x) \equiv 0 \pmod{p}$ has at most d incongruent solutions.

Proof. Let $f(x) = \sum_{j=0}^{d} a_j x^j$, and use induction on d. If $d = 1$, then $f(x) = a_1 x + a_0 \equiv 0 \pmod{p}$. Since $p \nmid a_1$, this congruence has exactly one solution modulo p by Theorem 2.3. This is the induction step. The induction hypothesis is that the result holds for any integral polynomial of degree less than d. Now let c be a root of f modulo p. Then

$$\frac{f(x) - f(c)}{x - c} = \sum_{j=1}^{d} a_j \frac{x^j - c^j}{x - c} = \sum_{j=1}^{d} a_j \sum_{i=1}^{j} c^{i-1} x^{j-i} = g(x)$$

where $g(x)$ is an integral polynomial of degree $d-1$. By the induction hypothesis, g has at most $d-1$ incongruent solutions modulo p. Since

$$f(x) \equiv (x - c)g(x) \pmod{p},$$

f has at most d incongruent solutions modulo p. $\square$

Biography 2.7 Joseph-Louis Lagrange (1736–1813) *was born on January 25, 1736, in Turin, Sardinia-Piedmont (now Italy). Although Lagrange's primary interests as a young student were in classical studies, his reading of an essay by Edmund Halley (1656–1743) on the calculus converted him to mathematics. While still in his teens, Lagrange became a professor at the Royal Artillery School in Turin in 1755 and remained there until 1766 when he succeeded Euler (see Biography 1.17 on page 56) as director of mathematics at the Berlin Academy of Science. In 1768, he published his result on polynomial congruences, Theorem 2.22 on the page before. He was also the first to prove Wilson's Theorem in 1770, (see Biography 1.16 on page 52). Lagrange left Berlin in 1787 to become a member of the Paris Academy of Science, where he remained for the rest of his professional life. In 1788 he published his masterpiece* Mécanique Analytique, *which may be viewed as both a summary of the entire field of mechanics to that time and an establishment of mechanics as a branch of analysis, mainly through the use of the theory of differential equations. When he was fifty-six, he married a young woman almost forty years younger than he, the daughter of the astronomer Lemonnier. She became his devoted companion until his death in the early morning of April 10, 1813, in Paris.*

Example 2.22 *Let p be a prime. If $d \in \mathbb{N}$ with $d \mid (p-1)$, then $p-1 = kd$ for some $k \in \mathbb{N}$, so by Lagrange's Theorem, $f_d(x) = x^{d(k-1)} + x^{d(k-2)} + \cdots x^d + 1$ has at most $d(k-1) = p-1-d$ incongruent solutions modulo p. But from Fermat's Little Theorem 2.7 on page 88,*

$$x^{p-1} \equiv 1 \pmod{p}$$

has exactly $p-1$ incongruent solutions, namely, $1, 2, \ldots, p-1$. Now if

$$c^{p-1} \equiv 1 \pmod{p}$$

such that $f(c) \not\equiv 0 \pmod{p}$, then $c^d \equiv 1 \pmod{p}$ since

$$(c^d - 1)f(c) \equiv c^{p-1} - 1 \equiv 0 \pmod{p}$$

with $p \nmid f(c)$. Hence, $x^d - 1 \equiv 0 \pmod{p}$ has a minimum of $p-1-(p-1-d) = d$ incongruent solutions. However, Lagrange's Theorem tells us that the latter congruence has no more than d solutions, so it must have precisely d solutions. We have shown the following.

For any $d > 0$ with $d \mid (p-1)$, $x^d - 1 \equiv 0 \pmod{p}$ has exactly d solutions.

$$(2.13)$$

It is essential that p be prime in Lagrange's result since, for instance, $x^2 - 1 \equiv 0 \pmod 8$ has four incongruent solutions $x = \pm 1, \pm 3$ modulo 8. However, it suffices to look at prime-power moduli by the following.

Theorem 2.23 | **The Chinese Remainder Theorem for Polynomials**

    If $n = p_1^{a_1} p_2^{a_2} \cdots p_d^{a_d}$ where the p_j are distinct primes for $j = 1, 2, \ldots, d$, then

$$f(x) \equiv 0 \pmod n \ \textit{if and only if } f(x) \equiv 0 \pmod {p_j^{a_j}} \textit{ for all } j = 1, 2, \ldots, d.$$

Proof. Since $f(x) \equiv 0 \pmod n$ if and only if $f(c) \equiv 0 \pmod n$ for some $c \in \mathbb{Z}$, then the result is immediate from the Chinese Remainder Theorem for integers given in Theorem 2.4 on page 85. □

    In view of Theorem 2.23, we need concentrate upon only prime-power moduli. Furthermore, there are methods to take solutions of $f(x) \equiv 0 \pmod {p^k}$ and use them to find solutions of $f(x) \equiv 0 \pmod {p^{k+1}}$. To do this we need some tools from elementary calculus.

Definition 2.17 | **Derivatives of Polynomials**

    If $f(x) = \sum_{j=0}^{d} a_j x^j$ where $a_j \in \mathbb{R}$ for $j = 0, 1, \ldots, d$, given nonnegative $d \in \mathbb{Z}$, then the derivative of $f(x)$, denoted by $f'(x)$, is given by

$$a_1 + a_2 x + \cdots + (d-1)a_{d-1}x^{d-2} + da_d x^{d-1},$$

when $d \in \mathbb{N}$ and $f'(x) = 0$ if $d = 0$. Furthermore, for any $n \in \mathbb{N}$, the nth derivative, denoted by $f^{(n)}(x)$ is given inductively by $(f^{(n-1)})'(x)$.

    The following type of result is often called a *lifting of solutions*, since we find solutions from a modulus, which is a lower power of a prime and "lift" them to a higher prime power modulus.

Theorem 2.24 | **Lifting Solutions Modulo Prime Powers**

    Let $f(x)$ be an integral polynomial, p a prime, and $k \in \mathbb{N}$. Suppose that $r_1, r_2, \ldots, r_m$ for some $m \in \mathbb{N}$ are all of the incongruent solutions of $f(x)$ modulo p^k, where $0 \le r_i < p^k$ for each $i = 1, 2, \ldots, m$. If $a \in \mathbb{Z}$ such that

$$f(a) \equiv 0 \pmod {p^{k+1}} \textit{ with } 0 \le a < p^{k+1}, \tag{2.14}$$

there exists $q \in \mathbb{Z}$ such that

  (a) *For some $i \in \{1, 2, \ldots, m\}$, $a = q p^k + r_i$ with $0 \le q < p$, and*

  (b) *$f(r_i) + q f'(r_i)p^k \equiv 0 \pmod {p^{k+1}}$.*

Additionally, if $f'(r_i) \not\equiv 0 \,(\mathrm{mod}\ p)$, then

$$f(qp^k + r_i) \equiv 0 \quad (\mathrm{mod}\ p^{k+1}) \tag{2.15}$$

has a unique solution for the value of q given by

$$q \equiv -\frac{f(r_i)}{p^k}(f'(r_i))^{-1} \quad (\mathrm{mod}\ p), \tag{2.16}$$

with $(f'(r_i))^{-1}$ being a multiplicative inverse of $f'(r_i)$ modulo p.
   If $f'(r_i) \equiv 0\,(\mathrm{mod}\ p)$ and $f(r_i) \equiv 0\,(\mathrm{mod}\ p^{k+1})$, then all values of $q = 0, 1, 2, \ldots, p-1$ yield incongruent solutions to (2.15).
   If $f'(r_i) \equiv 0\,(\mathrm{mod}\ p)$ and $f(r_i) \not\equiv 0\,(\mathrm{mod}\ p^{k+1})$, then $f(x) \equiv 0\,(\mathrm{mod}\ p^{k+1})$ has no solutions with $x \equiv r_i\,(\mathrm{mod}\ p^k)$.

Proof. Since the congruence (2.14) holds, then it follows that $f(a) \equiv 0\,(\mathrm{mod}\ p^k)$. Therefore, for some $i \in \{1, 2, \ldots, m\}$, $a \equiv r_i\,(\mathrm{mod}\ p^k)$. Hence, there exists $q \in \mathbb{Z}$ such that $a = qp^k + r_i$. Since $qp^k \leq r_i + qp^k = a < p^{k+1}$, then $q < p$. We need to verify that q is nonnegative to complete the proof of part (a). Assume to the contrary that $q < 0$. Then

$$a = qp^k + r_i < -p^k + p^k = 0,$$

contradicting the fact that $a \geq 0$. This completes the establishment of part (a).
   If $f(a) = \sum_{j=0}^d b_j a^j$, then by part (a),

$$f(a) \equiv \sum_{j=0}^d b_j(qp^k + r_i)^j \equiv \sum_{j=0}^d b_j(r_i^j + jr_i^{j-1}qp^k) \quad (\mathrm{mod}\ p^{k+1}),$$

where the last congruence follows from the Binomial Theorem. Hence,

$$0 \equiv f(a) \equiv \sum_{j=0}^d b_j r_i^j + qp^k \sum_{j=0}^d jr_i^{j-1} \equiv f(r_i) + qp^k f'(r_i) \quad (\mathrm{mod}\ p^{k+1}),$$

which secures part (b).
   If $f'(r_i) \not\equiv 0\,(\mathrm{mod}\ p)$, then $\gcd(f'(r_i), p) = 1$, so by Theorem 2.3 on page 84, the congruence (2.16) has the unique solution given by q. On the other hand, if $f'(r_i) \equiv 0\,(\mathrm{mod}\ p)$, then $\gcd(f'(r_i), p) = p$, so by Theorem 2.3, if $p \mid (f(r_i)/p^k)$ — which can occur if and only if $f(r_i) \equiv 0\,(\mathrm{mod}\ p^{k+1})$ — all values of $q = 0, 1, 2, \ldots, p-1$ are solutions of (2.15). Lastly, Theorem 2.3 says that if p does not divide $(f(r_i)/p^k)$, no values of q yield solutions. $\square$

Remark 2.17 *The result in Theorem 2.24, says that a solution to*

$$f(x) \equiv 0 \quad (\mathrm{mod}\ p^k)$$

lifts to a unique solution of

$$f(x) \equiv 0 \pmod{p^{k+1}}$$

when $f'(r_i) \not\equiv 0 \pmod{p}$, whereas such a solution lifts to p incongruent solutions modulo p^{k+1}, if $f'(r_i) \equiv 0 \pmod{p}$, and $f(r_i) \equiv 0 \pmod{p^{k+1}}$. Note that $f(r_i)/p^k$ is an integer since $f(r_i) \equiv 0 \pmod{p^k}$. Indeed, the congruence,

$$q f'(r_i) \equiv -\frac{f(r_i)}{p^k} \pmod{p}$$

holds in either case, so we merely need to solve this linear congruence for q to find possible solutions to $f(x) \equiv 0 \pmod{p^{k+1}}$ if the solutions modulo p^k are known.

Theorem 2.24 is a version of what is known as Hensel's Lemma, since Kurt Hensel was the first to prove the result that allows us to lift solutions of polynomial congruences — see Biography 2.8 on the following page.

Example 2.23 *Let $f(x) = x^3 + 2x^2 + 35$ and suppose we wish to solve*

$$f(x) \equiv 0 \pmod{3^3}. \tag{2.17}$$

We see by inspection that $f(8) \equiv 0 \pmod{9}$, and $f'(8) \equiv 2 \pmod{3}$, so by Theorem 2.24, we have a unique solution given by

$$q \equiv -f(8)/9 \cdot 2 \pmod{3}, \text{ where } 2 \equiv (f'(8))^{-1} \equiv 2^{-1} \pmod{3}.$$

However, $f(8)/9 \equiv 0 \pmod{3}$, so $x = 8 = 8 + 0 \cdot 9$ is the unique solution of $f(x) \equiv 0 \pmod{27}$ where x is of the form $x = 8 + 9q$. Furthermore, since the only congruence class of solutions of $f(x) \equiv 0 \pmod{9}$ is for $x \equiv 8 \pmod{9}$, there can be no more solutions to (2.17), other than $x \equiv 8 \pmod{27}$.

Example 2.24 *Let $f(x) = x^3 + 4x + 44$ and suppose we wish to solve*

$$f(x) \equiv 0 \pmod{49}. \tag{2.18}$$

By inspection, we see that $x = 1$ is a solution of $f(x) \equiv 0 \pmod{49}$, and since $f'(x) = 3x^2 + 4$, then $f'(1) \equiv 0 \pmod{7}$, Theorem 2.24 tells us that for $q = 0, 1, 2, 3, 4, 5, 6$, $x = 1 + 7q$ are incongruent solutions of $f(x) \equiv 0 \pmod{49}$, namely for for $x \equiv 1, 8, 15, 22, 29, 36, 43 \pmod{49}$.

However, we see that the only other congruence class for which

$$f(x) \equiv 0 \pmod{7}$$

is for $x \equiv 5 \pmod{7}$. Since $f'(5) \equiv 2 \pmod{7}$, then Theorem 2.24 says that the unique solution of (2.18) of the form $x = 5 + 7q$ is for

$$q \equiv -\frac{f(5)}{7}(f'(5))^{-1} \equiv -6 \cdot 4 \equiv 4 \pmod{7},$$

so for $x \equiv 33 \pmod{49}$, *(2.18) also has solutions. Hence, all incongruent solutions modulo* 49 *of* (2.18) *are given by*

$$x \equiv 1, 8, 15, 22, 29, 33, 36, 43 \pmod{49}.$$

Biography 2.8 Kurt Hensel (1861–1941) *was born in Prussia in a city then called Königsberg. When his family moved to Berlin, he began his formal mathematical training. Hensel was fortunate to have some of the greatest minds as his teachers. Among them were Weierstrass, Borchardt, Kirchhoff, Helmholz, and Kronecker. It was Kronecker who supervised his doctoral thesis completed in 1886. By 1897, Hensel had developed the foundations of what we now call* p-adic *numbers, which may be viewed as a completion of the rational number field in a different fashion than the usual completion which gives us the real field. This led to the development of the notion of a field with a* valuation *that had a deep influence on later mathematical development. In 1921, Hasse realized a great depth to p-adic numbers when he discovered his* local-global principle. *What this meant was that for quadratic forms, an equation has a rational solution if and only if it has a solution in p-adic numbers for each prime p, including a solution in the real field. Indeed, Hasse worked under Hensel at the University of Marburg, where Hensel was a professor until 1930. He was also editor, from 1901, of the highly well regarded and influential* Crelle's Journal, *which is its abbreviated name. Hensel died in Marburg, Germany on June 1, 1941.*

Exercises

2.66. Use the result established in Example 2.22 on page 114 to prove Wilson's Theorem.

2.67. Prove that if $n \in \mathbb{N}$, there exists a modulus $m \in \mathbb{N}$ such that $x^2 \equiv 1 \pmod{m}$ has more than n incongruent solutions.

2.68. Find all solutions of $x^4 + 3x^2 + 12 \equiv 0 \pmod{5^2}$.

2.69. Find all solutions of $x^3 + 3x^2 + 1 \equiv 0 \pmod{7^3}$.

2.70. Find all solutions of $x^3 + 2x^2 + 3 \equiv 0 \pmod{11^3}$.

2.71. Find all solutions of $x^3 + x^2 + x + 1 \equiv 0 \pmod{2^3}$.

2.72. Find all solutions of $x^3 + x^2 + x + 1 \equiv 0 \pmod{13^3}$.

2.73. Find all solutions of $x^3 + x^2 - 1 \equiv 0 \pmod{17^3}$.

2.7 Primality Testing

> *Probable impossibilities are to be preferred to improbable possibilities.*
> From Chapter 24 of **Poetics**
> **Aristotle (384–322 B.C.), Greek philosopher**

We now have the tools at our disposal to continue the study of *primality testing* begun in §1.8, where we looked at only *true* primality tests. We will look at not only more such tests, but also some *probabilistic primality tests*, which are primality tests that use *randomized algorithms*, namely those that make random decisions at certain points in their execution, so that the execution paths may differ each time the algorithm is invoked with the same input. Therefore, a probabilistic primality test will provide good, but not necessarily conclusive evidence that a given input is prime. For instance, the following probabilistic primality test will answer correctly if a prime is input, but might err with very small probability if the input is composite. (See Biographies 2.10 on page 122, 2.9 on page 121, and 2.11 on page 123.) The following is adapted from [32]. Note that we assume we have a method of generating "random numbers." In §3.4 we will learn how to do this in detail.

◆ The Miller-Selfridge-Rabin (MSR) Primality Test

Let $n - 1 = 2^t m$ where $m \in \mathbb{N}$ is odd and $t \in \mathbb{N}$. The value n is the input to be tested by executing the following steps, where all modular exponentiations are done using the repeated squaring method described on page 82.

(1) Choose a random integer a with $2 \le a \le n - 2$.

(2) Compute
$$x_0 \equiv a^m \pmod{n}.$$
If
$$x_0 \equiv \pm 1 \pmod{n},$$
then terminate the algorithm with

"*n* is probably prime."

If $x_0 \not\equiv \pm 1 \pmod{n}$ and $t = 1$, terminate the algorithm with

"*n* is definitely composite."

Otherwise, set $j = 1$ and go to step (3).

(3) Compute
$$x_j \equiv a^{2^j m} \pmod{n}.$$
If $x_j \equiv 1 \pmod{n}$, then terminate the algorithm with

"*n* is definitely composite."

If $x_j \equiv -1 \,(\mathrm{mod}\ n)$, terminate the algorithm with

"n is probably prime."

Otherwise set $j = j + 1$ and go to step (4).

(4) If $j = t - 1$, then go to step (5). Otherwise, go to step (3).

(5) Compute

$$x_{t-1} \equiv a^{2^{t-1}m} \pmod{n}.$$

If $x_{t-1} \not\equiv -1 \,(\mathrm{mod}\ n)$, then terminate the algorithm with

"n is definitely composite."

If $x_{t-1} \equiv -1 \,(\mathrm{mod}\ n)$, then terminate the algorithm with

"n is probably prime."

Example 2.25 Consider $n = 2821$. Since $n - 1 = 2^2 \cdot 705$, then $t = 2$ and $m = 705$. Select $a = 2$. Then

$$x_0 \equiv 2^{705} \equiv 2605 \pmod{n},$$

so we set $j = 1$ and compute

$$x_1 \equiv 2^{2 \cdot 705} \equiv 1520 \pmod{n},$$

so we set $j = 2$ and compute

$$x_2 \equiv 2^{4 \cdot 27} \equiv 1 \pmod{n}.$$

Thus, by step (3) of the MSR test we may conclude that n is definitely composite. This value of $n = 2821 = 7 \cdot 13 \cdot 31$ is an example of a Carmichael number introduced in Exercise 2.23 on page 101.

Remark 2.18 *If n is composite but declared to be "probably prime" with base a by the Miller-Selfridge-Rabin test, then*

n is said to be a strong pseudoprime to base a.

Thus, the MSR test is often called the strong pseudoprime test *in the literature. Strong pseudoprimes to base a are much sparser than composite n for which $a^{n-1} \equiv 1 \,(\mathrm{mod}\ n)$, called* pseudoprimes to base a. *An instance of the latter that is not an example of the former is given in Example 2.25, since $n = 2821$ is a pseudoprime to base 2 since it is a Carmichael number, but as the example demonstrates, is not a strong pseudoprime to base 2. Carmichael numbers are also called* absolute pseudoprimes, *since they are pseudoprimes to any base (including those bases a for which $\gcd(a, n) > 1$).*

▼ Analysis

Let us look a little closer at the MSR test to see why it is possible to declare that "n is definitely composite" in step (3). If $x \equiv 1 \pmod{n}$ in step (3), then for some j with $1 \leq j < t - 1$:

$$a^{2^j m} \equiv 1 \pmod{n}, \text{ but } a^{2^{j-1} m} \not\equiv \pm 1 \pmod{n}.$$

Thus, it can be shown that $\gcd(a^{2^{j-1} m} - 1, n)$ is a nontrivial factor of n. Hence, if the MSR test declares in step (3) that "n is definitely composite," then it is with 100% certainty. In other words, if n is prime, then MSR will declare it to be so. However, if n is composite, then it can be shown that the test fails to recognize n as composite with probability at most $(1/4)$.

This is why the most we can say is that "n is probably prime" at any step in the algorithm. However, if we perform the test r times for r large enough, the probability $(1/4)^r$ can be brought arbitrarily close to zero.

Also, in step (5), notice that we have not mentioned the possibility that

Biography 2.9 John Selfridge *was born in Ketchikan, Alaska, on February 17, 1927. He received his doctorate from U.C.L.A. in August of 1958, and became a professor at Pennsylvania State University six years later. He is a pioneer in computational number theory. The term "strong pseudoprime" was introduced by Selfridge in the mid-1970's, but he did not publish this reference. However, it did appear in a paper by Williams [52] in 1978. The MSR test is most often called the Miller-Rabin test. However, Selfridge was using the test in 1974 before the publication by Miller.*

$$a^{2^{t-1} m} \equiv 1 \pmod{n}$$

specifically. However, if this did occur, then that means that in step (3), we would have determined that

$$a^{2^{t-2} m} \not\equiv \pm 1 \pmod{n},$$

from which it follows that n cannot be prime. Furthermore, by the above method, we can factor n since $\gcd(a^{2^{t-2} m} - 1, n)$ is a nontrivial factor. This final step (4) is required since, if we get to $j = t - 1$, with $x \not\equiv \pm 1 \pmod{n}$ for any $j < t - 1$, then simply invoking step (3) again would dismiss those values of $x \not\equiv \pm 1 \pmod{n}$, and this would not allow us to claim that n is composite in those cases. Hence, it allows for more values of n to be deemed composite, with certainty, than if we merely performed step (3) as with previous values of j.

▼ How Pseudoprimes Pass MSR

We have mentioned that strong pseudoprimes are necessarily less likely to occur than pseudoprimes. We now present an example of a strong pseudoprime and explanation of the mechanism by which it escapes detection via MSR.

Consider $n = 1373653$ and $a = 2$. Since $n - 1 = 2^2 \cdot 343413 = 2^t \cdot m$, then

$$x_0 \equiv 2^m \equiv 890592 \pmod{n} \text{ and } x_1 = x_{t-1} \equiv 2^{2m} \equiv -1 \pmod{n},$$

then by step (3) of MSR, we declare that n is probably prime. However, the prime decomposition is $n = 829 \cdot 1657$. Hence, n is a strong pseudoprime. Now, we look at how this occurs in more detail.

From the above, we have that $x_0 \not\equiv 1 \pmod{q}$ for each of the prime divisors q of n, and $x_1 \equiv -1 \pmod{q}$ for each such q. But $x_2 \equiv 2^{n-1} \equiv 1 \pmod{n}$. In other words, the first time each of the $x_i \equiv 1 \pmod{q}$ for each prime q dividing n is at $i = 2$. It is rare to have the sequences $x_i \pmod{q}$ reach 1 at the same time for each prime dividing n. As an instance, we look to Example 2.25, which failed to pass the MSR even though it is an absolute pseudoprime. In that case,

$$x_0 \equiv 1 \pmod{7}, x_0 \equiv 5 \pmod{13}, x_0 \equiv 1 \pmod{31};$$

$$x_1 \equiv 1 \pmod{7}, x_1 \equiv -1 \pmod{13}, x_1 \equiv 1 \pmod{31};$$

$$x_2 \equiv 1 \pmod{7}, x_2 \equiv 1 \pmod{13}, x_2 \equiv 1 \pmod{31}.$$

Notice: the first time $x_i \equiv 1 \pmod{7}$ is for $i = 0$, the first time $x_i \equiv 1 \pmod{13}$ is for $i = 2$, and the first time $x_i \equiv 1 \pmod{31}$ is for $i = 0$. Hence, they do not all reach 1 at the same time. The scarcity of this phenomenon points to the effectiveness of the MSR test.

The MSR test is an example of a *Monte Carlo* algorithm, meaning a probabilistic algorithm that achieves a correct answer more than 50% of the time. More specifically, Miller-Selfridge-Rabin is a Monte Carlo algorithm for compositeness, since it provides a proof that a given input is composite but provides only some probabilistic evidence of primality. Furthermore, Miller-Selfridge-Rabin is a *yes-biased* Monte Carlo algorithm, meaning that a "yes" answer is always correct but a "no" answer may be incorrect. There are related algorithms that we have not discussed here, such as the Solovay-Strassen test, because the Miller-Selfridge-Rabin test is computationally less expensive, easier to implement, and at least as correct.

> **Biography 2.10** Gary Miller obtained his Ph.D. in computer science from U.C. Berkeley in 1974. He is currently a professor in computer science at Carnegie-Mellon University. His expertise lies in computer algorithms.

Outputs declared to be prime by probabilistic primality testing algorithms such as the MSR test are called *probable primes*. Sometimes, integers n satisfying $b^{n-1} \equiv 1 \pmod{n}$ are said to be *base-b probable primes*. The MSR test can be utilized as a vehicle for generating large probable primes as follows. We first need the following notions. If $B \in \mathbb{N}$, then a positive integer n is said to be a *B-smooth number*, if all primes dividing n are no larger than B, and B is called a *smoothness bound*.

> **Biography 2.11** Michael Rabin (1931–) *was born in Breslau, Germany (now Wroclaw, Poland), in 1931. In 1956, he obtained his Ph.D. from Princeton University where he later taught. In 1958, he moved to the Hebrew University in Jerusalem. He is known for his seminal work in establishing a rigorous mathematical foundation for finite automata theory. For such achievements, he was co-recipient of the 1976 Turing Award, along with Dana S. Scott. Both Rabin and Scott were doctoral students of Alonzo Church at Princeton. He now divides his time between positions at Harvard and the Hebrew University in Jerusalem.*

◆ Large (Probable) Prime Generation

We let b be the input bitlength of the desired prime and let B be the input smoothness bound (empirically determined). Execute the following steps.

(1) Randomly generate an odd b-bit integer n.

(2) Use trial division to test for divisibility of n by all odd primes no bigger than B. If n is so divisible, go to step (1). Otherwise go to step (3).

(3) Use the MSR to test n for primality. If it is declared to be a probable prime, then output n as such. Otherwise, go to step (1).

There is a mechanism for providing large *provable* primes, namely the positive output of a primality proving algorithm, or true primality test that we studied in §1.8. Before we state this result, we need to develop some more machinery. The first is a true primality test relying on knowledge of a partial factorization of $n - 1$ for a given $n \in \mathbb{N}$ — see Biography 2.12 on page 125.

Theorem 2.25 | Pocklington's Theorem

Let $n = ab + 1 \in \mathbb{N}$ with $a, b \in \mathbb{N}$, $b > 1$ and suppose that for every prime divisor q of b there exists an integer m such that $m^{n-1} \equiv 1 \pmod{n}$ and $\gcd(m^{(n-1)/q} - 1, n) = 1$. Then $p \equiv 1 \pmod{b}$ for every prime $p \mid n$. Furthermore, if $b > \sqrt{n} - 1$, then n is prime.

Proof. Let $p \mid n$ be prime and set $c = m^{(n-1)/q^e}$ where q is a prime and $e \in \mathbb{N}$ with $q^e \| b$. Therefore, since

$$\gcd(m^{(n-1)/q} - 1, n) = 1,$$

then $c^{q^e} \equiv 1 \pmod{p}$, but $c^r \not\equiv 1 \pmod{p}$ for any $r < q^e$. By Fermat's Little Theorem $q^e \leq p - 1$ so we let $p - 1 = q^e s + r$ where $0 \leq r < q^e$. Thus, by Fermat's Little Theorem again,

$$1 \equiv c^{p-1} \equiv c^{q^e s + r} \equiv (c^{q^e})^s c^r \equiv c^r \pmod{p},$$

so by the minimality of q^e we must have that $r = 0$. Since q was arbitrarily chosen, then $p \equiv 1 \pmod{b}$. For the last assertion of the theorem, assume that $b > \sqrt{n} - 1$ and that n is composite. Let p be the smallest prime dividing n. Then $p \leq \sqrt{n}$, so $\sqrt{n} \geq p > b \geq \sqrt{n}$, a contradiction. Hence, n is prime. $\qquad \square$

Example 2.26 *Suppose that we wish to test $n = 19079$ for primality using Pocklington's Theorem knowing that $n - 1 = 2 \cdot 9539$, where 9539 is prime, and if $b = 9539 = q$, with $a = m = 2$, then $m^{n-1} = 2^{n-1} \equiv 1 \pmod{n}$ but $\gcd(m^{(n-1)/q} - 1, n) = \gcd(3, 19079) = 1$, so n is prime.*

The following returns our attention to Fermat's Little Theorem 1.24 on page 50 as a true primality test. We will be able to use Pocklington's Theorem to verify the result.

| **Theorem 2.26** | **Testing via the Converse of Fermat's Little Theorem** |

Suppose that $n \in \mathbb{N}$ with $n \geq 3$. Then n is prime if and only if there exists an $m \in \mathbb{N}$ such that $m^{n-1} \equiv 1 \pmod{n}$, but $m^{(n-1)/q} \not\equiv 1 \pmod{n}$ for any prime $q \mid (n-1)$.

Proof. First suppose that n is prime. For each positive integer $d \leq n - 1$, we let $N(d)$ denote the number of those $m \in \{1, 2, \ldots, n - 1\}$ such that

$$m^d \equiv 1 \pmod{n} \text{ but } m^j \not\equiv 1 \pmod{n} \text{ for any positive integer } j < d. \quad (2.19)$$

By a similar argument to the proof of Pocklington's Theorem, each such d must divide $p - 1$. It follows that

$$\sum_{d \mid (n-1)} N(d) = n - 1.$$

However, from Theorem 2.11 on page 94,

$$\sum_{d \mid (n-1)} \phi(d) = n - 1.$$

Now if we can prove that $N(d) \leq \phi(d)$ for each d, then we will have that $N(d) = \phi(d)$ by virtue of the equality $\sum_{d \mid (n-1)} N(d) = \sum_{d \mid (n-1)} \phi(d)$.

If m is one of the $N(d)$ integers satisfying (2.19), then m^j for $j = 1, 2, \ldots, d$ are incongruent modulo n. Furthermore, each of the m^j are roots of $x^d - 1 \equiv 0 \pmod{n}$. Hence, from the result (2.13), established in Example 2.22 on page 114, namely that $x^d - 1$ has d incongruent roots modulo n, every one of those roots is congruent to one of those powers of m. If m^j is any one of those powers for $1 \leq j < d$ with $\gcd(j, d) = g$, then $(m^j)^{d/g} \equiv (m^d)^{j/g} \equiv 1 \pmod{n}$, so by (2.19), $g = 1$. There are $\phi(d)$ integers $j < d$ and relatively prime to d, so if there is one m satisfying (2.19), then there are exactly $\phi(d)$ values of m satisfying (2.19). This proves that $N(d) \leq \phi(d)$. Hence, by the above, $N(d) = \phi(d)$. In particular, if $d = n - 1$, then there are (for $n \geq 3$) $\phi(n - 1) \geq 1$ incongruent integers satisfying (2.19), so if we pick one of these as our value of m we have that $m^{n-1} \equiv 1 \,(\mathrm{mod}\ n)$, but $m^{(n-1)/q} \not\equiv 1 \,(\mathrm{mod}\ n)$ for any prime $q \mid (n-1)$.

Conversely assume that there is such an integer m. Let $n = ab + 1$. If both a and b are less than or equal to $\sqrt{n} - 1$, then

$$n = ab + 1 \le (\sqrt{n} - 1)^2 + 1 = n - 2\sqrt{n} + 2,$$

so $\sqrt{n} \le 1$ forcing $n = 1$, contradicting the hypothesis. Thus, without loss of generality, assume that $b > \sqrt{n}-1$, and the result now follows from Pocklington's Theorem. $\quad\square$

A major pitfall with the above primality test is that we must have knowledge of a factorization of $n-1$, so it works well on special numbers such as Fermat numbers, for instance. However, the above is a general "proof" that n is prime since the test finds an element of order $n - 1$ in $(\mathbb{Z}/n\mathbb{Z})^*$. Furthermore, it can be demonstrated that if we have a factorization of $n - 1$ and n is prime, then the above primality test can be employed to prove that n is prime in polynomial time; but if n is composite the algorithm will run without bound, or *diverge*.

> **Biography 2.12** Henry Cabourn Pocklington (1870–1952) *worked mainly in physics, the discoveries in which got him elected as a Fellow of the Royal Society. His professional career was spent as a physics teacher at Leeds Central Higher Grade School in England up to his retirement in 1926. Nevertheless, his six papers in number theory were practical and innovative. See* [44] *for more detail.*

There is one more observation worth making before we leave the discussion of Fermat's Little Theorem and primality testing. The following is immediate from that result.

◆ **Compositeness Test Via Fermat's Little Theorem**

If $n \in \mathbb{N}$, $a \in \mathbb{Z}$, and $\gcd(a, n) = 1$, such that

$$a^{n-1} \not\equiv 1 \pmod{n}, \tag{2.20}$$

then n is composite.

It is clear that if (2.20) holds, then n is composite, but if it fails, we cannot conclude that n is prime. Carmichael numbers provide an infinite number of counterexamples to that conclusion since they are absolute pseudoprimes, given that they satisfy $a^{n-1} \equiv 1 \,(\mathrm{mod}\ n)$ for any base a prime to n.

We have now seen three important types of tests for *recognizing primes*:

(1) The test has a condition for compositeness. If n satisfies the condition, then n *must* be composite. If n fails the test, it might still be composite (with low probability). Therefore, a successful completion of the test always guarantees that n is composite, but an unsuccessful completion of the test does *not* prove that n is prime. For instance, the above test for compositeness using Fermat's

Little Theorem with condition (2.20) is one such test. Such tests are known as *compositeness tests*.

(2) The test has a condition for primality. If n satisfies the condition, then n *must* be prime and if n fails the condition, then n *must* be composite. Theorem 2.26 on page 124 is an instance of this type of test. This type of test is known as a *deterministic primality test*.

(3) The test has a condition for primality. If n passes the test, then n is *probably* prime (with high probability). Such tests are known as *probabilistic primality tests*. For example, the MSR test in this section is such a test. In fact, the MSR test is an instance of a randomized algorithm that provides a proof of compositeness, but only good evidence of primality. If such tests are run a sufficient number of times, the evidence that n is prime becomes overwhelming, meaning that the probability of error is brought to negligible levels.

Now we provide the promised method for generating provable primes.

◆ Large (Provable) Prime Generation

Begin with a prime p_1, and execute the following steps until you have a prime of the desired size. Initialize the variable counter $j = 1$.

(1)   Randomly generate a small odd integer m and form $n = 2mp_j + 1$.

(2)   If $2^{n-1} \not\equiv 1 \pmod{n}$, then go to step (1). Otherwise, go to step (3).

(3)   Using the primality test given in Theorem 2.26 on page 124, with prime bases $2 \leq a \leq 23$, if for any such a,

$$a^{(n-1)/p} \not\equiv 1 \pmod{n}$$

for any prime p dividing $n - 1$, then n is prime. If n is large enough, terminate the algorithm with output n as the provable prime. Otherwise, set $n = p_{j+1}$, $j = j + 1$, and go to step (1). If the test fails, go to step (1).

Note that since we have a known factorization of $n-1$ in the above algorithm, and a small value of m to check, then the test is simple and efficient.

It is important in cryptographic applications to have an adequate supply of large random primes and the above method is one mechanism for so doing. We will return to these issues as we explore these applications in more depth later on.

Exercises

2.74. Use the MSR test to determine if $n = 9547$ is prime.

2.75. Use the MSR test to determine if $n = 9221$ is prime.

2.76. Prove that if n is a base-2 pseudoprime, then $2^n - 1$ is a strong pseudoprime to base 2.

2.77. For $n \in \mathbb{N}$ prove that $2^{2^n} + 1$ is a strong pseudoprime to base 2.

2.8 Cryptology

> *A secret in the Oxford sense: you may tell it to only one person at a*
> *time.*   In the Sunday Telegraph. January 30, 1977
> **Lord Franks (1905–1992), British philosopher and**
> **administrator**

One of the most important applications of number theory is to the area of secret communication, which has been of interest since antiquity. In this section, we look at applications of modular arithmetic to secret communication. Some of the following is adapted from [34].

◆ Terminology

Whether communication is in the military, commerce, diplomacy, or the strictly personal, the goal is to send a message so that only the intended recipient can read it. The study of methods for sending messages in *secret* (namely, in *enciphered* or *disguised* form) so that only the intended recipient can remove the disguise and read the message (or *decipher* it) is called *cryptography*. Cryptography has, as its etymology, *kryptos* from the Greek, meaning *hidden*, and *graphein*, meaning *to write*. The original message is called the *plaintext*, and the disguised message is called the *ciphertext*. The final message, encapsulated and sent, is called a *cryptogram*.

The process of transforming plaintext into ciphertext is called *encryption* or *enciphering*. The reverse process of turning ciphertext into plaintext, which is accomplished by the recipient who has the knowledge to remove the disguise, is called *decryption* or *deciphering*. Anyone who engages in cryptography is called a *cryptographer*. On the other hand, the study of mathematical techniques for attempting to defeat cryptographic methods is called *cryptanalysis*. Those practicing cryptanalysis (usually termed the "enemy") are called *cryptanalysts*.

The term *cryptology* is used to embody the study of both cryptography and cryptanalysis, and the practitioners of cryptology are *cryptologists*. The etymology of cryptology is the Greek *kryptos* meaning *hidden* and *logos* meaning *word*. Also, the term *cipher* (which we will use interchangeably with the term *cryptosystem*) is a method for enciphering and deciphering. We now formalize the above discussion in mathematical terms.

Definition 2.18 | Enciphering and Deciphering Transformations

An enciphering transformation (*also called an* enciphering function) *is a bijective function*

$$E_e : \mathcal{M} \mapsto \mathcal{C},$$

where the key $e \in \mathcal{K}$ uniquely determines E_e acting upon plaintext message units $m \in \mathcal{M}$ to get ciphertext message units

$$E_e(m) = c \in \mathcal{C}.$$

A deciphering transformation (*or* deciphering function) *is a bijective function*

$$D_d : \mathcal{C} \mapsto \mathcal{M},$$

which is uniquely determined by a given key $d \in \mathcal{K}$, *acting upon ciphertext message units* $c \in \mathcal{C}$ *to get plaintext message units*

$$D_d(c) = m.$$

The application of E_e *to* m, *namely the operation* $E_e(m)$, *is called* enciphering, encoding, *or* encrypting $m \in \mathcal{M}$, *whereas the application of* D_d *to* c *is called* deciphering, decoding, *or* decrypting $c \in \mathcal{C}$.

Definition 2.19 | Cryptosystems/Ciphers |

A cryptosystem *is composed of a set*

$$\{E_e : e \in \mathcal{K}\}$$

consisting of enciphering transformations and the corresponding set

$$\{E_e^{-1} : e \in \mathcal{K}\} = \{D_d : d \in \mathcal{K}\}$$

of deciphering transformations. In other words, for each $e \in \mathcal{K}$, *there exists a unique* $d \in \mathcal{K}$ *such that* $D_d = E_e^{-1}$, *so that* $D_d(E_e(m)) = m$ *for all* $m \in \mathcal{M}$.

The keys (e, d) *are called a key* pair *where possibly* $e = d$. *A cryptosystem is also called a* cipher. *We reserve the term* Cipher Table *for the pairs of plaintext symbols and their ciphertext equivalents*

$$\{(m, E_e(m)) : m \in \mathcal{M}\}.$$

The case where $e = d$ *or where one of them may be "easily" determined from the other in the key pair is called a* symmetric-key *cipher, which is the simplest of the possibilities for cryptosystems, and so has the longest history. Such ciphers are also called* single-key, one-key, *and* conventional

The simplest examples are *monographic*, *character*, or *substitution* ciphers, which replace individual letters with other letters by a *substitution*. As a simple example, the plaintext might be *palace*, and the ciphertext might be *QZYZXW* when a,c,e,l,p are replaced by Z,X,W,Y,Q, respectively. (The cryptographic convention is to use *lower-case* letters for *plaintext* and *UPPER-CASE* letters for *CIPHERTEXT*.)

The following is a famous substitution cipher.

Example 2.27 *Julius Caesar invented a cipher based upon a simple shift of the letters three places to the right in the given alphabet. The following table gives the Caesar cipher for the English alphabet.*

Table 2.1

Plaintext	a	b	c	d	e	f	g	h	i	j	k	l	m
Cipher	D	E	F	G	H	I	J	K	L	M	N	O	P
Plaintext	n	o	p	q	r	s	t	u	v	w	x	y	z
Ciphertext	Q	R	S	T	U	V	W	X	Y	Z	A	B	C

Now by assigning numbers to each letter, we have the following.

Table 2.2

Plaintext	a	b	c	d	e	f	g	h	i	j	k	l	m
Ciphertext	0	1	2	3	4	5	6	7	8	9	10	11	12
Plaintext	n	o	p	q	r	s	t	u	v	w	x	y	z
Ciphertext	13	14	15	16	17	18	19	20	21	22	23	24	25

Putting these together, we get the full Caesar cipher equivalents both numerically and alphabetically.

Table 2.3

Plaintext	a	b	c	d	e	f	g	h	i	j	k	l	m
	0	1	2	3	4	5	6	7	8	9	10	11	12
Ciphertext	D	E	F	G	H	I	J	K	L	M	N	O	P
	3	4	5	6	7	8	9	10	11	12	13	14	15
Plaintext	n	o	p	q	r	s	t	u	v	w	x	y	z
	13	14	15	16	17	18	19	20	21	22	23	24	25
Ciphertext	Q	R	S	T	U	V	W	X	Y	Z	A	B	C
	16	17	18	19	20	21	22	23	24	25	0	1	2

The mathematical interpretation of the Caesar cipher may be defined as that transformation E_e uniquely determined by the key $e = 3$, which is addition of 3 modulo 26. Thus,

$$E_3(m) = c \equiv m + 3 \pmod{26},$$

or simply

$$E_3(m) = c = m + 3 \in \mathcal{C} = \mathbb{Z}/26\mathbb{Z}.$$

Also, $m \in \mathcal{M} = \mathbb{Z}/26\mathbb{Z}$ is the numerical equivalent of the plaintext letter as described above. Similarly, $D_3(c)$ is that deciphering transformation uniquely defined by the key $d = 3$, which is modular subtraction of 3 modulo 26. In other words,

$$D_3(c) = m \equiv c - 3 \pmod{26},$$

or simply

$$D_3(c) = m = c - 3 \in \mathbb{Z}/26\mathbb{Z},$$

and $c \in \mathcal{C} = \mathbb{Z}/26\mathbb{Z}$ is the numerical equivalent of the ciphertext letter. Notice that $D_3(E_3(m)) = m$ for each $m \in \mathcal{M}$.

An example of a cryptogram made with the Caesar cipher is: brutus, via modular arithmetic on $1, 17, 20, 19, 20, 18$, becomes $4, 20, 23, 22, 23, 21$, that yields *the ciphertext* EUXWXV, *which the reader may check via Table 2.3 on the preceding page.*

Suppose that we are given the following ciphertext accomplished via the Caesar cipher: **WKH GLH LV FDVW**. *To decipher it, we translate to numerical values via Table 2.3 to get:* $22, 10, 7, \quad 6, 11, 7, \quad 11, 21, \quad 5, 3, 21, 22$. *Then we perform* $D_3(c) \equiv c - 3 \,(\mathrm{mod}\ 26)$ *on each value to get,* $19, 7, 4, \quad 3, 8, 4, \quad 8, 18, \quad 2, 0, 18, 19$, *the plaintext equivalent of* **the die is cast***, which is actually a quote by Julius Caesar himself made when crossing the River Rubicon, which delineated the frontier between Gaul and Italy proper. The quote indicates the fact that he was virtually declaring war on Rome since his military power was limited to Gaul.*

The Caesar cipher is a member of a family described by *shift transformations*, $E_e(m) \equiv m + b \,(\mathrm{mod}\ 26)$ where b is the key describing the magnitude of the shift of the letters in our alphabet. More generally, let $a, b, n \in \mathbb{N}$ and for $m \in \mathbb{Z}$ define

$$E_e(m) \equiv am + b \pmod{n},$$

where transformation key e is the ordered pair (a, b). Notice that for $a = 1$ we are back to the shift transformation where the key is b. Such a transformation is called an *affine function*. In order to guarantee that the deciphering transformation exists, we need to know that the inverse of the affine function exists. This means that $f^{-1}(c) \equiv a^{-1}(c - b) \,(\mathrm{mod}\ n)$ must exist and this can happen only if $\gcd(a, n) = 1$. Also, we know that there are $\phi(n)$ natural numbers less than n and relatively prime to it. Hence, since b can be any of the choices of natural numbers less than n, we have shown that there are exactly $n\phi(n)$ possible affine ciphers, the product of the possible choices for a with the number for b, since this is the total number of possible keys. We have motivated the following.

◆ Affine Ciphers

Let $\mathcal{M} = \mathcal{C} = \mathbb{Z}/n\mathbb{Z}$, $n \in \mathbb{N}$, $\mathcal{K} = \{(a, b) : a, b \in \mathbb{Z}/n\mathbb{Z} \text{ and } \gcd(a, n) = 1\}$, and for $e, d \in \mathcal{K}$, and $m, c \in \mathbb{Z}/n\mathbb{Z}$, set $E_e(m) \equiv am + b \,(\mathrm{mod}\ n)$, and $D_d(c) \equiv a^{-1}(c - b) \,(\mathrm{mod}\ n)$.

Thus, as with the shift transformation of which the affine cipher is a generalization, $e = (a, b)$ since e is multiplication by a followed by addition of b modulo n, and $d = (a^{-1}, -b)$ is subtraction of b followed by multiplication with a^{-1}. In the case of the shift transformation, the inverse is additive and in the case of the affine cipher, the inverse is multiplicative. Of course, these coincide precisely when $a = 1$. In either case, knowing e or d allows us to easily determine the other, so they are symmetric-key cryptosystems.

Example 2.28 *Let* $n = 26$, *and let* $\mathcal{M} = \mathcal{C} = \mathbb{Z}/26\mathbb{Z}$. *Define an Affine Cipher as follows.*

$$E_e(m) = 5m + 9 = c \in \mathbb{Z}/26\mathbb{Z},$$

and since $5^{-1} \equiv 21 \,(\mathrm{mod}\ 26)$,

$$D_d(c) = 21(c-9) = 21c - 7 \in \mathbb{Z}/26\mathbb{Z} = \mathcal{M}.$$

Table 2.2 on page 129 provides the numerical equivalents for each element in
$\mathcal{M} = \mathcal{C}$.

Using the above, we wish to decipher the following message and provide plain-text:

JIIXWD

*To do this, we first translate each letter into the numerical equivalent in the
alphabet of definition, via Table 2.2 on page 129 as follows.*

$$9 \quad 8 \quad 8 \quad 23 \quad 22 \quad 3.$$

Then we apply $D_d(m)$ *to each of these numerical equivalents m to get the
following.*

$$0 \quad 5 \quad 5 \quad 8 \quad 13 \quad 4,$$

whose letter equivalents are

affine

Monoalphabetic ciphers suffer from the weakness that they can be cryptanalyzed via a frequency count of the letters in the ciphertext. For instance,
if a letter occurs most frequently in ciphertext, we might guess the plaintext
equivalent to be the letter E since E is the most commonly occurring letter in
the English alphabet. If correct, this would lead to other decryptions and the
cipher would be broken in this manner. For instance if the second most commonly occurring letter is guessed to be T, the second most commonly occurring
letter in English, then we have more decryptions.

Table 2.4 provides the letter frequencies for the English alphabet.

Relative Letter Frequencies for English
Table 2.4

a	b	c	d	e	f	g	h	i
8.167	1.492	2.782	4.253	12.702	2.228	2.015	6.094	6.966
j	k	l	m	n	o	p	q	r
0.153	0.772	4.025	2.406	6.749	7.507	1.929	0.095	5.987
s	t	u	v	w	x	y	z	
6.327	9.056	2.758	0.978	2.360	0.150	1.974	0.074	

To prevent cryptanalysis via frequency analysis as described above, we may
use ciphers that operate on blocks of plaintext rather than individual letters.

Definition 2.20 | **Block/Polygraphic Ciphers**

A block cipher, also know as a polygraphic cipher *is a cryptosystem that separates the plaintext message into strings, called* blocks, *of fixed length* $k \in \mathbb{N}$, *called the* blocklength, *and enciphers one block at a time.*

An illustration of a polygraphic cipher is the following due to Vigenère (see Biography 2.13 on page 133). He employed the idea that others had invented of using the plaintext as its own key. However, he added something new, a *priming key*, which is a single letter (known only to the sender and the legitimate receiver) that is used to decipher the first plaintext letter, which would, in turn, be used to decipher the second plaintext letter, and so on. The following is an example of an *autokey cipher*, which is a cryptosystem wherein the plaintext itself (in whole or in part) serves as the key (usually after employing an initial priming key).

◆ **The Autokey Vigenère Cipher**

Let $n \in \mathbb{N}$ and call $k_1 k_2 \cdots k_r$ for $1 \leq r \leq n$ a *priming key*. Then given a plaintext message unit $m = (m_1, m_2, \ldots, m_s)$ where $s > r$, we generate a keystream as follows:

$$k = k_1 k_2 \cdots k_r m_1 m_2 \cdots m_{s-r}.$$

Then we encipher via

$$E_{k_j}(m_j) = m_j + k_j \pmod{n} = c_j \text{ for } j = 1, 2, \ldots, r,$$

and

$$E_{k_j}(m_j) = m_j + m_{j-r} \pmod{n} = c_j \text{ for } j > r,$$

and decipher via

$$D_{k_j}(c_j) = c_j - k_j \pmod{n} = m_j \text{ for } j = 1, 2, \ldots, r,$$

and

$$D_{k_j}(c_j) = c_j - m_{j-r} \pmod{n} = m_j \text{ for } j > r.$$

Here is a simple example, where $n = 26$, which is the most commonly used value.

Example 2.29 *Given a priming key* $k = k_1 k_2 k_3 = 273$ *and* $n = 26$ *in the autokey Vigenère cipher, suppose we want to decrypt the Vigenère ciphertext*

CAWAVDNQG,

using Table 2.2 on page 129. Converting ciphertext to numerical equivalents, we have

$$2, 0, 22, 0, 21, 3, 13, 16, 6.$$

Thus, we compute the following:

$$m_1 = c_1 - k_1 = 2 - 2 = 0 \pmod{26}, \quad m_2 = c_2 - k_2 = 0 - 7 \equiv 19 \pmod{26},$$

$$m_3 = c_3 - k_3 = 22 - 3 = 19 \pmod{26}, \quad m_4 = c_4 - m_1 = 0 - 0 \equiv 0 \pmod{26},$$

$$m_5 = c_5 - m_2 = 21 - 19 \equiv 2 \pmod{26}, \quad m_6 = c_6 - m_3 = 3 - 19 \equiv 10 \pmod{26},$$

$$m_7 = c_7 - m_4 = 13 - 0 \equiv 13 \pmod{26}, \quad m_8 = c_8 - m_5 = 16 - 2 \equiv 14 \pmod{26},$$

and

$$m_9 = c_9 - m_6 = 6 - 10 \equiv 22 \pmod{26}.$$

Via Table 2.2, the letter equivalents give us

attack now

Biography 2.13 Blaise de Vigenère (1523–1596) *was born in Saint-Pourçain, France. He had his first contact with cryptography at age twenty-six when he went to Rome on a two-year diplomatic mission. He read cryptographic books, and discussed the subject with the experts there. In 1570, he retired from court, married, and settled down to a life of writing. He authored over 20 books, including his masterpiece,* Traicté des Chiffres, *published in 1585, containing his contributions to cryptography. Vigenère discussed a variety of cryptographic ideas, including the idea for an autokey polyalphabetic substitution cipher. Moreover, therein he discusses such subjects as magic and alchemy.*

In an effort to inject more mathematical security into block ciphers, Lester Hill invented a block cipher in 1929 that uses some matrix theory — see Biography 2.14 on the following page.

◆ The Hill Cipher

Let $\mathcal{K} = \{e \in \mathcal{M}_{r \times r}(\mathbb{Z}/n\mathbb{Z}) : e \text{ is invertible}\}$, for fixed $r, n \in \mathbb{N}$, and set $\mathcal{M} = \mathcal{C} = (\mathbb{Z}/n\mathbb{Z})^r$. Then for $m \in \mathcal{M}$, $e \in \mathcal{K}$, $E_e(m) = me$, and $D_d(c) = ce^{-1}$, where $c \in \mathcal{C}$. (Note that e is invertible if and only if $\gcd(\det(e), n) = 1$. See Theorem A.5 on page 297.) This cryptosystem is known as the Hill cipher. The most common usage is for $r = 2$ and $n = 26$ as illustrated below, which is an illustration of a *digraph cipher*, which encrypts pairs of plaintext letters to produce pairs of ciphertext letters.

Example 2.30 Let $r = 2$ and $n = 26$ where Table 2.2 on page 129 gives the numerical equivalents of plaintext letters. Thus, $\mathcal{M} = \mathcal{C} = (\mathbb{Z}/26\mathbb{Z})^2$, and $\mathcal{K}$ consists of all invertible two-by-two matrices with entries from $\mathbb{Z}/26\mathbb{Z}$, so if $e \in \mathcal{K}$, then $\gcd(\det(e), 26) = 1$. Let us take

$$e = \begin{pmatrix} 7 & 2 \\ 5 & 3 \end{pmatrix}$$

for which $\det(e) = 11$. Suppose that we want to encipher *money*. First we get the numerical equivalents from Table 2.2: 12, 14, 13, 4, 24. Thus, we may set $m_1 = (12, 14)$, $m_2 = (13, 4)$, and $m_3 = (24, 25)$, where z, with numerical equivalent of 25, is used to complete the last pair. Now use the enciphering transformation defined in the Hill cipher.

$$E_e(m_1) = (12, 14) \begin{pmatrix} 7 & 2 \\ 5 & 3 \end{pmatrix} = (24, 14),$$

$$E_e(m_2) = (13, 4) \begin{pmatrix} 7 & 2 \\ 5 & 3 \end{pmatrix} = (7, 12),$$

and

$$E_e(m_3) = (24, 25) \begin{pmatrix} 7 & 2 \\ 5 & 3 \end{pmatrix} = (7, 19).$$

Now we use Table 2.2 to get the ciphertext letter equivalents and send *YO HM HT* as the cryptogram of pairs.

Now we show how decryption works. Once the cryptogram is received, we must calculate the inverse of e, which is

$$e^{-1} = \begin{pmatrix} 5 & 14 \\ 9 & 3 \end{pmatrix}.$$

Now apply the deciphering transformation to the numerical equivalents of the ciphertext as follows. Given $c_1 = (24, 14)$, $c_2 = (7, 12)$, $c_3 = (7, 19)$, we have

$$D_d(c_1) = D_{e^{-1}}(24, 14) = (24, 14) \begin{pmatrix} 5 & 14 \\ 9 & 3 \end{pmatrix} = (12, 14),$$

$$D_d(c_2) = D_{e^{-1}}(7, 12) = (7, 12) \begin{pmatrix} 5 & 14 \\ 9 & 3 \end{pmatrix} = (13, 4),$$

and

$$D_d(c_3) = D_{e^{-1}}(7, 19) = (7, 19) \begin{pmatrix} 5 & 14 \\ 9 & 3 \end{pmatrix} = (24, 25).$$

The letter equivalents now give us back the original plaintext message *money* after discarding the letter z at the end.

Biography 2.14 Lester S. Hill *devised this cryptosystem in 1929. His only published papers in the area of cryptography appeared in 1929 and 1931. Thereafter, he kept working on cryptographic ideas but turned all of his work over to the Navy in which he had served as a lieutenant in World War I. He taught mathematics at Hunter College in New York from 1927 until his retirement in 1960. He died in Lawrence Hospital in Bronxville, New York, after suffering through a lengthy illness. Hill's rigorous mathematical approach may be said to be one of the factors which has helped foster today's solid grounding of cryptography in mathematics.*

We have learned about one type of symmetric-key cryptosystem, block ciphers. Now we look at the other type of symmetric-key cryptosystem. First we need the following notions.

Definition 2.21 | **Keystreams, Seeds, and Generators**

If $\mathcal{K}$ is the keyspace for a set of enciphering transformations, then a sequence $k_1 k_2 \cdots \in \mathcal{K}$ is called a keystream. *A keystream is either randomly chosen or generated by an algorithm, called a* keystream generator, *which generates the keystream from an initial small input keystream called a* seed. *Keystream generators that eventually repeat their output are called* periodic.

Definition 2.22 | **Stream Ciphers**

Let $\mathcal{K}$ be a keyspace for a cryptosystem and let $k_1 k_2 \cdots \in \mathcal{K}$ be a keystream. This cryptosystem is called a stream cipher *if encryption upon plaintext strings $m_1 m_2 \cdots$ is achieved by repeated application of the enciphering transformation on plaintext message units, $E_{k_j}(m_j) = c_j$, and if d_j is the inverse of k_j, then deciphering occurs as $D_{d_j}(c_j) = m_j$ for $j \geq 1$. If there exists an $\ell \in \mathbb{N}$ such that $k_{j+\ell} = k_j$ for all $j \in \mathbb{N}$, then we say that the stream cipher is* periodic *with period ℓ.*

The following is the simplest flow chart for a stream cipher.

Diagram 2.1 A Stream Cipher

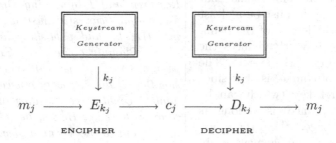

Generally speaking, stream ciphers are faster than block ciphers from the perspective of hardware. The reason is that stream ciphers encrypt individual plaintext message units, usually, but not always, one binary digit at a time. In practice, the stream ciphers used are most often those that do indeed encipher one bit at a time.

One of the simplest stream ciphers is the following — see Biography 2.15 on the next page.

◆ **The Vernam Cipher**

The Vernam cipher is a stream cipher that enciphers in the following fashion. Given a bitstring

$$m_1 m_2 \cdots m_n \in \mathcal{M},$$

and a keystream

$$k_1 k_2 \cdots k_n \in \mathcal{K},$$

the enciphering transformation is given by

$$E_{k_j}(m_j) = m_j + k_j = c_j \in \mathcal{C},$$

and the deciphering transformation is given by

$$D_{k_j}(c_j) = c_j + k_j = m_j,$$

where $+$ is addition modulo 2. The keystream is randomly chosen and never used again.

For this reason, the Vernam cipher is also called the *one-time pad* that can be shown to be unbreakable. This means that since the key is used only once then discarded, a cryptanalyst with access to the ciphertext $c_1 c_2 \cdots c_n$ can only guess at the plaintext $m_1 m_2 \cdots m_n$, since both are equally likely. Conversely, it has been shown that to have a theoretically unbreakable system means that the keylength must be at least that of the length of the plaintext. This vastly reduces the practicality of the system. The reason, of course, is that since the secret key (which can be used only once) is as long as the message, then there are serious key-management problems. Today, one-time pads are in use for military and diplomatic purposes when unconditional security is of the utmost importance. See [32, Chapter 11] for a detailed insight into the theory behind the proof of the security of the one-time pad, known as *information theory*.

Biography 2.15 Gilbert S. Vernam, (1890–1960) *a cryptologist working for the American Telephone and Telegraph (AT&T) Company, came to the realization that if the Vigenère cipher were used with a truly random key, with keylength the size of the plaintext, called a* running key, *then attacks would fail. At this time, AT&T was working closely with the armed forces, so the company reported this to the Army. It came to the attention of Major Mauborgne, head of the Signal Corps' Research and Engineering Division. (When Mauborgne was still just a first lieutenant in 1914, he had published the first solution of the Playfair cipher, see Exercise 2.90 on page 138.) He played with Vernam's idea and saw that if the key were reused, then a cryptanalyst could piece together information and recover the key. Hence, he added the second component to the Vernam idea. The key must be used once, and only once, then destroyed. Now, the idea was complete. Use the Vigenère cipher with a truly random running key that is* used *exactly once, then destroyed. The system is called the* one-time pad *and sometimes, perhaps inappropriately in view of Mauborgne's contribution, the* Vernam cipher.

We will return to one more symmetric-key cryptosystem in §3.5, where *public-key cryptography* is discussed. We require the notions in Chapter 3 to present the cipher, known as an *exponentiation cipher*, which will be valuable in setting the stage for public-key cryptography in general, and RSA in particular.

Exercises

2.78. Use the Caesar cipher to encrypt the plaintext

launch the attack.

2.79. Use the Caesar cipher to encrypt the plaintext

find the weapons.

2.80. Use the Caesar cipher to decrypt the ciphertext

QHYHU VDB DQBWKLQJ.

2.81. Use the Caesar cipher to decrypt the ciphertext

DVVXPH WKH ZRUVW.

2.82. Use the affine cipher given in Example 2.28 on page 130 to encrypt the plaintext

the banks will fail.

2.83. Use the affine cipher given in Example 2.28 on page 130 to encrypt the plaintext

follow the money.

2.84. Use the affine cipher given in Example 2.28 to decrypt the ciphertext

ABG VDTQDA.

2.85. Use the affine cipher given in Example 2.28 to decrypt the ciphertext

VDTFQXAZ JNDWTZ.

2.86. Use the affine cipher given in Example 2.30 on page 133 to encrypt the plaintext

fire all torpedos.

2.87. Use the affine cipher given in Example 2.30 to encrypt the plaintext

drop all bombs.

2.88. Use the affine cipher given in Example 2.30 to decrypt the ciphertext

PS RL HZ IW HZ.

2.89. Use the Vigenère cipher given in Example 2.29 on page 132 to encrypt the plaintext

Caesar cipher.

2.90. Consider the following digraph cipher, where the letters W and X are considered as a single entity.

A	Z	I	WX	D
E	U	T	G	Y
O	N	K	Q	M
H	F	J	L	S
V	R	P	B	C

Pairs of letters are enciphered according to the following rules.

(a) If two letters are in the same row, then their ciphertext equivalents are immediately to their right. For instance, VC in plaintext is RV in ciphertext. (This means that if one is at the right or bottom edge of the table, then one "wraps around" as indicated in the example.)

(b) If two letters are in the same column, then their cipher equivalents are the letters immediately below them. For example, ZF in plaintext is UR in ciphertext, and XB in plaintext is GW in ciphertext.

(c) If two letters are on the corners of a diagonal of a rectangle formed by them, then their cipher equivalents are the letters in the opposite corners and same row as the plaintext letter. For instance, UL in plaintext becomes GF in ciphertext and SZ in plaintext is FD in ciphertext.

(d) If the same letter occurs as a pair in plaintext, then we agree by convention to put a Z between them and encipher.

(e) If a single letter remains at the end of the plaintext, then a Z is added to it to complete the digraph.

Decipher the following message, which was enciphered using the above digraph cipher.

UP TG JA HY GU ZF WH

(*The idea behind the above digraph cipher was conceived by Sir Charles Wheatstone, and was sponsored at the British Foreign Office by Lord Lyon Playfair. Thus, it has become known as the Playfair cipher.*)

2.91. Using the Playfair cipher, described in in Exercise 2.90, decipher the following

AY PD VJ UV.

Chapter 3

Primitive Roots

3.1   Order

In this chapter we look at the multiplicative structure of $\mathbb{Z}/n\mathbb{Z}$ introduced
in Definition 2.4 on page 79. The topic of this chapter, primitive roots, defined
in this section, may be used to simplify the calculations in $\mathbb{Z}/n\mathbb{Z}$. The results
we develop will allow us to look at further applications to primality testing and
to random number generation, both of which are important in cryptographic
applications, such as in §3.5. We need to develop the tools to do so. First we
need the following concept related to Euler's Theorem 2.10 on page 93, which
tells us that for $m \in \mathbb{Z}$ and $n \in \mathbb{N}$ with $\gcd(m,n) = 1$, we have $m^{\phi(n)} \equiv 1$
(mod n). One may naturally ask for the *smallest* exponent $e \in \mathbb{N}$ such that
$m^e \equiv 1 \,(\text{mod } n)$.

Definition 3.1 | **Modular Order of an Integer**

Let $m \in \mathbb{Z}$, $n \in \mathbb{N}$, and $\gcd(m,n) = 1$. Then the order of m modulo n is the
smallest $e \in \mathbb{N}$ such that $m^e \equiv 1\,(\text{mod } n)$, denoted by $e = \operatorname{ord}_n(m)$, and we say
that m belongs to the exponent e modulo n.

Note that the modular order of an integer given in Definition 3.1 is the same
as the element order in the multiplicative group $(\mathbb{Z}/n\mathbb{Z})^*$, which we discussed
on page 81, see Equation (2.2).

Example 3.1 *We calculate that* $3^{12} \equiv 1 \,(\mathrm{mod}\ 35)$, *but* $3^j \not\equiv 1 \,(\mathrm{mod}\ 35)$ *for any natural number* $j < 12$, *so* $\mathrm{ord}_{35}(3) = 12$.

Example 3.2 *Since* $2^3 \equiv 1 \,(\mathrm{mod}\ 7)$ *but* $2^j \not\equiv 1 \,(\mathrm{mod}\ 7)$ *for* $j = 1, 2$, *then* $\mathrm{ord}_7(2) = 3$.

Example 3.3 *If we look at powers of* 2 *modulo* 11, *and see that*

$$2^{10} \equiv 1 \quad (\mathrm{mod}\ 11),$$

which we know by Fermat's Little Theorem. However, $2^d \not\equiv 1 \,(\mathrm{mod}\ 11)$ *for any positive integer* $d < 10$. *Hence,* $\mathrm{ord}_{11}(2) = 10$. *There is a name for integers which satisfy that* $\mathrm{ord}_n(m) = \phi(n)$, *namely the topic of this chapter,* primitive roots, *introduced by Euler in 1773.*

Definition 3.2 | **Primitive Roots**

If $m \in \mathbb{Z}$, $n \in \mathbb{N}$ *and*

$$\mathrm{ord}_n(m) = \phi(n),$$

then m *is called a* primitive root modulo n. *In other words,* m *is a primitive root if it belongs to the exponent* $\phi(n)$ *modulo* n.

Example 3.4 *In Example 3.3, we see that* 2 *is a primitive root modulo* 11. *We also have that* 2 *is a primitive root modulo the prime* 9547. *An example for* 3 *is given when we calculate that*

$$\mathrm{ord}_{2962}(3) = 1480,$$

so 3 *is a primitive root modulo the* $2962 = 2 \cdot 1481$, *where* 1481 *is prime, since* $\phi(2962) = 1480$. *However,* 35, *for instance, has no primitive roots. Later in this chapter, we will determine exactly those moduli that have primitive roots* — *see Theorem 3.7 on page 151.*

In Examples 3.1—3.2, we see that the order of an integer modulo n divides $\phi(n)$, and this is no coincidence.

Proposition 3.1 | **Divisibility by the Order of an Integer**

If $m \in \mathbb{Z}$, $d, n \in \mathbb{N}$ *such that* $\gcd(m, n) = 1$, *then* $m^d \equiv 1 \,(\mathrm{mod}\ n)$ *if and only if* $\mathrm{ord}_n(m) \mid d$.

Proof. If $\text{ord}_n(m) \mid d$, then $d = \text{ord}_n(m)x$ for some $x \in \mathbb{N}$, so

$$m^d = (m^{\text{ord}_n(m)})^x \equiv 1 \pmod{n}.$$

Conversely, if $m^d \equiv 1 \pmod{n}$, then $d \geq \text{ord}_n(m)$ so there exist integers q and r with $d = q \cdot \text{ord}_n(m) + r$ where $0 \leq r < \text{ord}_n(m)$ by the Division Algorithm. Thus, $1 \equiv m^d \equiv (m^{\text{ord}_n(m)})^q m^r \equiv m^r \pmod{n}$, so by the minimality of $\text{ord}_n(m)$, $r = 0$. In other words, $\text{ord}_n(m) \mid d$.

(Also, see the solution of Exercise 2.55 on page 107.) $\square$

Corollary 3.1 *If* $\gcd(m, n) = 1$, *where* $m \in \mathbb{Z}$ *and* $n \in \mathbb{N}$, *then*

$$\text{ord}_n(m) \mid \phi(n).$$

Proof. Given $\gcd(m, n) = 1$, Euler's Theorem says that $m^{\phi(n)} \equiv 1 \pmod{n}$. Therefore, by Proposition 3.1, $\text{ord}_n(m) \mid \phi(n)$. $\square$

Example 3.5 *Since the only possible orders modulo n are divisors of $\phi(n)$ by Corollary 3.1, then this reduces the search. For instance, to find the order of 3 modulo 25, we need only look at divisors of 20, namely $1, 2, 4, 5, 10, 20$. Since*

$$3^1 \equiv 3 \pmod{25}, \quad 3^2 \equiv 9 \pmod{25}, \quad 3^4 \equiv 6 \pmod{25}, \quad 3^5 \equiv 18 \pmod{25},$$

$$3^{10} \equiv 24 \pmod{25}, \quad 3^{20} \equiv 1 \pmod{25},$$

then we conclude that 3 is a primitive root modulo 25, without having to try all exponents $1 \leq j \leq 20$.

Note that we may rephrase Proposition 3.1 in terms of the group-theoretic language surrounding $(\mathbb{Z}/n\mathbb{Z})^*$, namely that if $\mathfrak{d}$ is the order of an element $m \in (\mathbb{Z}/n\mathbb{Z})^*$, then for any $d \in \mathbb{N}$, if $m^d = 1 \in (\mathbb{Z}/n\mathbb{Z})^*$, d must be a multiple of $\mathfrak{d}$. We use this language to prove the next fact.

Corollary 3.2 *If* $d, n \in \mathbb{N}$, *and* $m \in \mathbb{Z}$ *with* $\gcd(m, n) = 1$, *then*

$$\text{ord}_n(m^d) = \frac{\text{ord}_n(m)}{\gcd(d, \text{ord}_n(m))}.$$

Proof. Set $f = \text{ord}_n(m^d)$ (the order of m^d in $(\mathbb{Z}/n\mathbb{Z})^*$) and $g = \gcd(d, \text{ord}_n(m))$. Thus, by Proposition 3.1, $\text{ord}_n(m) \mid df$, so $(\text{ord}_n(m)/g) \mid fd/g$. Therefore, by Claim 1.3 on page 26, $(\text{ord}_n(m)/g) \mid f$. Also, since

$$(m^{\text{ord}_n(m)})^{d/g} = (m^d)^{\text{ord}_n(m)/g} = 1 \in (\mathbb{Z}/n\mathbb{Z})^*,$$

then by Proposition 3.1, applied to m^d this time, $f \mid (\text{ord}_n(m)/g)$. Hence, $f = (\text{ord}_n(m)/g)$, which is the intended result. $\square$

Example 3.6 *Since* $\text{ord}_{25}(3) = 20$ *by Example 3.5, and* $\gcd(5, 20) = 5$, *then by Corollary 3.2,* $\text{ord}_{25}(3^5) = \text{ord}_{25}(3)/\gcd(5, 20) = 4$.

Corollary 3.3 *Let* $m \in \mathbb{Z}$, e, $n \in \mathbb{N}$, *and* $\gcd(m, n) = 1$. *Then*

$$\text{ord}_n(m^e) = \text{ord}_n(m)$$

if and only if

$$\gcd(e, \text{ord}_n(m)) = 1.$$

Proof. By Corollary 3.2 on page 141,

$$\text{ord}_n(m^e) = \text{ord}_n(m)/\gcd(e, \text{ord}_n(m)).$$

Therefore, $\text{ord}_n(m^e) = \text{ord}_n(m)$ if and only if $\gcd(e, \text{ord}_n(m)) = 1$.   □

Example 3.7 *By Example 3.2 on page 140, we know that* $\text{ord}_7(2) = 3$, *so* $\text{ord}_7(2^2) = \text{ord}_7(2) = 3$ *by Corollary 3.3.*

Corollary 3.4 *If* m *is a primitive root modulo* n, *then* m^e *is a primitive root modulo* n *if and only if* $\gcd(e, \phi(n)) = 1$.

Proof. By Corollary 3.3, if m is a primitive root modulo n, then $\text{ord}_n(m^e) = \text{ord}_n(m)$ if and only if $\gcd(e, \phi(n)) = 1$.   □

We prove a useful result as a segue into a result taking us back to reduced residue systems introduced in Definition 2.8 on page 91.

Lemma 3.1 *If* $m \in \mathbb{Z}$ *and* $n \in \mathbb{N}$ *with* $\gcd(m, n) = 1$, *then* $m^i \equiv m^j \pmod{n}$ *for nonnegative integers* i, j *if and only if* $i \equiv j \pmod{\text{ord}_n(m)}$.

Proof. If $m^i \equiv m^j \pmod{n}$ for $0 \leq i \leq j \leq \phi(n)$, then since $\gcd(m, n) = 1$, we have $m^{j-i} \equiv 1 \pmod{n}$, by Proposition 2.3 on page 75. Therefore, by Proposition 3.1 on page 140, $\text{ord}_n(m) \mid (j - i)$, namely $i \equiv j \pmod{\text{ord}_n(m)}$.

Conversely, if $i \equiv j \pmod{\text{ord}_n(m)}$ for $0 \leq i \leq j$, then $j = i + q \cdot \text{ord}_n(m)$ where $q \geq 0$. Thus,

$$m^j \equiv m^{i + q\,\text{ord}_n(m)} \equiv m^i (m^{\text{ord}_n(m)})^q \equiv m^i \cdot 1^q \equiv m^i \pmod{n},$$

which secures the result.   □

Theorem 3.1 | **Primitive Roots and Reduced Residues** |

Let $m \in \mathbb{Z}$ and $n \in \mathbb{N}$ relatively prime to m. If m is a primitive root modulo n, then $\{m^j\}_{j=1}^{\phi(n)}$ is a complete set of reduced residues modulo n.

Proof. By Definition 2.8, we need to show both that $\gcd(m^j, n) = 1$, and that $m^i \equiv m^j \pmod{n}$ if and only if $i = j$. Since $\gcd(m, n) = 1$, then $\gcd(m^j, n) = 1$, which is the first part. If $m^i \equiv m^j \pmod{n}$, then by Lemma 3.1, this occurs if and only if $i \equiv j \pmod{\mathrm{ord}_n(m)}$. However, for $1 \le i, j \le \phi(n)$, this occurs if and only if $i = j$, which is the second part. $\qquad\square$

Example 3.8 *By Example 3.3 on page 140, we know that 2 is a primitive root modulo 11, so*
$$\{2, 2^2, 2^3, 2^4, 2^5, 2^6, 2^7, 2^8, 2^9, 2^{10}\},$$
is a reduced residue system modulo 11. Also, by Example 3.5 on page 141, 3 is a primitive root modulo 25, so $\{3^j\}_{j=1}^{20}$ is a reduced residue system modulo 25.

Theorem 3.1 leads to the following.

Theorem 3.2 | **The Number of Primitive Roots** |

If $n \in \mathbb{N}$ has a primitive root, then it has $\phi(\phi(n))$ incongruent primitive roots.

Proof. Let m be a primitive root modulo n. By Theorem 3.1, another primitive root must be of the form m^e with $1 \le e \le \phi(n)$. Thus, by Corollary 3.3 on the preceding page, $\mathrm{ord}_n(m) = \mathrm{ord}_n(m^e)$ if and only if $\gcd(e, \phi(n)) = 1$, and there are precisely $\phi(\phi(n))$ such integers e. $\qquad\square$

Example 3.9 *By Example 3.4 on page 140, 3 is a primitive root modulo 2962, so by Theorem 3.2, 2962 has exactly $\phi(\phi(2962)) = 576$ primitive roots. A simpler illustration also comes from Example 3.4, namely 2 is a primitive root modulo 11, so 11 has exactly $\phi(\phi(11)) = 4$ primitive roots.*

We close this section with an application of order to Diophantine analysis — see Biography 1.15 on page 48.

Example 3.10 *Suppose that we want to find all the solutions of the equation*
$$3^a + 1 = 2^b$$
for nonnegative integers a, b. We may use the order of an integer for this problem. Suppose that $a > 1$. Then $2^b \equiv 1 \pmod 9$. However, $\mathrm{ord}_9(2) = 6$, so $6 \mid b$ by Proposition 3.1 on page 140. Thus, there exists an integer m such that $b = 6m$. Hence, by Fermat's Little Theorem,
$$2^b \equiv (2^6)^m \equiv 1 \pmod 7,$$
so $7 \mid (2^b - 1) = 3^a$, a contradiction. Therefore, $a = 0, 1$ for which $b = 1, 2$, respectively, are the only solutions.
The reader may try Exercise 3.6 on the next page to test understanding of this methodology.

Exercises

3.1. Let $\gcd(m, n) = 1$ for $m \in \mathbb{Z}$, $n \in \mathbb{N}$. Prove that if $\mathrm{ord}_n(m) = ab$, then $\mathrm{ord}_n(m^b) = a$.

3.2. Find all Fermat primes $\mathcal{F}_n = 2^{2^n} + 1$ such that 2 is a primitive root modulo $\mathcal{F}_n$.

3.3. Let g be a primitive root modulo a prime $p > 2$. Prove that $p - g$ is a primitive root modulo p if and only if $p \equiv 1 \pmod 4$.

3.4. Let $q = 2p + 1$ where p and q are odd primes. Is 2 is a primitive root modulo q?

3.5. Prove that if g is a primitive root modulo a prime $p > 2$, then so is g_1 where $gg_1 \equiv 1 \pmod p$.

3.6. Use the methodology in Example 3.10 on the preceding page to find all solutions of $2^b + 1 = 3^a$ for nonnegative integers a, b.

   (*Hint: Prove that $b < 4$.*)

   (*The equations in Example 3.10 and Exercise 3.6 are related to a problem of Catalan, who proved in 1885 that $a^b - b^a = 1$, with $a > 1$, $b > 1$, only has solutions for $(a, b) \in \{(3, 2), (2, 3)\}$. He conjectured, more generally, that $a^b - c^d = 1$, with a, b, c, d, all bigger than 1, has solutions for only $a = 3, b = 2, c = 2, d = 3$.*)

3.7. Use Theorem 3.1 on page 142 to prove Wilson's Theorem.

Biography 3.1 Eugène Charles Catalan (1814–1894) *was born on May 30, 1814 in Brugge, Belgium. He obtained his degree in mathematics from École Polytechnique in 1841. He taught descriptive geometry at Charlemagne College, and was appointed, in 1865, to chair of analysis at the University of Liège. By 1883 he was working for the Belgian Academy of Science in number theory. He died in Liège on February 14, 1894.*
The conjecture bearing his name, cited above, was posed in 1844, and proved in 2002 by Preda Mihăilescu. The proof employs wide use of cyclotomic fields and Galois modules, and was published in Crelle's Journal, formally known as Journal für die reine und angewandte Mathematik.

Biography 3.2 Preda Mihăilescu (1955–) *is a German mathematician, born in Bucharest, Romania on May 23, 1955. He left Romania in 1973 to settle in Switzerland. He finished his doctorate, entitled,* Cyclotomy of rings and primality testing, *from ETH Zürich in 1997, under the direction of Erwin Engeler and Hendrik Lenstra. He was a researcher at the University of Paderborn, Germany until 2005 when he became a professor at the Georg-August University of Göttingen. The proof of Catalan's conjecture has now earned him the honour of having the result named* Mihăilescu's Theorem.

3.2 Existence

Existence precedes and rules essence.
From **L'etre et le néant (1943)** part 4, chapter 1
Jean-Paul Sartre (1905–1980), French philosopher, novelist, dramatist, and critic

In §3.1, we determined, among other things, the number of primitive roots modulo n, given that a primitive root exists. We now show that for primes, a primitve root always exists.

Theorem 3.3 | Primitive Roots Modulo a Prime |

Let p be a prime and let $e \in \mathbb{N}$ such that $e \mid (p-1)$, then there exist exactly $\phi(e)$ incongruent $m \in \mathbb{Z}$, with $\text{ord}_p(m) = e$.

Proof. For any $e \mid (p-1)$, let $r(e)$ be the number of incongruent natural numbers $m < p$, that belong to the exponent e modulo p. Since every natural number $m < p$ must belong to *some* exponent modulo p, then

$$\sum_{e \mid (p-1)} r(e) = p - 1.$$

Also, by Theorem 2.11 on page 94,

$$\sum_{d \mid (p-1)} \phi(d) = p - 1,$$

so

$$\sum_{e \mid (p-1)} r(e) = \sum_{d \mid (p-1)} \phi(d). \tag{3.1}$$

Claim 3.1 $r(e) \leq \phi(e)$ *for any natural number* $e \mid (p-1)$.

Certainly if $r(e) = 0$, then the result is true. If $r(e) > 0$, then there exists an integer m of order e modulo p. Thus, the integers $m, m^2, \ldots, m^e$ are incongruent modulo p, and each of them satisfies the polynomial congruence

$$x^e - 1 \equiv 0 \pmod{p}.$$

By Example 2.22 on page 114, there are no more solutions to Equation (3.1). Hence, any integer having order e modulo p must be one of the m^j for $j \in \{1, 2, \ldots, e\}$. However, there are only $\phi(e)$ of the m^j having order e, namely those such that $\gcd(j, e) = 1$. Thus, if there is one element of order e modulo p, there must be exactly $\phi(e)$ such positive integers less than e. Hence, $r(e) \leq \phi(e)$, which is Claim 3.1.

By Equation (3.1) and Claim 3.1, we must have $r(e) = \phi(e)$ for any positive $e \mid (p-1)$. Thus, there are exactly $\phi(e)$ incongruent integers m having $\text{ord}_p(m) = e$. $\square$

Corollary 3.5 *If p is prime, then there exist exactly $\phi(p-1)$ incongruent primitive roots modulo p.*

Proof. Let $e = p - 1$ in Theorem 3.3.   □

Example 3.11 *If $p = 11$, then 1 has order 1; 10 has order 2; $3, 4, 5, 9$ have order 5; and $2, 6, 7, 8$ have order 10; so*

$$\sum_{e \mid 10} r(e) = r(1) + r(2) + r(5) + r(10) = 1 + 1 + 4 + 4 = 10 = \sum_{e \mid 10} \phi(e).$$

Remark 3.1 *If we look at the Möbius inversion formula, Theorem 2.16 on page 98, we get that*

$$r(e) = \sum_{d \mid e} \mu(d)e/d = \phi(e),$$

where $e = \sum_{d \mid e} r(d)$, by Example 2.22 on page 114.

Example 3.12 *If $p \equiv 1 \pmod 4$ is prime then by Theorem 3.3, there exists an element e of order $4 \mid (p-1)$ modulo p. Thus,*

$$e^4 - 1 \equiv (e^2 - 1)(e^2 + 1) \equiv 0 \pmod p.$$

Therefore, $p \mid (e^2 - 1)$ or $p \mid (e^2 + 1)$. If the former occurs, then this contradicts that e has order 4 modulo p, so the latter must occur, and we have that there is a solution to the congruence $x^2 \equiv -1 \pmod p$, which we saw in Example 2.12 on page 88 by different methods.

Although Theorem 3.3 verifies the *existence* of primitive roots modulo a prime, it does not provide us with a *method* for finding them since it is merely an existence result. However, Gauss developed a methodology for computing primitive roots in [16, Articles 73–74, pp. 47–49], as follows.

◆ **Gauss' Algorithm for Computing Primitive Roots Modulo p**

(1) Let $m \in \mathbb{N}$ such that $1 < m < p$ and compute m^t for $t = 1, 2, \ldots$, until $m^t \equiv 1 \pmod p$. In other words, compute powers until $\mathrm{ord}_p(m)$ is achieved. If $t = \mathrm{ord}_p(m) = p - 1$, then m is a primitive root and the algorithm terminates. Otherwise, go to step (2).

(2) Choose $b \in \mathbb{N}$ such that $1 < b < p$ and $b \not\equiv m^j \pmod p$ for any $j = 1, 2, \ldots, t$. Let $u = \mathrm{ord}_p(b)$.[3.1] If $u \neq p - 1$, then let $v = \mathrm{lcm}(t, u)$.

[3.1]Observe if $u|t$, then $b^t \equiv 1 \pmod p$. However, it follows from (1) and Example 2.22 on page 114 that m^j for $0 \le j \le t - 1$ are all the incongruent solutions of $x^t \equiv 1 \pmod p$, so $b \equiv m^j \pmod p$ for some such j, a contradiction to the choice of b. Hence, $u \nmid t$.

Therefore, $v = ac$ where $a \mid t$ and $c \mid u$ with $\gcd(a,c) = 1$. Let m_1 and b_1 be the least nonnegative residues of $m^{t/a}$ and $b^{u/c}$ modulo p, respectively. Thus, $g = m_1 b_1$ has order $ac = v$ modulo p. If $v = p - 1$, then g is a primitive root and the algorithm is terminated. Otherwise, go to step (3).

(3) Repeat step (2) with v taking the role of t and $m_1 b_1$ taking the role of m. (Since $v > t$ at each step, the algorithm terminates after a finite number of steps with a primitive root modulo p.)

Gauss used the following to illustrate his algorithm.

Example 3.13 Let $p = 73$. Choose $m = 2$ in step (1), and we compute $t = \mathrm{ord}_p(m) = 9$ with

$$m^j \equiv 1, 2, 4, 8, 16, 32, 64, 55, 37, 1 \pmod{p}$$

for $j = 0, 1, 2, 3, 4, 5, 6, 7, 8, 9 = t = \mathrm{ord}_p(m)$, respectively. Now we go to step (2) since $m = 2$ is not a primitive root modulo $p = 73$. Since $3 \not\equiv 2^j \pmod{73}$ for any natural number $j \leq 9$, we choose $b = 3$. Compute b^j for $j = 1, 2, \ldots u$, where $3^u = 3^{12} \equiv 1 \pmod{73}$, where

$$3^j \equiv 3, 9, 27, 8, 24, 72, 70, 64, 46, 65, 49, 1 \pmod{p}$$

for $j = 1, 2, 3, 4, 5, 6, 7, 8, 9, 10, 11, 12 = u = \mathrm{ord}_p(b) = \mathrm{ord}_{73}(3)$, respectively. Since $u \neq p - 1$, then set $v = \mathrm{lcm}(t, u) = 36 = ac = 9 \cdot 4$. Then $m_1 = 2^{t/a} = 2$ and $b_1 = 3^{u/c} = 3^3 = 27$, so $m_1 b_1 = 54$, but $v = \mathrm{ord}_{73}(54) = 36 \neq p - 1$. Thus, we repeat step (2) with $v = 36$ replacing t and choose a value of b not equivalent to any power of the new $m = 54 = m_1 b_1$ modulo 73. Since $b = 5$ qualifies for the role and it is a primitive root modulo 73, the algorithm terminates.

Gauss also conjectured that 10 is a primitive root modulo infinitely many primes. This conjecture was generalized in the early twentieth century by Emil Artin — see Biography 3.3 on page 151.

Conjecture 3.1 | Artin's Conjecture

Every nonsquare integer $m \neq -1$ is a primitive root modulo infinitely many primes.

Although this conjecture remains open, Heath-Brown proved in 1986 that, with the possible exception of at most two primes, it is true that for each prime p there are infinitely primes q such that p is a primitive root modulo q. For example, there are infinitely many primes q such that one of 2, 3, or 5 is a primitive root modulo q (see [18, p. 249]).

Now we wish to conclude this section on existence with the result that tells us exactly those moduli possessing primitive roots. We know from the above that prime moduli do, and we now complete the answer in what follows. We begin with powers of odd primes.

Theorem 3.4 | Primitive Roots Modulo Odd Prime Powers |

If $p > 2$ is prime, then there exists a primitive root modulo p^n for all $n \in \mathbb{N}$. Furthermore, if g is a primitive root modulo p^2, then g is a primitive root modulo p^n for all $n \in \mathbb{N}$.

Proof. We know that p has a primitive root g by Theorem 3.3 on page 145.

Claim 3.2 *Either g or $p + g$ is a primitive root modulo p^2.*

Since $\operatorname{ord}_p(g) = p - 1$, we may set $d = \operatorname{ord}_{p^2}(g)$ where $(p - 1) \mid d$ by Proposition 3.1 on page 140. Also, by Corollary 3.1 on page 141, $d \mid \phi(p^2)$, whence,

$$(p - 1) \mid d \mid p(p - 1),$$

so either $d = p - 1$ or $d = p(p - 1)$. In the latter case we have that g is a primitive root modulo p^2, and in the former case we have that

$$g^{p-1} \equiv 1 \pmod{p^2}. \tag{3.2}$$

In this instance, set $g_1 = g + p$. Moreover, by the Binomial Theorem and (1.26) established in the proof of Fermat's Little Theorem on page 50, we have that

$$g_1^{p-1} \equiv (g + p)^{p-1} \equiv g^{p-1} + (p - 1)g^{p-2}p + \sum_{j=2}^{p-1} \binom{p-1}{j} g^{p-1-j} p^j \equiv$$

$$g^{p-1} + (p - 1)g^{p-2}p \equiv g^{p-1} - g^{p-2}p \pmod{p^2}.$$

Now, if $g_1^{p-1} \equiv 1 \pmod{p^2}$, then by the latter congruence and the congruence assumed in (3.2),

$$1 \equiv g^{p-1} - g^{p-2}p \equiv 1 - g^{p-2}p \pmod{p^2},$$

and it follows that $g^{p-2}p \equiv 0 \pmod{p^2}$, so $g^{p-2} \equiv 0 \pmod{p}$, which is impossible since the fact that g is a primitive root modulo p implies $\gcd(p, g) = 1$. This establishes Claim 3.2.

By Claim 3.2, we may choose g to be a primitive root modulo p, which is also a primitive root modulo p^2.

Claim 3.3 *For any natural number $n > 1$, $g^{p^{n-2}(p-1)} \not\equiv 1 \pmod{p^n}$.*

We use induction on n. If $n = 2$, we know that the result holds since g is a primitive root modulo p^2. Now assume that the result holds for n, namely,

$$g^{p^{n-2}(p-1)} \not\equiv 1 \pmod{p^n} \tag{3.3}$$

and we will prove that it holds for $n + 1$. By Euler's Theorem 2.10 on page 93, we have,

$$g^{p^{n-2}(p-1)} \equiv g^{\phi(p^{n-1})} \equiv 1 \pmod{p^{n-1}},$$

so there is a $z \in \mathbb{N}$ such that

$$g^{p^{n-2}(p-1)} = 1 + zp^{n-1}, \tag{3.4}$$

and by (3.3), $g^{p^{n-2}(p-1)} \not\equiv 1 \,(\text{mod } p^n)$, so $p \nmid z$. Raising both sides of (3.4) to the power p, and using the Binomial Theorem and (1.26) again, we get,

$$g^{p^{n-1}(p-1)} = (1 + zp^{n-1})^p \equiv 1 + zp^n + \sum_{j=2}^{p} \binom{p}{j} (zp^{n-1})^j \equiv 1 + zp^n \ (\text{mod } p^{n+1}).$$

Since $p \nmid z$, it is not possible for $g^{p^{n-1}(p-1)} \equiv 1 \,(\text{mod } p^{n+1})$, which secures Claim 3.3 via induction.

Claim 3.3 also secures the theorem. $\square$

Corollary 3.6 | Primitive Roots for $2p^n$

For any odd prime p and $n \in \mathbb{N}$ there exists a primitive root modulo $2p^n$.

Proof. Let g be a primitive root modulo p^n by Theorem 3.4. Without loss of generality, we may assume that g is odd since if it were even then we could select $g + p^n$ which is also a primitive root modulo p^n. Thus, $\gcd(2p^n, g) = 1$ and

$$\phi(2p^n) = \phi(2)\phi(p^n) = \phi(p^n),$$

so if $g^d \equiv 1 \,(\text{mod } 2p^n)$, then $g^d \equiv 1 \,(\text{mod } p^n)$. Hence, $\phi(p^n) \mid d$. Also, since g is a primitive root modulo p^n, $d = \phi(p^n) = \phi(2p^n)$, which secures the result. $\square$

Now we look at powers of 2.

Theorem 3.5 | Primitive Roots and Moduli 2^n

If g is a primitive root modulo 2^n, then $n = 1$ or $n = 2$.

Proof. The proof hinges on the following critical result.

Claim 3.4 *If n is a natural number with $n \geq 3$, and m is any odd integer, then*

$$m^{2^{n-2}} \equiv 1 \ (\text{mod } 2^n).$$

We prove the result by induction on n. If $n = 3$, then $m^2 \equiv 1 \,(\text{mod } 8)$ by Exercise 2.2 on page 83. Assume the induction hypothesis,

$$m^{2^{n-2}} \equiv 1 \ (\text{mod } 2^n).$$

Therefore, there exists an integer z such that

$$m^{2^{n-2}} = 1 + z2^n.$$

Squaring both sides, we get,

$$m^{2^{n-1}} = 1 + z2^{n+1} + z^2 2^{2n},$$

which implies that,

$$m^{2^{n-1}} \equiv 1 \pmod{2^{n+1}},$$

establishing Claim 3.4.

By Claim 3.4, if $n > 2$, then for any odd integer m,

$$m^{\phi(2^n)/2} \equiv 1 \pmod{2^n},$$

so m cannot be a primitive root modulo 2^n since $\mathrm{ord}_{2^n}(m) \neq \phi(2^n)$. Hence, the only powers of 2 that have primitive roots are 2, with primitive root 1, and 4 with primitive root 3. □

Remark 3.2 *Although Theorem 3.5 says there can be no primitive roots for powers of 2 other than 2 or 4, Exercise 3.8 on page 152 tells us that there is always an element of maximum possible order modulo 2^n, namely 5 with $\mathrm{ord}_{2^n}(5) = \phi(2^n)/2 = 2^{n-2}$.*

Now we complete the search by showing that no other moduli than those we have seen above have primitive roots.

Theorem 3.6 | **Moduli with No Primitive Roots**

If m, n are relatively prime natural numbers with $m > 2$ and $n > 2$, then the modulus mn has no primitive root.

Proof. Let $\ell = \mathrm{lcm}(\phi(m), \phi(n))$, and $g = \gcd(\phi(m), \phi(n))$. Since $m > 2$ and $n > 2$, then $2 \mid \phi(m)$ and $2 \mid \phi(n)$, by Exercise 2.22 on page 100. Hence, $g \geq 2$. By part (b) of Theorem 1.13 on page 26,

$$\ell = \frac{\phi(m)\phi(n)}{g} \leq \frac{\phi(m)\phi(n)}{2}.$$

Now if we take any integer z relatively prime to mn, then

$$z^\ell = (z^{\phi(m)})^{\phi(n)/g} \equiv 1 \pmod{m},$$

and similarly $z^\ell \equiv 1 \pmod{n}$. Hence,

$$z^\ell \equiv 1 \pmod{mn},$$

so

$$\mathrm{ord}_{mn}(z) < \phi(mn) = \phi(m)\phi(n) \text{ for any } z \text{ with } \gcd(z, mn) = 1. \qquad (3.5)$$

This secures the result since (3.5) says there can be no primitive root modulo mn. □

We may summarize our results into a single result as follows.

Theorem 3.7 | **The Primitive Root Theorem**

The natural number $n > 1$ has a primitive root if and only if

$$n \subset \{2, 4, p^a, 2p^a\} \text{ where } p \text{ is an odd prime.}$$

Theorem 3.7 was proved by Gauss in 1801. This corrected and completed earlier contributions by Euler, Legendre, and Lagrange.

Biography 3.3 Emil Artin (1898–1962) *was born in Vienna, Austria, in 1898. In World War I, he served in the Austrian army. In 1921, after the war, he obtained his Ph.D. from the University of Leipzig. He also attended the University of Göttingen in 1922–1923, the latter year being when he was appointed to a position at the University of Hamburg. However, by 1937 he emigrated to the U.S.A. to escape the Nazi restrictions, since his wife was Jewish, although he was not. He taught at the University of Notre Dame for one year. Then he spent eight years at Indiana, and in 1946 went to Princeton where he remained for the next twelve years. In 1958, he returned to Germany where he remained for the rest of his life. He was reappointed to the University of Hamburg, which he had left two decades before. Artin contributed to finite group theory, the theory of associative algebras, as well as number theory. His name is attached to numerous deep mathematical entities. For instance, there are the Artin Reciprocity Law, Artin L-functions, and Artinian Rings (see [30]). Furthermore, he invented the notion of braid structures, used today by topologists. Among Artin's students were Serge Lang, John Tate, and Max Zorn. Artin had interests outside of mathematics, including astronomy, biology, chemistry, and music. In the latter, he excelled as an accomplished musician in his own right, playing not only the flute but also the harpsichord and the clavichord. He died in Hamburg on December 20, 1962.*

We close this section by presenting a result that links solutions of congruences with the existence of primitive roots.

Theorem 3.8 | **Quadratic Congruences and Primitive Roots**

A positive integer n has a primitive root if and only if the only solutions to

$$x^2 \equiv 1 \pmod{n} \tag{3.6}$$

are $x \equiv \pm 1 \pmod{n}$.

Proof. Assume that g is a primitive root modulo n, and that $x^2 \equiv 1 \pmod{n}$ for some integer x. Then by Theorem 3.1 on page 142, $x \equiv g^j \pmod{n}$ for some nonnegative integer $j \leq \phi(n)$. Hence, $g^{2j} \equiv 1 \pmod{n}$. Thus, $\phi(n) \mid 2j$ since g is a primitive root, so there is an integer z such that $2j = \phi(n)z$, namely, $j = \phi(n)z/2$. Therefore,

$$x \equiv g^j \equiv g^{\phi(n)z/2} \equiv (g^{\phi(n)/2})^z \equiv (-1)^z \equiv \pm 1 \pmod{n}.$$

Conversely, we prove the contrapositive, namely assume that n has no primitive root and prove that there is a solution to (3.6) that is not congruent to ± 1
modulo n. By Theorem 3.7, $n \notin \{2, 4, p^a, 2p^a\}$ for any odd prime p. Thus, if p
is an odd prime with $p^a || n$, then by the Chinese Remainder Theorem, there is
a solution to the system of congruences,

$$x \equiv 1 \pmod{p^a} \quad x \equiv -1 \pmod{n/p^a}, \tag{3.7}$$

where $n/p^a > 2$. Hence, $x \not\equiv \pm 1 \pmod n$, but $x^2 \equiv 1 \pmod n$, with the latter
congruence following again from the Chinese Remainder Theorem upon squaring
the congruences in (3.7). If no odd prime divides n, then $n = 2^a$ where $a > 2$.
By Exercise 3.8, there are more than two solutions to the congruence (3.6), so
one of them is not congruent to ± 1 modulo n. $\qquad\square$

Exercises

3.8. Let $n \geq 3$ be a natural number. Prove that $\text{ord}_{2^n}(5) = 2^{n-2}$.

   (*Hint: Use induction to verify that $5^{2^{n-3}} \equiv 1 + 2^{n-1} \pmod{2^n}$.*)

3.9. Which of the following have primitive roots? Provide an example if such
   a root exists, and provide a reason that it does not otherwise.
   (a) 169.   (b) 55.
   (c) 26.   (d) 34.
   (e) 206.   (f) 118.

3.10. Which of the following have primitive roots? Provide an example if such
   a root exists, and provide a reason that it does not otherwise.
   (a) 25.   (b) 122.
   (c) 52.   (d) 38.
   (e) 222.   (f) 226.

3.11. If p is prime, f is called a *Fibonacci primitive root modulo p* if

$$f^2 \equiv f + 1 \pmod p.$$

   Prove that if f is a Fibonacci primitive root modulo p, then

$$f^{j+1} \equiv F_{j+1} f + F_j \pmod p,$$

   where F_j is the jth Fibonacci number for $j \in \mathbb{N}$, defined on page 3.

3.12. Find all incongruent primitive roots modulo 14.

3.13. Find all incongruent primitive roots modulo 26.

3.14. Prove that if g is a primitive root modulo p^2 for an odd prime p, then x
   is a solution of $x^{p-1} \equiv 1 \pmod{p^2}$ if and only if $x \equiv g^k \pmod{p^2}$ where
   $p \mid k$.

3.15. Prove that if p is an odd prime, $n \in \mathbb{N}$ then both $2p^n$ and p^n have the
   same number of primitive roots. (Exercises 3.12–3.13 are examples of this
   fact.)

3.3 Indices

> If all the arts aspire to the condition of music, all the sciences aspire
> to the condition of mathematics.
>   From **Some turns of thought in modern philosophy (1933)**
>   **George Santayana (1863–1952), Spanish-born philosopher**
> and critic

The concept in the header of this section was developed by Gauss in his *Disquistiones Arithmeticae* — see Biography 1.7 on page 33. If $n \in \mathbb{N}$ has a primitive root m, then by Theorem 3.1 on page 142, the values $1, m, m^2, \ldots, m^{\phi(n)-1}$ form a complete set of reduced residues modulo n. Thus, if $b \in \mathbb{N}$ with $\gcd(b, n) = 1$, there is exactly one nonnegative integer $e < \phi(n)$ for which $b \equiv m^e \pmod{n}$. This value has a distinguished name.

Definition 3.3 Index

Let $n \in \mathbb{N}$ with primitive root m, and $b \in \mathbb{N}$ with $\gcd(b, n) = 1$. Then for exactly one of the values $e \in \{0, 1, \ldots, \phi(n) - 1\}$, $b \equiv m^e \pmod{n}$ holds. This unique value e modulo $\phi(n)$ is the index *of b to the base m modulo n, denoted by $\operatorname{ind}_m(b)$.*

Note that the notation $\operatorname{ind}_m(b)$ makes no mention of the modulus n, which will be understood in context since it is fixed.

Example 3.14 If $n = 53$, then $m = 2$ is a primitive root modulo n. Also, $\operatorname{ind}_2(5) = 31$, since $5 \equiv 2^{31} \pmod{53}$, so 5 has index 31 to base 2 modulo 53.

Example 3.15 *If $n = 11$, then 2 is a primitive root modulo 11 and*

$$2^1 \equiv 2 \pmod{11}, 2^2 \equiv 4 \pmod{11}, 2^3 \equiv 8 \pmod{11}, 2^4 \equiv 5 \pmod{11},$$

$$2^5 \equiv 10 \pmod{11}, 2^6 \equiv 9 \pmod{11}, 2^7 \equiv 7 \pmod{11}, 2^8 \equiv 3 \pmod{11},$$

$$2^9 \equiv 6 \pmod{11}, \text{ and } 2^{10} \equiv 1 \pmod{11}.$$

Therefore,

$$\operatorname{ind}_2(1) = 0, \operatorname{ind}_2(2) = 1, \operatorname{ind}_2(3) = 8, \operatorname{ind}_2(4) = 2, \operatorname{ind}_2(5) = 4,$$

$$\operatorname{ind}_2(6) = 9, \operatorname{ind}_2(7) = 7, \operatorname{ind}_2(8) = 3, \operatorname{ind}_2(9) = 6, \text{ and } \operatorname{ind}_2(10) = 5.$$

With different primitive roots come some different indices. For instance, since 6 is a primitive root modulo 11, then we calculate,

$$\operatorname{ind}_6(1) = 0, \operatorname{ind}_6(2) = 9, \operatorname{ind}_6(3) = 2, \operatorname{ind}_6(4) = 8, \operatorname{ind}_6(5) = 6,$$

$$\operatorname{ind}_6(6) = 1, \operatorname{ind}_6(7) = 3, \operatorname{ind}_6(8) = 7, \operatorname{ind}_6(9) = 4, \text{ and } \operatorname{ind}_6(10) = 5.$$

Definition 3.3 gives rise to an arithmetic of its own, the *index calculus*. The following are some of the properties.

Theorem 3.9 | Index Calculus |

If $n \in \mathbb{N}$ and m is a primitive root modulo n, then for any $c, d \in \mathbb{Z}$ each of the following holds.

(a) $\mathrm{ind}_m(1) \equiv 0 \,(\mathrm{mod}\ \phi(n))$.

(b) $\mathrm{ind}_m(cd) \equiv \mathrm{ind}_m(c) + \mathrm{ind}_m(d) \,(\mathrm{mod}\ \phi(n))$.

(c) For any $t \in \mathbb{N}$, $\mathrm{ind}_m(c^t) \equiv t \cdot \mathrm{ind}_m(c) \,(\mathrm{mod}\ \phi(n))$.

Proof. For part (a), let $\mathrm{ind}_m(1) = w$. Then $1 \equiv m^w \,(\mathrm{mod}\ n)$. Since m is a primitive root modulo n, then $w \equiv 0 \,(\mathrm{mod}\ \phi(n))$, by Proposition 3.1 on page 140.

For part (b), let $x = \mathrm{ind}_m(cd)$, $y = \mathrm{ind}_m(c)$, and $z = \mathrm{ind}_m(d)$. Since $cd \equiv m^x \,(\mathrm{mod}\ n)$, $c \equiv m^y \,(\mathrm{mod}\ n)$ and $d \equiv m^z \,(\mathrm{mod}\ n)$, then

$$m^{y+z} \equiv cd \equiv m^x \quad (\mathrm{mod}\ n).$$

Therefore,
$$m^{y+z-x} \equiv 1 \quad (\mathrm{mod}\ n),$$

so since m is a primitive root modulo n, then

$$y + z - x \equiv 0 \quad (\mathrm{mod}\ \phi(n)),$$

by Proposition 3.1 again, and this secures part (b).

For part (c), we maintain the notation from part (b) to get, $c \equiv m^y \,(\mathrm{mod}\ n)$. Therefore,
$$c^t \equiv m^{yt} \quad (\mathrm{mod}\ n),$$

so $\mathrm{ind}_m(c^t) \equiv ty \equiv t \cdot \mathrm{ind}_m(c) \,(\mathrm{mod}\ \phi(n))$. $\qquad\square$

The reader will recognize that the properties of the index mimic those of logarithms from elementary calculus. Hence, if $n = p$ is prime, the index of b to the base p is often called the *discrete logarithm* of b to the base p. (See the discussion on page 167.) For another property, see Exercise 3.28 on page 159. Moreover, Theorem 3.9 provides us with a tool for finding indices by solving linear congruences

$$cx \equiv b \quad (\mathrm{mod}\ n)$$

for $x \in \mathbb{Z}$. To see this note that when this congruence holds, then

$$\mathrm{ind}_m(c) + \mathrm{ind}_m(x) \equiv \mathrm{ind}_m(b) \quad (\mathrm{mod}\ \phi(n)),$$

for any primitive root m modulo n. The process is illustrated as follows.

Example 3.16 *Suppose that we wish to solve*

$$3x^3 \equiv 7 \pmod{11}. \tag{3.8}$$

Then by the index calculus, $\operatorname{ind}_2(3x^3) \equiv \operatorname{ind}_2(7) \pmod{10}$, *so*

$$\operatorname{ind}_2(3) + 3\operatorname{ind}_2(x) \equiv 7 \pmod{10},$$

where the latter congruence is known from Example 3.15 on page 153. Therefore, since we also know that $\operatorname{ind}_2(3) = 8$ *from Example 3.15,*

$$\operatorname{ind}_2(x) \equiv 3^{-1}(7 - 8) \equiv -7 \equiv 3 \pmod{10}.$$

Hence,

$$x \equiv 2^3 \equiv 8 \pmod{11},$$

and we have a congruence class of solutions for (3.8).

The above is related to the following notion.

Definition 3.4 | **Modular Roots and Power Residues**

If $m, n \in \mathbb{N}$, $b \in \mathbb{Z}$, $\gcd(b, n) = 1$, *then* b *is called an* m^{th} *power residue modulo* n *if* $x^m \equiv b \pmod{n}$ *for some* $x \in \mathbb{Z}$, *and* x *is called an* m^{th} *root modulo* n.

For instance, if $m = 2$, then x is called a *square root modulo* n, and b is called a *quadratic residue modulo* n; if $m = 3$, then b is called a *cubic residue modulo* n, and x is called a *cube root modulo* n, and so on.

It is valuable to have a criterion for the solvability of such congruences.

Theorem 3.10 | **Criterion for Power Residue Congruences**

Let $e, n \in \mathbb{N}$ *such that* n *has a primitive root, let* $b \in \mathbb{Z}$, *such that* $\gcd(b, n) = 1$, *and set* $g = \gcd(e, \phi(n))$. *Then the congruence*

$$x^e \equiv b \pmod{n} \tag{3.9}$$

is solvable if and only if

$$b^{\phi(n)/g} \equiv 1 \pmod{n}.$$

Moreover, if there are solutions to (3.9), then there are exactly g *incongruent solutions* x *modulo* n

Proof. Let a be a primitive root modulo n. Then (3.9) holds if and only if

$$e \cdot \operatorname{ind}_a(x) \equiv \operatorname{ind}_a(b) \pmod{\phi(n)}. \tag{3.10}$$

By Theorem 2.3 on page 84, (3.10) has solutions if and only if $g \mid \mathrm{ind}_a(b)$. If $g \mid \mathrm{ind}_a(b)$, then there are exactly g incongruent solutions modulo $\phi(n)$ such that (3.10) holds, so there are exactly g integers x incongruent modulo n such that (3.9) holds. Since $g \mid \mathrm{ind}_a(b)$ if and only if

$$\mathrm{ind}_a(b)\phi(n)/g \equiv 0 \pmod{\phi(n)},$$

which holds if and only if

$$(a^{\mathrm{ind}_a(b)})^{\phi(n)/g} \equiv b^{\phi(n)/g} \equiv 1 \equiv a^0 \pmod{n},$$

we have secured the result.   □

Corollary 3.7 *If p is an odd prime, $c, e \in \mathbb{N}$, and $b \in \mathbb{Z}$ with $\gcd(p, b) = 1$, then $x^e \equiv b \pmod{p^c}$ if and only if*

$$b^{p^{c-1}(p-1)/g} \equiv 1 \pmod{p^c},$$

where $g = \gcd(e, p^{c-1}(p-1))$. Moreover, if it has a solution, then it has exactly g solutions.

Example 3.17 *Suppose that we seek solutions to*

$$x^5 \equiv 5 \pmod{27}. \tag{3.11}$$

In the notation of Corollary 3.7, $g = \gcd(e, \phi(p^c)) = \gcd(5, 18) = 1$. Since $5^{18} \equiv 1 \pmod{27}$, Theorem 3.10 tells us that there are solutions to (3.11), and that there is only one congruence class of solutions since $g = 1$. The index calculus allows us to find this class as follows. Since $\mathrm{ind}_2(5) = 5$, and

$$5 \cdot \mathrm{ind}_2(x) \equiv \mathrm{ind}_2(5) \equiv 5 \pmod{18},$$

we have that $m = 2$ is a primitive root mod 3^3. Therefore,

$$\mathrm{ind}_2(x) \equiv 1 \pmod{18},$$

so either $\mathrm{ind}_2(x) \equiv 1 \pmod{27}$, or $\mathrm{ind}_2(x) \equiv 19 \pmod{27}$. Thus, the only distinct congruence class that satisfies the given congruence is $x \equiv 2 \pmod{27}$, since $2^{19} \equiv 2 \pmod{27}$.

Example 3.18 *Consider $x^3 \equiv 4 \pmod{27}$. Then since $4^6 \equiv 19 \pmod{27}$, where $\phi(27)/\gcd(\phi(27), 3) = 18/3 = 6$, then Theorem 3.10 says that there are no solutions to this congruence.*

We observe that Corollary 3.7 is a generalization of a well-known result by Euler on quadratic congruences that we isolate here for its elegance and historical value.

Corollary 3.8 | **Euler's Criterion for Quadratic Congruences**

If $p > 2$ is prime and $b \in \mathbb{Z}$ is relatively prime to p, then

$$x^2 \equiv b \pmod{p} \tag{3.12}$$

is solvable if and only if $b^{(p-1)/2} \equiv 1 \pmod{p}$.

Proof. Take $c = 1$, $e = 2$, with $g = 2$ in Corollary 3.7. ☐

Note that another way of stating Corollary 3.8 is that (3.12) is solvable if $b^{(p-1)/2} \equiv 1 \pmod{p}$, and is not solvable if $b^{(p-1)/2} \equiv -1 \pmod{p}$. Later when we study quadratic reciprocity, we will return to this result.

Another immediate consequence of Corollary 3.7 is (2.13) in Example 2.22 on page 114, which we now state to illustrate the applicability of our above development.

Corollary 3.9 *If $p > 2$ is prime, then for any positive integer $d \mid (p-1)$, $x^d - 1 \equiv 0 \pmod{p}$ has exactly d solutions.*

Proof. Take $b = c = 1$ and $e = g = d$ in Corollary 3.7.

A consequence of Euler's criterion that links quadratic congruences and primitive roots is the following.

Corollary 3.10 *If a is a primitive root modulo an odd prime p, and $b \in \mathbb{Z}$, then $b \equiv x^2 \pmod{p}$ is solvable if and only if $2 \mid \mathrm{ind}_a(b)$.*

Proof. Let $b \equiv a^j \pmod{p}$ where $j = \mathrm{ind}_a(b)$, so $a \equiv b \cdot a^{-j+1} \pmod{p}$. By Corollary 3.8, $x^2 \equiv 1 \pmod{p}$ is solvable if and only if $b^{(p-1)/2} \equiv 1 \pmod{p}$. If j is odd, then this contradicts that a is a primitive root modulo p since $a^{(p-1)/2} \equiv b^{(p-1)/2}(a^{(1-j)/2})^{p-1} \equiv 1 \pmod{p}$. Hence, j is even. ☐

The above covers power residues for odd prime power moduli, but Theorem 3.10 on page 155 fails to say anything about the case where $n = 2^c$ for $c > 2$ by the Primitive Root Theorem. See Exercises 3.24–3.27 for solutions of power residue congruences when the moduli are higher powers of 2.

Remark 3.3 *Theorem 3.10 tells us when solutions to (3.9) exist, and how many there are. A related question is how many e-th power residues exist? In other words, Theorem 3.10 gives us g incongruent solutions x of (3.9) for a given fixed b, but the latter question asks for the number of values of b for which there are solutions x to (3.9). The next result answers that query.*

Corollary 3.11 | **The Number of Power Residues**

Suppose that $n \in \mathbb{N}$ possesses a primitive root and $b \in \mathbb{Z}$ relatively prime to n. Then b is an e-th power residue modulo n if and only if $b^{\phi(n)/g} \equiv 1 \pmod{n}$, where $g = \gcd(e, \phi(n))$, and there exist $\phi(n)/g$ such b, each of which is the e-th power of exactly g integers modulo n.

Proof. By Theorem 3.10, we need only find the number of incongruent solutions of $x^{\phi(n)/g} \equiv 1 \pmod{p^c}$. If a is a primitive root modulo n, then the $\phi(n)/g$ values $a^g, a^{2g}, \ldots, a^{(\phi(n)/g)g}$ are incongruent modulo n. Hence, each of these values serves as an e-th power residue, which completes the task.   □

Example 3.19 *Suppose that we want to determine the number of incongruent fourth power residues modulo 27, namely the number of incongruent $b \in \mathbb{N}$ such that $x^4 \equiv b \pmod{27}$.*

By Corollary 3.11, there must be $\phi(p^c)/g = 18/2 = 9$ incongruent such solutions. Since 2 is a primitive root modulo 27 and $g = \gcd(e, \phi(n)) = \gcd(4, 18) = 2$, then by the proof of Corollary 3.11, we may find these values of b via reduction of powers of $2^g = 2^2 = 4$ modulo 27. If we let $\overline{4^j}$ denote the reduction modulo 27 of 4^j for $j = 1, 2, \ldots, 9$ then our values of b are given by

$$b \in \{1, 4, 7, 10, 13, 16, 19, 22, 25\} = \{\overline{4^9}, \overline{4^1}, \overline{4^8}, \overline{4^3}, \overline{4^4}, \overline{4^2}, \overline{4^6}, \overline{4^7}, \overline{4^5}\},$$

and each of these values of b is the fourth power of $g = 2$ integers modulo $n = 27$. For instance, $b = 4$ is the fourth power of exactly $x = 5, 22$ modulo 27.

▼ Some Comments on Applications

An important consideration in complexity theory (see Appendix B) is the search for an efficient algorithm which, given a prime p and a primitive root m modulo p, computes $\text{ind}_m(x)$ for any given $x \in \mathbb{F}_p^*$. These algorithms have significant ramifications for the construction of secure pseudorandom number generators, which we will study as an application in §3.4.

Exercises

3.16. Find each of the following.

    (a) $\text{ind}_2(11) \pmod{13}$ (b) $\text{ind}_3(13) \pmod{17}$

    (c) $\text{ind}_5(15) \pmod{23}$ (d) $\text{ind}_6(16) \pmod{41}$

    (e) $\text{ind}_3(19) \pmod{43}$ (f) $\text{ind}_5(21) \pmod{47}$

3.17. Find each of the following.

    (a) $\text{ind}_2(11) \pmod{19}$ (b) $\text{ind}_5(13) \pmod{23}$

    (c) $\text{ind}_7(25) \pmod{71}$ (d) $\text{ind}_5(26) \pmod{73}$

    (e) $\text{ind}_5(29) \pmod{97}$ (f) $\text{ind}_5(22) \pmod{103}$

3.18. Using the index calculus, find solutions to each of the following.

    (a) $4x^3 \equiv 3 \pmod{11}$ (b) $3x^4 - 5 \equiv 0 \pmod{17}$

    (c) $3x^5 \equiv 4 \pmod{19}$ (d) $2x^7 \equiv 3 \pmod{23}$

    (e) $5x^8 \equiv 3 \pmod{29}$ (f) $3x^7 \equiv 4 \pmod{31}$

3.19. Using the index calculus, find solutions to each of the following.
    (a) $3x^3 \equiv 4 \,(\mathrm{mod}\ 5)$ (b) $4x^4 \equiv 5 \,(\mathrm{mod}\ 7)$
    (c) $3x^4 \equiv 5 \,(\mathrm{mod}\ 11)$ (d) $2x^5 \equiv 9 \,(\mathrm{mod}\ 11)$
    (e) $4x^3 \equiv 3 \,(\mathrm{mod}\ 17)$ (f) $3x^5 \equiv 4 \,(\mathrm{mod}\ 19)$

3.20. Prove that $\mathrm{ind}_m(m) \equiv 1 \,(\mathrm{mod}\ \phi(n))$ for any primitive root m of the modulus n.

3.21. Prove that for a modulus $n > 2$, $\mathrm{ind}_m(-1) \equiv \phi(n)/2 \,(\mathrm{mod}\ \phi(n))$.

3.22. Prove that for a modulus n,

$$\mathrm{ind}_m(-c) \equiv \phi(n)/2 + \mathrm{ind}_m(c) \ \ (\mathrm{mod}\ \phi(n)).$$

3.23. Suppose that p is an odd prime, $b \in \mathbb{Z}$, $p \nmid b$ and $e \in \mathbb{N}$ with $p \nmid e$. Prove that if $x^e \equiv b \,(\mathrm{mod}\ p)$ has a solution, then so does $x^e \equiv b \,(\mathrm{mod}\ p^c)$ for all $c \in \mathbb{N}$.

3.24. Given an integer $c \geq 2$, prove that every odd integer b satisfies the following congruence for some nonnegative integer $a < 2^c$,

$$b \equiv \pm 5^a \ \ (\mathrm{mod}\ 2^c).$$

(*Hint: Use Exercise 3.8 on page 152.*)

☆ 3.25. Let $c \in \mathbb{N}$ with $c > 1$. Prove that if $e \in \mathbb{N}$, is an odd positive integer, then

$$b \equiv x^e \ \ (\mathrm{mod}\ 2^c)$$

has a solution $x \in \mathbb{Z}$ for any odd integer $b \in \mathbb{N}$.

☆ 3.26. Prove that if $c \geq 2$ is an integer and e is an even positive integer then b is an e-th power residue modulo 2^c if and only if $b \equiv 1 \,(\mathrm{mod}\ \gcd(4e, 2^c))$.
(*Hint: Use the criterion* (S5) *developed in the solution of Exercise 3.25 on page 336.*)

☆ 3.27. Let $c \geq 2$ be an integer and $e \in \mathbb{N}$. Prove that the number of incongruent e-th power residues modulo 2^c is

$$\frac{2^{c-1}}{\gcd(2, e)\gcd(e, 2^{c-2})}.$$

(*Hint: Use Exercises 3.24–3.26.*)

3.28. Let p be a prime with primitive roots a and b. Also, let c be an integer relatively prime to p. Prove that

$$\mathrm{ind}_a(c) \equiv \mathrm{ind}_b(c) \cdot \mathrm{ind}_a(b) \ \ (\mathrm{mod}\ p-1).$$

(This property mimics the change of base formula for logarithms.)

3.4   Random Number Generation

> *Anyone who considers arithmetic methods of producing random digits is, of course, in a state of sin.*
> **John von Neumann (1903–1957), Mathematician**

"Random numbers" are employed in many areas such as cryptography, to generate keys, for instance. They are also employed in programming slot machines, testing computer chips for flaws, and testing the performance of computer algorithms, to mention a few. Now we have to decide upon what we mean by the notion of *randomness*.

A truly random number is a sequence of natural numbers where each element in the sequence is selected by chance without any dependence on the previously chosen numbers. These are difficult, if not impossible to achieve, in most cases. We could, for instance, use the time between output tics from a Geiger counter exposed to a radioactive element, but for any reasonable application this is infeasible.

When using a computer, the notion of a randomly generated sequence can only be approximated. In practice, we use a computer program that generates a sequence of digits in a fashion that appears to be random, called a *pseudorandom number generator* (PRNG). Here we say "appears to be random" since computers are *finite state* devices, so any random-number generator on a computer must be periodic, which means it is predictable, so it cannot be truly random. The most that one can expect, therefore, from a computer is *pseudorandomness*, meaning that the numbers pass at least one statistical test for randomness.

Biography 3.4 John von Neumann (1903–1957) *was a Hungarian-born American mathematician. He received his Ph.D. from the University of Budapest when he was 23, and was a private lecturer in Berlin, Germany from 1926 to 1930. In that year he moved to Princeton, New Jersey. There he was among those selected for the first faculty of the Institute for Advanced Study, with Albert Einstein and Kurt Gödel being two of the others. He was a major figure in twentieth century mathematics and physics. Indeed, he was involved in the A-bomb development in the Manhattan Project, with its scientific research directed by the American physicist J. Robert Oppenheimer. It is believed that when von Neumann died of pancreatic cancer in 1957, it may have been due to his exposure to radioactivity when observing A-bomb tests in the Pacific. His contributions were to the areas of logic, quantum physics, optimization theory and game theory, of which he was co-creator, as he was with the concepts of cellular automata, and the working out of the key steps involved in thermonuclear reactions and the hydrogen bomb.*

Of course, these pseudorandom number generators are periodic, but if the periods are large enough, then they can be used for cryptographic applications, for instance.

We begin with a method first developed by von Neumann in 1946.

◆ Von Neumann's Middle-Square Method

The following generates $m \in \mathbb{N}$ random numbers with at most $2n$ ($n \in \mathbb{N}$) digits each.

(1) Set $j = 0$, and randomly select a $2n$-digit *seed* number n_0.

(2) Square n_j to get an intermediate number M possessing at most $4n$ digits. (We pad zeros to the left of M to create $4n$ digits if necessary.)

(3) Set $j = j + 1$ and select the middle $2n$ digits of M as the new random number n_i.

(4) If $j < m$, go to step (2). If $j = m$, then terminate the algorithm.

Example 3.20 *We give a very small illustration, $2n = 4$ and $m = 10$, for pedagogical purposes, whereas typically we would choose very large m and n in practice.*

We randomly select $n_0 = 1211$ and apply the middle-square method as follows.

j	n_j	$n_j^2 = M$	n_{j+1}
0	1211	01466521	4665
1	4665	21762225	7622
2	7622	58094884	948
3	948	00898704	8987
4	8987	80766169	7661
5	7661	58690921	6909
6	6909	47734281	7342
7	7342	53904964	9049
8	9049	81884401	8844
9	8844	78216336	2163
$10 = m$	2163		

Table 3.1

We have a set of ten randomly generated 4-digit integers,

$$\{4665, 7622, 0948, 8987, 7661, 6909, 7342, 9049, 8844, 2163\}.$$

Remark 3.4 *There is a serious problem with the middle-square method, namely that for some choices of the seed, the algorithm becomes quickly periodic, producing the same numbers repeatedly. For instance, for $n_0 = 6100$, then*

$$n_1 = 2100, n_2 = 4100, n_3 = 8100, n_4 = 6100, n_5 = 2100 \ldots.$$

Also, if an initial seed leads to zero, then all subsequent numbers are zero. For instance, if $n_0 = 1010$, then application of the algorithm yields,

	j	n_j	$n_j^2 = M$	n_{j+1}
	0	1010	1020100	201
	1	201	00040401	404
	2	404	00163216	1632
	3	1632	02663424	6634
	4	6634	44009956	99
	5	99	00009801	98
	6	98	00009604	96
Table 3.2	7	96	00009216	92
	8	92	00008464	84
	9	84	00007056	70
	10	70	00004900	49
	11	49	00002401	24
	12	24	00000576	5
	13	5	00000025	0
	14	0	00000000	0
	15	0	$\cdots$	$\cdots$

Sequences produced by the middle-square method are not really randomly chosen since, once the seed is chosen all subsequent integers are determined. Nevertheless, the sequence output by this method *appears* to be random, and such sequences are readily employed for computer simulations. Thus, these are truly *pseudorandom* sequences.

The most popular method for generating pseudorandom sequences is the following developed in 1949 by D.H. Lehmer — see Biography 1.19 on page 64.

◆ **Linear Congruential Generator**

Let $a, n \in \mathbb{N}$, with $n \geq 2$, $a \leq n - 1$, and $b \leq n - 1$ a nonnegative integer. Then the method known as the *linear congruential method* is described as follows. Choose a nonnegative seed $s_0 \in \mathbb{Z}$ with $s_0 \leq n - 1$, and define

$$s_j \equiv as_{j-1} + b \pmod{n},$$

for $1 \leq j \leq \ell$, where $\ell \in \mathbb{N}$ is the least value such that $s_{\ell+1} = s_j$ for some natural number $j \leq \ell$. Then

$$f(s_0) = (s_1, s_2, \ldots, s_\ell)$$

is a *linear congruential pseudo-random number generator*. We call ℓ the *period length* of f, a the *multiplier*, and b the *increment*.

Example 3.21 *Let $n = 23$, $a = 5$, $b = 7$, and $s_0 = 9$. We calculate,*

j	s_j	$a \cdot s_j + b$	s_{j+1}
0	9	$5 \cdot 9 + 7$	6
1	6	$5 \cdot 6 + 7$	14
2	14	$5 \cdot 14 + 7$	8
3	8	$5 \cdot 8 + 7$	1
4	1	$5 \cdot 1 + 7$	12
5	12	$5 \cdot 12 + 7$	21
6	21	$5 \cdot 21 + 7$	20
7	20	$5 \cdot 20 + 7$	15
8	15	$5 \cdot 15 + 7$	13
9	13	$5 \cdot 13 + 7$	3
10	3	$5 \cdot 3 + 7$	22
11	22	$5 \cdot 22 + 7$	2
12	2	$5 \cdot 2 + 7$	17
13	17	$5 \cdot 17 + 7$	0
14	0	$5 \cdot 0 + 7$	7
15	7	$5 \cdot 7 + 7$	19
16	19	$5 \cdot 19 + 7$	10
17	10	$5 \cdot 10 + 7$	11
18	11	$5 \cdot 11 + 7$	16
19	16	$5 \cdot 16 + 7$	18
20	18	$5 \cdot 18 + 7$	5
21	5	$5 \cdot 5 + 7$	9
22	9	$5 \cdot 9 + 7$	6

Table 3.3

Hence,

$$f(5) = (6, 14, 8, 1, 12, 21, 20, 15, 13, 3, 22, 2, 17, 0, 7, 19, 10, 11, 16, 18, 5, 9)$$

is a linear congruential generator.

Notice in Example 3.21, the period length is $\ell = 22$. Naturally, the best random number generator will be the one that generates numbers that do not repeat *early on*. We note that the maximum length of distinct numbers generated by the linear congruential generator is the modulus n. Establishment of criteria for achievement of that maximum period length is beyond the scope of this book, but may be found in [22]. Nevertheless, we may illustrate it here.

Example 3.22 *Let $n = 16$, $a = 5$, $b = 3$, and $s_0 = 7$. Then we have the following application.*

j	s_j	$a \cdot s_j + b$	s_{j+1}
0	7	$5 \cdot 7 + 3$	6
1	6	$5 \cdot 6 + 3$	1
2	1	$5 \cdot 1 + 3$	8
3	8	$5 \cdot 8 + 3$	11
4	11	$5 \cdot 1 + 3$	10
5	10	$5 \cdot 10 + 3$	5
6	5	$5 \cdot 5 + 3$	12
7	12	$5 \cdot 12 + 3$	15
8	15	$5 \cdot 15 + 3$	14
9	14	$5 \cdot 14 + 3$	9
10	9	$5 \cdot 9 + 3$	0
11	0	$5 \cdot 0 + 3$	3
12	3	$5 \cdot 3 + 3$	2
13	2	$5 \cdot 2 + 3$	13
14	13	$5 \cdot 3 + 3$	4
15	4	$5 \cdot 4 + 3$	7
16	7	$5 \cdot 7 + 3$	6

Table 3.4

Thus, we have the linear congruential generator

$$f(7) = (6, 1, 8, 11, 10, 5, 12, 15, 14, 9, 0, 3, 2, 13, 4, 7)$$

of maximum period length $\ell = 16 = n$.

The reason that we have achieved it here and only barely missed achieving it in Example 3.21 is that we satisfied Knuth's conditions here but not in the previous example. Knuth's conditions may be succinctly stated as follows. A linear congruential generator f will have period length $\ell = n$ if and only if $\gcd(b, n) = 1$, $a \equiv 1 \pmod{p}$ for all primes $p|n$, and $a \equiv 1 \pmod 4$ if $4|n$. The reader may verify that these conditions are satisfied in this example. However, in Example 3.21, $a = 5 \not\equiv 1 \pmod{23}$.

Another another pseudo-random number generator that is more recent than the linear congruential generator, and relates to *public key cryptography*, which we will study in §3.5, is the following.

◆ The RSA Generator

Let $n = pq$, where p, q are primes. Choose $a \in \mathbb{N}$ such that $\gcd(a, \phi(n)) = 1$, and select a *seed* $s_0 \in \mathbb{N}$ with $1 \leq s_0 \leq n - 1$. Define

$$s_j \equiv s_{j-1}^a \pmod{n},$$

for $1 \leq j < \ell$, where ℓ is the least integer such that $s_{\ell+1} = s_j$ for some natural number $j \leq \ell$. Then

$$r(s_0) = (s_1, s_2, \ldots, s_\ell)$$

is an *RSA pseudo-random number generator*. The value ℓ is the *period length* of r and a is the *exponent*.

For simplicity of illustration, we choose two small primes p and q. However, as we will see in §3.5, we need much bigger primes to ensure *security*.

Example 3.23 *Let* $p = 5$, $q = 17$, $n = 85$, $a = 5$, *and* $s_0 = 3$. *Then we have the following.*

Table 3.5

j	s_j	$s_j^a \pmod{n} = s_{j+1}$
0	3	73
1	73	48
2	48	63
3	63	3
4	3	73

Thus, $\ell = 4$, *and*

$$r(3) = (73, 48, 63, 3)$$

is an RSA pseudo-random number generator.

Exercises

3.29. Find the sequence of $m = 8$ (at most) 4-digit pseudorandom numbers generated using von Neumann's middle-square method with seed 6666.

3.30. Find the sequence of $m = 8$ (at most) 4-digit pseudorandom numbers generated using von Neumann's middle-square method with seed 8888.

3.31. Find the linear congruential generator with parameters $n = 26$, $a = 2$, $b = 3$, and seed $s_0 = 15$.

3.32. Find the linear congruential generator with parameters $n = 29$, $a = b = 1$, and seed $s_0 = 2$.

3.33. Find the RSA generator with parameters $n = 133$, $a = 5$, and seed $s_0 = 6$.

3.34. Find the RSA generator with parameters $n = 145$, $a = 11$, and seed $s_0 = 2$.

3.35. Prove that the j-th term produced by the linear congruential generator, as described on page 162, is given by

$$s_j \equiv a^j s_0 + b\frac{(a^j - 1)}{(a - 1)} \pmod{n}.$$

3.36. Let $n \in \mathbb{N}$, and $s_0 \in \mathbb{N}$, with $\mathrm{ord}_n(s_0) = 2^w t$, where $t \in \mathbb{N}$ is odd and $w \geq 0$. Define $s_j \equiv s_{j-1}^2 \pmod{n}$ for all $j \in \mathbb{N}$, $0 \leq s_j \leq n-1$. Prove that the smallest $\ell \in \mathbb{N}$ such that $s_{\ell+j} = s_j$ for some nonnegative integer j is $\ell = \mathrm{ord}_t(2)$.

(Hint: Use Proposition 3.1 on page 140 and Lemma 3.1 on page 142.)

3.5   Public-Key Cryptography

> *We must plan for freedom, and not only for security, if for no other*
> *reason than that only freedom can make security secure.*
>   From Volume 2, Chapter 21 of **The Open Society and its**
> **Enemies (1945)**
>   **Karl Popper (1902–1994), Austrian-born philosopher**

In §2.8, we looked at classical symmetric-key cryptography. At the end of that section we spoke of one more symmetric-key cipher to set the stage for the discussion of public-key cryptography. The following was developed in 1978 — see [39], and Biography 3.5 on the facing page.

◆ **The Pohlig-Hellman Symmetric-Key Exponentiation Cipher**

(a)   A secret prime p is chosen and a secret enciphering key $e \in \mathbb{N}$ with $e \le p-2$ and $\gcd(e, p-1) = 1$.

(b)   A secret deciphering key d is computed via $ed \equiv 1 \pmod{p-1}$.

(c)   Encryption of plaintext message units m is: $c \equiv m^e \pmod{p}$.

(d)   Decryption is achieved via $m \equiv c^d \pmod{p}$.

Example 3.24 *Let $p = 347$, and set $e = 69$, with plaintext*

$$18, 19, 20, 3, 24, 7, 0, 17, 3.$$

Then we encipher each by exponentiating as follows, where all congruences are modulo 347.

$$18^{69} \equiv 239; \quad 19^{69} \equiv 164; \quad 20^{69} \equiv 70; \quad 3^{69} \equiv 267;$$

$$24^{69} \equiv 41; \quad 7^{69} \equiv 191; \quad 0^{69} \equiv 0; \quad 17^{69} \equiv 260; \quad 3^{69} \equiv 267.$$

Then we send off the ciphertext. To decipher, we need the inverse of e modulo $346 = p - 1$, and this is achieved by using the Euclidean algorithm to solve

$$69d + 346x = 1,$$

which has a solution $d = 341$ for $x = -68$, and this is the least positive such value of d. So we may decipher via $239^{341} \equiv 18$ and so on to retrieve the plaintext.

Since knowledge of e and p would allow a cryptanalyst to obtain d, then both p and e must be kept secret. The security of this cipher is based on the difficulty of solving the following problem.

Discrete Log Problem (DLP):

Given a prime p, a generator m of $\mathbb{F}_p^*$, and an element $c \in \mathbb{F}_p^*$, find the unique integer e with $0 \leq e \leq p - 2$ such that

$$c \equiv m^e \pmod{p}. \tag{3.13}$$

The DLP is often called simply *discrete log*. Here $e \equiv \text{ind}_m(c) \pmod{p-1}$. It can be shown (see [37], for instance) that the complexity of finding e in (3.13), when p has n digits, is roughly the same as factoring an n digit number. Therefore, computing discrete logs is virtually the same degree of difficulty as factoring, and since there are no known tractable factoring algorithms, we assume that the *integer factoring problem* (IFP) is intrinsically difficult. We will return to this important problem in §4.3 where we formally define the IFP and look at several factoring algorithms.

In the Pohlig-Hellman cipher, the enciphering key is (e, p) where p is prime, and the deciphering key is (d, p) where d is an inverse of e modulo $p - 1$. This is symmetric-key since, as shown above, it is relatively easy to find d if we know e. However, suppose that we want an exponentiation cipher where the enciphering key may be made *public*, while the decrypting key is kept *private* and known only to the intended recipient. In other words, we seek a *public-key* cryptosystem, where an encrypting key can be made public, because the cryptosystem will be set up to ensure that an impossibly large amount of computer time would be required to find the decrypting key from it.

In 1978, a paper [42] was published by R. Rivest, A. Shamir, and L. Adleman (see Biographies 3.6 on the next page, 3.7 on page 169, and 3.8 on page 170). In this paper they describe a public-key cryptosystem, including key generation and a public-key cipher, whose security rests upon the IFP. This cryptosystem, which has

> **Biography 3.5** Martin E. Hellman *was born on October 2, 1945. He obtained all his academic degrees in electrical engineering: his bachelor's degree from New York University in 1966; his master's degree in 1967; and his Ph.D. in 1969, the latter two from Stanford. He was employed at IBM and at MIT, but returned to Stanford in 1971. He remained there until 1996, when he received his Professor Emeritus status. We already learned above that he was one of the pioneers of PKC. He has been involved in computer privacy issues going back to the debate over the DES keylength in 1975. He has not only demonstrated his scholarship with numerous publications, but also has excelled in teaching. He was recognized with four teaching awards; three of these were from minority-student organizations. He is now retired from research and teaching. He and Dorothie, his wife of some thirty-five years, live on campus at Stanford.*

come to be known by the acronym from the authors' names, the *RSA cryptosystem*, has stood the test of time to this day, where it is used in cryptographic applications from banking and in e-mail security to e-commerce on the Internet. Now we mathematically formalize what we mean by a public-key cipher.

◆ Public-Key Cryptosystems (PKCs)

A cryptosystem consisting of a set of enciphering transformations $\{E_e\}$ and a set of deciphering transformations $\{D_d\}$ is called a *public-key cryptosystem* or an *asymmetric cryptosystem* if for each key pair (e, d) the enciphering key e, called the *public key*, is made publicly available, whereas the deciphering key d, called the *private key*, is kept secret. The cryptosystem satisfies the property that it is infeasible to compute d from e, given the prohibitive amount of computer time it would take to do so.

The term *private* is used in PKC rather than the term *secret* used in symmetric-key cryptography (SKC) since it takes two or more entities to share a secret (such as the symmetric secret key), whereas it is truly *private* when only one entity knows about it, such as with the asymmetric private key. We now look at a means of viewing PKC from a nonmathematical viewpoint that will shed more light on the concept from a different angle.

Biography 3.6 Ronald L. Rivest *received a B.A. in mathematics from Yale University in 1969 and a Ph.D. in computer science from Stanford University in 1974. He is a co-inventor of the RSA public-key cryptosystem and co-founder of RSA Data Security (now called RSA Security after having been bought by Security Dynamics, the company that holds all the patents on the RSA cryptosystem). Among his numerous, outstanding honours and positions are Fellow of the American Academy of Arts and Science, Fellow of the Association for Computing Machinery, Director of the Financial Cryptographic Association, Director of the International Association for Cryptologic Research, and Fellow of the World Technology Network. He, together with Adleman and Shamir, was awarded the 2000 IEEE Koji Kobayashi Computers and Communications Award, as well as the Secure Computing Lifetime Achievement Award. He is widely respected as an expert in cryptographic design and cryptanalysis, as well as the areas of machine learning and VLSI design.*

◆ PKC Analogy

In the paper [42] by Rivest, Shamir, and Adleman, the cryptographic characters Alice and Bob were introduced. We will use these characters along with a growing cast of cryptographic players to illustrate the various concepts. We begin with an analogy for PKC itself. Suppose that Bob has a *public* wall safe with a *private* combination known only to him. Moreover, suppose that the safe is left open and made available to passers-by. Then anyone, including Alice can put messages in the safe and lock it. However, only Bob can retrieve the message because, even Alice, who left the message in the safe, has no way of retrieving it.

Biography 3.7 Adi Shamir *is an Israeli cryptographer who is, at the time of this writing, the Borman Professor in the Applied Mathematics Department of the Weizman Institute of Science in Israel. He obtained his Ph.D. from Stanford in 1977 after which he did postdoctoral work at Warwick University in England. Shamir's name is attached to a wide variety of cryptographic schemes, including the Fiat-Shamir identification protocol, RSA, DC, and his polynomial secret-sharing scheme, to mention only a few. On April 14, 2003, the ACM formally announced that the A.M. Turing Award (essentially the "Nobel Prize of computer science") would go to Adleman, Shamir, and Rivest for their developmental work on PKC.*

We now wish to view the typical interaction between PKC and SKC in practice, say on the Internet for e-commerce. The reason for not using PKC *exclusively* for general-purpose encryption is that PKCs are slower than SKCs, meaning that they take longer to run on a computer. For instance, the RSA cryptosystem, which we will study in this section, is a thousand times slower than the symmetric-key cryptosystem known as the *Data Encryption Standard* (DES), which was the first commercially available algorithm put in use and standardized by the US federal government in 1977 — see [34, Chapter 3] for a complete description of the algorithm including a scaled-down version, known as S-DES, presented for pedagogical purposes. We do not present DES (or S-DES) in this text since its description is very lengthy. DES was replaced in 2000 by the *Advanced Encryption Standard* (AES), *Rijndael* that is also described in detail in [34].

What is done in practice, and is mathematically described in what follows, is that a PKC is used to encipher keys for SKC use, thereby securely transmitting the keys, which are then used with the SKC to transmit the bulk of the data. Some of the following is adapted from [34].

◆ **Hybrid Cryptosystems — Digital Envelopes**

Alice and Bob have access to an SKC, which we will call S. Also, Bob has a public-private key pair (e, d). In order to send a message m to Bob, Alice first generates a symmetric key, called a *session key* or *data encryption key*, k to be used only once. (The property of producing a new session key each time a pair of users wants to communicate is called *key freshness*.) Alice enciphers m using k and S obtaining ciphertext $E_k(m) = c$. Using Bob's public key e, Alice encrypts k to get $E_e(k) = k'$. Both of these encryptions are fast since S is efficient in the first enciphering, and the session key is small in the second enciphering. Then Alice sends c and k' to Bob, who deciphers k with his private key d, via $D_d(k') = k$. Then Bob easily deduces the symmetric deciphering key k^{-1}, which he uses to decipher

$$D_{k^{-1}}(c) = D_{k^{-1}}(E_k(m)) = m.$$

Hence, the PKC is used only for the sending of the session key, which provides a digital envelope that is both secure and efficient.

The following illustrates the above discussion.

Diagram 3.1 Digital Envelope — Hybrid Cryptosystem

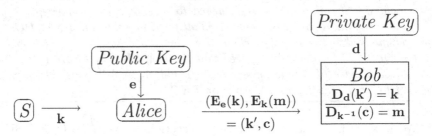

Now we are ready to look at our first PKC, invented in 1978, and patented in 1983.

Biography 3.8 Leonard Adleman *was born on December 31, 1945, in San Francisco, California. He received his B.Sc. in mathematics from the University of California at Berkeley in 1972 and his Ph.D. there in 1976. His doctoral thesis was done under the guidance of Manuel Blum and was titled* Number Theoretic Aspects of Computational Complexity. *In 1980, he was hired in the computer science department of the University of Southern California. His professional interests are algorithms, computational complexity, computer viruses — a term he coined — cryptography, immunology, molecular biology, number theory, and quantum computing. He also works on computing using DNA. He observed that a protein, called* polymerase, *which produces complementary strands of DNA, resembles the operation of a Turing machine. Adleman reached the conclusion that DNA formation essentially functions in a fashion similar to a computer, so he is interested in constructing a viable DNA computer that would have the potential for vastly faster computation in the future.*

◆ **The RSA Public-Key Cryptosystem**

This cipher is a PKC based on modular exponentiation. Suppose that Alice wants to send a message to Bob.

She first obtains Bob's public key, the pair (n, e), where $n = pq$ is the modulus consisting of the product of two large primes p and q, and $e \in \mathbb{N}$ is the *enciphering exponent*, where $\gcd(e, \phi(n)) = 1$. To encipher a message, the plaintext is converted from letters to numerical equivalents, and put into blocks of m-digit base-26 plaintext, that get encrypted to ℓ-digit base-26 integers, where $m < \ell$ and the primes p, q are selected so that

$$26^m < n < 26^\ell.$$

In this way, each plaintext message m gets transformed to a unique ℓ-digit base-26 integer via the encryption,

$$E_e(m) \equiv m^e \equiv c \pmod{n}, \text{ where } 0 \le c < n.$$

To decipher, Bob has knowledge of the inverse d — called the *RSA deciphering exponent* — of e modulo $\phi(n)$. Given the stipulation that $\gcd(e, \phi(n)) = 1$, d must exist via

$$ed = x\phi(n) + 1 \text{ for some } x \in \mathbb{Z}. \tag{3.14}$$

Bob deciphers C via

$$D_d(c) \equiv c^d \equiv (m^e)^d \equiv m^{ed} \equiv m^{x\phi(n)+1} \equiv (m^{\phi(n)})^x m \equiv m \pmod{n},$$

where the penultimate congruence follows from Euler's Theorem 2.10 on page 93. The pair (d, n) is the *RSA deciphering key*.

Remark 3.5 *In order to invoke Euler's Theorem in the above description of the RSA cipher, we have to assume that $\gcd(m, n) = 1$. In the very highly unlikely event that $\gcd(m, n) > 1$, then we still recover m as follows. If $p \mid m$, then $m \equiv 0 \pmod{p}$. Hence, $m^{ed} \equiv 0 \pmod{p}$. Also, $m^{ed} \equiv m \pmod{q}$. Since $p \ne q$, $m^{ed} \equiv m \pmod{n}$. Thus, the Chinese Remainder Theorem uniquely determines m^{ed} modulo n,*

$$c^d \equiv (m^e)^d \equiv m \pmod{n}.$$

Example 3.25 *Suppose that Bob chooses $(p, q) = (29, 43)$. Then $n = 1247$ and $\phi(n) = 1176$. If Bob selects $e = 5$, then solving*

$$1 = 5d + \phi(n)x = 5d + 1176x,$$

we get $x = -4$, so $d = 941$. Also, $(e, n) = (5, 1247)$ is his public key. Alice obtains Bob's public key and wishes to send the message **send money**. *The numerical equivalents are given by Table 2.2:*

$$\{m_1, m_2, m_3, m_4, m_5, m_6, m_7, m_8, m_9\} = \{18, 04, 13, 03, 12, 14, 13, 04, 24\},$$

so she picks blocks of size $m = 2$ for plaintext and $\ell = 4$ for ciphertext. Note that

$$26^m = 26^2 = 676 < n < 26^4 = 456976 = 26^\ell.$$

She enciphers using Bob's public key to get

$$m_1^5 \equiv 18^5 \equiv 363 \equiv c_1 \pmod{n}, \quad m_2^5 \equiv 4^5 \equiv 1024 \equiv c_2 \pmod{n},$$

$$m_3^5 \equiv 13^5 \equiv 934 \equiv c_3 \pmod{n}, \quad m_4^5 \equiv 3^5 \equiv 243 \equiv c_4 \pmod{n},$$

$$m_5^5 \equiv 12^5 \equiv 679 \equiv c_5 \pmod{n}, \quad m_6^5 \equiv 14^5 \equiv 367 \equiv c_6 \pmod{n},$$

$$m_7^5 \equiv 13^5 \equiv 934 \equiv c_7 \pmod{n}, \quad m_8^5 \equiv 4^5 \equiv 1024 \equiv c_8 \pmod{n},$$

$$m_9^5 \equiv 24^5 \equiv 529 \equiv c_9 \pmod{n},$$

and she sends the ciphertext

$$\{c_1, c_2, c_3, c_4, c_5, c_6, c_7, c_8, c_9\} = \{363, 1024, 934, 243, 679, 367, 934, 1024, 529\}$$

to Bob. He uses his private key d to decipher via

$$c_1^d \equiv 363^{941} \equiv 18 \equiv m_1 \pmod{n}, \quad c_2^d \equiv 1024^{941} \equiv 4 \equiv m_2 \pmod{n},$$

and so on to recover the plaintext.

▼ Security

The above description of RSA, known as *Plain* RSA (namely, *without* any preprocessing of plaintext message units) is insecure in the following sense. The attack against plain RSA given in [7] shows that even though an m-bit key is used in plain RSA, the *effective security* is $m/2$ bits. Hence, it is essential that, before encryption, a preprocessing step be implemented that uses the *Optimal Asymmetric Encryption Padding* (OAEP) (introduced in [5]) such as [38], a recent standard from RSA Labs. In order to obtain a *secure* RSA cryptosystem from a *plain* RSA cryptosystem, there should be an application of a preprocessing function to the plaintext *before enciphering*. In [5], there are new standards for "padding" plain RSA so that it is secure against certain *chosen ciphertext attacks*, which are attacks where the cryptanalyst chooses the ciphertext and is given the corresponding plaintext. This attack is most effective against public-key cryptosystems, but sometimes is effective against symmetric-key ciphers as well.

Another important point about security that involves a bad implementation of RSA is the following. Suppose that Alice and Bob use the RSA cryptosystem but they choose the same RSA modulus n, and enciphering (public) keys e_A and e_B, respectively, with $\gcd(e_A, e_B) = 1$. Suppose that Eve intercepts two cryptograms m^{e_A} and m^{e_B}, enciphering the same message m, sent by a third entity to Alice and Bob, respectively. Given that $\gcd(e_A, e_B) = 1$, the Euclidean algorithm allows Eve to solve

$$e_A x + e_B y = 1$$

for some $x, y \in \mathbb{Z}$. Then, Eve calculates:

$$c_A^x c_B^y \equiv (m^{e_A})^x (m^{e_B})^y \equiv m^{e_A x + e_B y} \equiv m \pmod{n},$$

and this is done without knowledge of a factorization of n or of knowledge of the private keys. This is called *common modulus protocol failure* (CMPF). This is not a failure of the RSA cryptosystem, rather a very bad implementation of it. In fact, the CMPF shows, in no uncertain terms, that an RSA modulus should *never* be used by more than one entity. The CMPF illustrates the fact that no matter how strong a cipher might be, a bad implementation of it will render the scheme to be insecure, and useless. The true security of RSA requires a proper implementation. For instance, even the RSA modulus size of 2048 bits suggested above, is useless in the face of a bad implementation such as the CMPF.

The security of RSA is based on the following, as-yet-unproved conjecture.

> **The RSA Conjecture**
> Cryptanalyzing RSA must be as difficult as factoring.

A good reason for believing that this conjecture is valid is that the only known method for finding d given e is the Euclidean algorithm applied to e and $\phi(n)$. Yet, to compute $\phi(n)$, we need to know p and q, namely, we need to know how to factor n.

There are numerous cryptosystems that are called *equivalent to the difficulty of factoring*. For instance, there are RSA-like cryptosystems whose difficulty to break is as hard as factoring the modulus. It can be shown that any cryptosystem for which there is a constructive proof of equivalence to the difficulty of factoring, is vulnerable to a chosen-ciphertext attack — see the discussion of security on page 172. We have already seen that factoring an RSA modulus allows the breaking of the cryptosystem, but the converse is not known. In other words, it is not known if there are other methods of breaking RSA, but some new attacks presented concerns.

◆ Attacks on RSA

In what follows, the attacks on RSA must really be seen as attacks on particular *implementations* of RSA. Hence, taken together, the following present a cogent argument and criteria for secure implementations of RSA.

In 1995, Paul Kocher, then a Stanford undergraduate, discovered that RSA could be cryptanalyzed by recovering the decryption exponent through a careful timing of the computation times for a sequence of decryptions. This weakness was a surprising and unexpected discovery, and although there are means of thwarting the attack, it was another wake-up call. An analogue for Kocher's timing attack is for a thief to watch someone turning the lock on a safe and measuring the time it takes to go to each combination number in order to guess the combination, quite clever. For a description of this attack, and means to thwart it, see [32, Chapter 4, pp. 177–178].

Another outstanding idea from Kocher, called *power cryptanalysis*, involves a very careful measurement of the computer's *power consumption*. This works during decryption when Eve, a passive eavesdropper, could recover the secret key. This works since during multiprecision multiplications the computer's power consumption is necessarily higher than it would normally be. Hence, if Eve measures the length of these high consumption episodes, she can easily decide when the computer is performing one or two multiplications, and this gives away the bits of d. Protection against this attack is some kind of physical shielding of the power output. See

http://www.cryptography.com/resources/whitepapers/DPA.html.

Another attack is known as the *low public exponent attack*. For the sake of efficiency, one would like to use small public RSA exponents. However, one has to be careful not to compromise security in so doing. One typically used public exponent is, surprisingly, $e = 3$. However, if the same message m, in a single block, is sent to three different entities, having pairwise relatively prime RSA moduli n_j, with $m < n_j$ for $j = 1, 2, 3$, this allows recovery of the plaintext. Here is how it is done. Using the Chinese remainder theorem, there is a solution to

$$x \equiv c_i \equiv m^3 \pmod{n_i} \text{ for each } i = 1, 2, 3.$$

Since $m^3 < n_1 n_2 n_3$, then $x = m^3$. By computing the cube root of the integer x, we retrieve m. Furthermore, this attack can be generalized to show that a plaintext m can be recovered if e is the RSA enciphering exponent and m is sent to $k \geq e$ recipients with pairwise relatively prime RSA moduli n_i such that $m < n_i$ for $i = 1, 2, \ldots, k$.

The attack described in the above paragraph works because the messages are linearly related, allowing use of the Chinese Remainder Theorem. In fact, a generalization of results by Coppersmith (see [12]) were given by Hastad in [21], called the *Strong Hastad Broadcast Attack*. He proved that *any* fixed polynomial applied as padding is insecure. Therefore, a defense against his attack is to pad with a *randomized* polynomial, not a fixed one.

Another powerful attack developed by Coppersmith is his *partial key exposure attack* (see [12]), which can be described as follows. Given an RSA modulus $n = pq$ of bitlength ℓ, the bitlength of either p or q is about $\ell/2$. Knowledge of either the $\ell/4$ most significant bits of p or the $\ell/4$ least significant bits of p, can be shown to allow one to efficiently factor n.

Another attack, called the *low secret exponent attack*, was developed by Weiner [51]. Again, as with the reason for choosing small public exponents, we want increased efficiency in the decryption process, so we choose small secret exponents. For instance, given a 1024-bit RSA modulus, the decryption process can have efficiency increased ten-fold with the choice of small d. However, Weiner's attack yields a total break of the RSA cryptosystem. (A *total break*, means that a cryptanalyst can recover d, hence retrieve all plaintext from ciphertext.) Succinctly, his attack says that if $n = pq$ where p and q are primes such that

$$q < p < 2q \text{ and } d < n^{1/4}/3,$$

then given a public key e with

$$ed \equiv 1 \pmod{\phi(n)},$$

d can be efficiently calculated. We conclude that use of small decryption exponents, in the sense given by Weiner above, leads to a total loss of security in the RSA cryptosystem. Weiner's method was improved in [8] by Boneh and Durfee who showed that RSA is insecure if $d < n^{0.292}$.

Exercises

3.37. Prove that the DLP presented on page 167 is independent of the generator m of $\mathbb{F}_p^*$. (This means you must demonstrate that any algorithm that computes logs to base m can be used to compute logs to any other base m' that is a generator of $\mathbb{F}_p^*$.)

3.38. The *generalized discrete log problem* is described as follows. Given a finite cyclic group G of order $n \in \mathbb{N}$, a generator α of G, and an element $\beta \in G$, find that unique nonnegative integer $x \leq n - 1$ such that

$$\alpha^x = \beta.$$

Given the fact that such a group G is isomorphic to $\mathbb{Z}/n\mathbb{Z}$ (see page 303), one would expect that an efficient algorithm for computing discrete logs in one group would imply an efficient algorithm for the other group. Explain why this is *not* the case.

3.39. Explain why selecting an enciphering exponent $e = \phi(n)/2 + 1$ in the RSA cipher is a bad choice leading to a very undesirable outcome.

3.40. Suppose that *Mallory*, a malicious active attacker, wants to decipher

$$c \equiv m^e \pmod{n}$$

to recover plaintext m enciphered using RSA and sent by Alice to Bob. Furthermore, suppose that Mallory can intercept and disguise c by selecting a random $x \in (\mathbb{Z}/n\mathbb{Z})^*$ and computing

$$\bar{c} \equiv cx^e \pmod{n}.$$

Not knowing this, Bob computes

$$\bar{m} \equiv \bar{c}^d \pmod{n}$$

and sends it to Alice. Explain how Mallory may now recover the plaintext if he intercepts $\bar{m}$.

Exercises 3.41–3.44 refer to the Pohlig-Hellman exponentiation cipher presented on page 166. In each exercise, use the data to decipher the given cryptogram and produce the plaintext via Table 2.2 on page 129.

3.41. $p = 173$, $e = 3$, and

$$c = (125, 166, 43, 170, 112, 112, 149, 112, 170, 64, 27, 64, 0, 112, 170).$$

3.42. $p = 281$, $e = 23$, and

$$c = (260, 0, 39, 200, 54, 69, 44, 66, 141, 144, 69, 54, 200, 66, 144).$$

3.43. $p = 389$, $e = 29$, and

$$c = (264, 124, 0, 342, 298, 0, 376, 264, 294, 285, 182).$$

3.44. $p = 1151$, $e = 51$, and

$$c = (769, 11, 0, 257, 506, 242, 11, 400, 242, 0, 966, 361, 11).$$

Exercises 3.45–3.48 pertain to the RSA public-key cryptosystem described on page 170. Find the plaintext numerical value of m from the parameters given. You will first have to determine the private key d from the given data via the methodology illustrated in Example 3.25 on page 171. If repeated squaring is not employed (see page 82), then a computer utilizing a mathematical software package will be required for these calculations.

3.45. $(p, q) = (379, 911)$, $n = 345269$, $e = 11$, and $c \equiv 69234 \pmod{n}$.

3.46. $(p, q) = (173, 337)$, $n = 58301$, $e = 11$, and $c \equiv 2191 \pmod{n}$.

3.47. $(p, q) = (367, 911)$, $n = 334337$, $e = 17$, and $c \equiv 226756 \pmod{n}$.

3.48. $(p, q) = (3677, 6323)$, $n = 23249671$, $e = 13$, and $c \equiv 234432 \pmod{n}$.

In the RSA cipher, if we know $\phi(n)$ and n, then we can factor n. The reason is that we can find p and q by successively computing

$$p + q = n - (p-1)(q-1) + 1 \ and \ p - q = \sqrt{(p+q)^2 - 4n},$$

so we get

$$p = \frac{1}{2}\left[(p+q) + (p-q)\right] \ and \ q = \frac{1}{2}\left[(p+q) - (p-q)\right].$$

In Exercises 3.49–3.54, use the above to find the primes p and q.

3.49. $pq = 632119$ and $\phi(pq) = 630496$.

3.50. $pq = 1009427$ and $\phi(pq) = 1007400$.

3.51. $pq = 5396749$ and $\phi(pq) = 5391936$.

3.52. $pq = 13554791$ and $\phi(pq) = 13547400$.

3.53. $pq = 43179137$ and $\phi(pq) = 43165956$.

3.54. $pq = 62091299$ and $\phi(pq) = 62075520$.

Suddenly Christopher Robin began to tell Pooh about some of the things People called Kings and Queens and something called Factors...
from **The House on Pooh Corner**
by **A.A. Milne (1882–1956)**

Chapter 4

Quadratic Residues

> *As with everything else, so with mathematical theory, beauty can be perceived, but not explained.*
>
> **Arthur Cayley (1821–1895), British mathematician**

We have encountered the notion in the title of this chapter already. For instance, when we introduced power residues in Definition 3.4 on page 155, quadratic residues were given as an illustration. Also, in Examples 2.8–2.12 in Chapter 2, quadratic congruences were illustrated. Indeed, quadratic congruences are the next simplest after linear congruences the solutions for which we classified in Theorem 2.3 on page 84. As we have seen, it is from quadratic congruences that the notion of quadratic residues arise. For instance, such questions arise as: given a prime p and an integer a, when does there exist an integer x such that $x^2 \equiv a \,(\mathrm{mod}\ p)$? Such queries were studied by Euler, Gauss, and Legendre, the latter having his name attached to the symbol that we study in §4.1, a symbol which provides a mechanism for answering the above question.

4.1   The Legendre Symbol

We begin by formally defining the name in the title of this chapter.

Definition 4.1 | Quadratic Residues and Nonresidues |

If $n \in \mathbb{N}$ and $a \in \mathbb{Z}$ with $\gcd(a, n) = 1$, then a is said to be a quadratic residue modulo n *if there exists an integer x such that*

$$x^2 \equiv a \pmod{n}, \tag{4.1}$$

and if (4.1) has no such solution, then a is a quadratic nonresidue modulo n.

Remark 4.1 *Although $0^2 \equiv 0 \,(\mathrm{mod}\ n)$, we typically consider positive values for the purposes of quadratic residues. We are only interested in the values relatively prime to the modulus as given in Definition 4.1. The following discussion shows that by confining ourselves to these relatively prime values, we also achieve some nice symmetry.*

Example 4.1 *To determine the quadratic residues modulo 7, we consider the squares of all positive integers less than 7. They are*

$$1^2 \equiv 6^2 \equiv 1 \pmod 7, 2^2 \equiv 5^2 \equiv 4 \pmod 7, \text{ and } 3^2 \equiv 4^2 \equiv 2 \pmod 7.$$

Hence, the quadratic residues modulo 7 are $1, 2, 4$, so the quadratic nonresidues modulo 7 are $3, 5, 6$.

Notice that the number of quadratic residues modulo 7 in Example 4.1 is three, which is exactly the same as the number of quadratic nonresidues, namely $(p-1)/2$ each. This is a general fact.

Theorem 4.1 | **The Number of Quadratic Residues and Nonresidues**

If $p > 2$ is prime, then, in the set $\mathcal{S} = \{1, 2, 3, \ldots, p-1\}$, there are exactly $(p-1)/2$ quadratic residues and $(p-1)/2$ quadratic nonresidues.

Proof. By Theorem 3.1 on page 142, if a is a primitive root modulo p, the set $\mathcal{S}$ is identical to the set of least positive residues of the integers

$$a, a^2, a^3, \ldots, a^{p-1} \text{ modulo } p.$$

Also, by Corollary 3.10 on page 157, $x^2 \equiv b \,(\mathrm{mod}\ p)$ has a solution if and only if $2 \mid \mathrm{ind}_a(b)$. Hence, for $b \in \mathcal{S}$, $x^2 \equiv b \,(\mathrm{mod}\ p)$ has a solution if and only if

$$b \equiv a^{2j} \pmod p$$

for some $j = 1, 2, \ldots, (p-1)/2$. Hence there are exactly $(p-1)/2$ quadratic residues, leaving exactly $(p-1)/2$ quadratic nonresidues in $\mathcal{S}$.   □

Immediate from the above proof is the following result.

Corollary 4.1 *If p is prime and g is a primitive root modulo p, then $a \in \mathbb{Z}$ is a quadratic residue modulo p if and only if $\mathrm{ind}_g(a)$ is even. Equivalently, a is a quadratic nonresidue modulo p if and only if $\mathrm{ind}_g(a)$ is odd.*

Remark 4.2 *Euler's criterion stated in Corollary 3.8 on page 157 may be re-stated as follows. The integer b is a quadratic residue modulo p if and only if*

$$b^{(p-1)/2} \equiv 1 \pmod{p}.$$

Therefore, quadratic residues have order at most $(p-1)/2$ modulo p, so all primitive roots of a prime $p > 2$ are quadratic nonresidues modulo p. (Note that the Euler criterion is an immediate consequence of the proof of Theorem 4.1. However, we could not state it as a corollary thereof since we used that criterion to prove Corollary 3.10, which was used in the proof of Theorem 4.1. Therefore, to do so would have amounted to an (invalid) circular argument.)

Study of quadratic residues and nonresidues is simplified by the following notation that is the header for this section.

Definition 4.2 | **Legendre's Symbol** |

If $c \in \mathbb{Z}$ and $p > 2$ is prime, then

$$\left(\frac{c}{p}\right) = \begin{cases} 0 & \text{if } p \mid c, \\ 1 & \text{if } c \text{ is a quadratic residue modulo } p, \\ -1 & \text{otherwise,} \end{cases}$$

and $\left(\frac{c}{p}\right)$ is called the Legendre Symbol *of c with respect to p.*

The following is a fundamental result on *quadratic residuacity* modulo n. This term means the determination of whether an integer a quadratic residue or a nonresidue modulo n.

Theorem 4.2 | **Euler's Criterion for Quadratic Residuacity** |

If $p > 2$ is prime, then

$$\left(\frac{c}{p}\right) \equiv c^{(p-1)/2} \pmod{p}.$$

Proof. This is merely a restatement of Corollary 3.8 on page 157 in terms of the Legendre symbol.   □

An application of Euler's criterion is the promised proof of the converse of Theorem 1.28 on page 61. We state the full result here together with a reminder, for the reader, of the sequence upon which the primality test is based.

Theorem 4.3 | **The Lucas-Lehmer Test for Mersenne Primes** |

Let p be an odd prime, set $s_1 = 4$ and recursively define for $j \geq 2$, $s_j = s_{j-1}^2 - 2$. Then $M_p = 2^p - 1$ is prime if and only if

$$s_{p-1} \equiv 0 \pmod{M_p}. \tag{4.2}$$

Proof. For a proof that (4.2) implies that M_p is prime see Theorem 1.28. Now assume that M_p is prime. Let $\tau = (1 + \sqrt{3})/\sqrt{2}$ and $\bar{\tau} = (1 - \sqrt{3})/\sqrt{2}$. In what follows, all congruences are assumed to take place in $\mathbb{Z}[\tau] = \{a + b\tau : a, b \in \mathbb{Z}\}$. First we need the following result.

Claim 4.1 $\tau^{M_p+1} \equiv -1 \,(\mathrm{mod}\ M_p)$.

We look at $\sqrt{2}\tau = 1 + \sqrt{3}$ raised to the power M_p modulo M_p.

$$(\sqrt{2}\tau)^{M_p} \equiv \tau^{M_p} 2^{(M_p-1)/2}\sqrt{2} \equiv (1 + \sqrt{3})^{M_p} \equiv 1 + 3^{(M_p-1)/2}\sqrt{3} \ \ (\mathrm{mod}\ M_p),$$

where the last congruence follows from the Binomial Theorem.

Since $M_p \equiv -1 \,(\mathrm{mod}\ 8)$, then by Euler's criterion,

$$2^{(M_p-1)/2} \equiv \left(\frac{2}{M_p}\right) \equiv 1 \ \ (\mathrm{mod}\ M_p),$$

and since $M_p \equiv 1 \,(\mathrm{mod}\ 3)$, then by Euler's criterion,

$$3^{(M_p-1)/2} \equiv \left(\frac{3}{M_p}\right) \equiv -1 \ \ (\mathrm{mod}\ M_p).$$

Hence,

$$\tau^{M_p} \equiv \bar{\tau} \ \ (\mathrm{mod}\ M_p) \quad \text{and} \quad \tau^{M_p+1} \equiv \tau\bar{\tau} \equiv -1 \ (\mathrm{mod}\ M_p),$$

which is Claim 4.1.

From Claim 4.1, we have that

$$\tau^{2^p} + 1 \equiv 0 \ \ (\mathrm{mod}\ M_p).$$

Since $\tau^2 = 2 + \sqrt{3}$, then

$$(2 + \sqrt{3})^{2^{p-1}} + 1 \equiv 0 \ \ (\mathrm{mod}\ M_p). \tag{4.3}$$

Multiplying both sides of (4.3) by $(2 - \sqrt{3})^{2^{p-2}}$ we get

$$(2 + \sqrt{3})^{2^{p-1}}(2 - \sqrt{3})^{2^{p-2}} + (2 - \sqrt{3})^{2^{p-2}} \equiv 0 \ \ (\mathrm{mod}\ M_p). \tag{4.4}$$

However,

$$(2 + \sqrt{3})^{2^{p-1}}(2 - \sqrt{3})^{2^{p-2}} = (2 + \sqrt{3})^{2^{p-2}}(2 + \sqrt{3})^{2^{p-2}}(2 - \sqrt{3})^{2^{p-2}} =$$

$$(2 + \sqrt{3})^{2^{p-2}}\left[(2 + \sqrt{3})(2 - \sqrt{3})\right]^{2^{p-2}} = (2 + \sqrt{3})^{2^{p-2}},$$

since

$$(2 - \sqrt{3})(2 + \sqrt{3}) = 1.$$

Thus, (4.4) becomes

$$(2 + \sqrt{3})^{2^{p-2}} + (2 - \sqrt{3})^{2^{p-2}} \equiv 0 \ \ (\mathrm{mod}\ M_p).$$

The result will now follow from the following claim.

Claim 4.2 *For any* $j \in \mathbb{N}$, $s_j = (2+\sqrt{3})^{2^{j-1}} + (2-\sqrt{3})^{2^{j-1}}$.

Let $\omega_j = (2+\sqrt{3})^{2^{j-1}} + (2-\sqrt{3})^{2^{j-1}}$. Then

$$\omega_1 = (2+\sqrt{3}) + (2-\sqrt{3}) = 4,$$

and

$$\omega_{j+1} = \left[(2+\sqrt{3})^{2^{j-1}} + (2-\sqrt{3})^{2^{j-1}}\right]^2 - 2(2+\sqrt{3})(2-\sqrt{3}) = \omega_j^2 - 2,$$

so $\omega_j = s_j$ for all $j \in \mathbb{N}$, securing both Claim 4.2, and the theorem.   □

Remark 4.3 *The proof of the necessity of Condition 4.2 is due to Rosen [43], and is one of the most elementary of the proofs of the result in the literature. Typically in introductory number theory texts, a weaker result is given or no proof at all since most are lengthy and detailed by comparison. In any case, this completes the primality test for Mersenne primes begun in §1.8, and illustrates the applicability of Euler's criterion.*

Biography 4.1 Adrien-Marie Legendre (1752–1833) *was educated at the Collège Mazarin in Paris. During the half decade 1775–1780, he taught along with Laplace (1749–1827) at Ecole Militaire. He also took a position at the Académie des Sciences, becoming first adjoint in 1783, then associé in 1785, and his work finally resulted in his election to the Royal Society of London in 1787. In 1794 Legendre published his phenomenally successful book* Eléments de Géométrie, *which remained the leading introductory text in the subject for over a century. In 1795, he was appointed professor at the École Normale. In 1808, Legendre published his second edition of* Théorie des Nombres, *which included Gauss' proof of the Quadratic Reciprocity Law (about which we will learn in §4.2). Legendre also published his three-volume work* Exercises du Calcul Intégral *during 1811–1819. Then his three-volume work* Traité des Fonctions Elliptiques *was published during the period 1825–1832. Therein he introduced the name "Eulerian Integrals" for beta and gamma functions. This work also provided the fundamental analytic tools for mathematical physics, and today some of these tools bear his name, such as* Legendre Functions. *He also made fundamental contributions in the areas of mathematical astronomy and geodesy.*

Example 4.2 *Let* $p = 7$ *and* $c = 5$. *Then*

$$\left(\frac{c}{p}\right) \equiv \left(\frac{5}{7}\right) \equiv 5^3 \equiv -1 \equiv c^{(p-1)/2} \pmod{p},$$

so 5 is a quadratic nonresidue modulo 7 by Theorem 4.2.

Now we establish some fundamental facts about Legendre's symbol.

Theorem 4.4 $\boxed{\textbf{Properties of the Legendre Symbol}}$

If $p > 2$ is prime and $b, c \in \mathbb{Z}$, then

(1) If $b \equiv c \,(\mathrm{mod}\ p)$, then $\left(\dfrac{b}{p}\right) = \left(\dfrac{c}{p}\right)$.

(2) $\left(\dfrac{b}{p}\right)\left(\dfrac{c}{p}\right) = \left(\dfrac{bc}{p}\right)$.

(3) $\left(\dfrac{-1}{p}\right) = (-1)^{(p-1)/2}$.

Proof. Each part is a consequence of Euler's criterion in Theorem 4.2 on page 179. For instance, for part (2),

$$\left(\frac{b}{p}\right)\left(\frac{c}{p}\right) \equiv b^{(p-1)/2} c^{(p-1)/2} \equiv (bc)^{(p-1)/2} \equiv \left(\frac{bc}{p}\right) \quad (\mathrm{mod}\ p),$$

and parts (1) and (3) clearly follow as well.   □

Example 4.3 *Since* $3 \equiv 11\,(\mathrm{mod}\ 7)$ *then*

$$\left(\frac{3}{7}\right) = \left(\frac{11}{7}\right) = 1,$$

which illustrates part (1) of Theorem 4.4.

Example 4.4 *A special case of part (2) of Theorem 4.4 is*

$$\left(\frac{b^2}{p}\right) \equiv \left(\frac{b}{p}\right)^2 \equiv 1 \ (\mathrm{mod}\ p).$$

Example 4.5 *A reinterpretation of part (3) of Theorem 4.4 is the following. Since* $(p-1)/2$ *is even when* $p \equiv 1\,(\mathrm{mod}\ 4)$ *and it is odd when* $p \equiv 3\,(\mathrm{mod}\ 4)$, *then*

$$\left(\frac{-1}{p}\right) = \begin{cases} 1 & \textit{if } p \equiv 1 \ (\mathrm{mod}\ 4), \\ -1 & \textit{if } p \equiv -1 \ (\mathrm{mod}\ 4). \end{cases}$$

In other words, we have established the quadratic residuacity of -1 *modulo any odd prime* p.

Now we look to establish the quadratic residuacity of 2 modulo any prime $p > 2$. First we need the following result due to Gauss.

Theorem 4.5 | Gauss' Lemma |

Let $p > 2$ be prime with $c \in \mathbb{Z}$ relatively prime to p. Suppose that $\mathcal{R}$ is a set consisting of the least positive residues r_j of the integers cj modulo p for $j = 1, 2, \ldots, (p-1)/2$, and $\mathcal{S} = \{r_j \in \mathcal{R} : r_j > p/2\}$, with $|\mathcal{S}| = s$. Then

$$\left(\frac{c}{p}\right) = (-1)^s.$$

Proof. If $r_i, r_j \in \mathcal{R}$ and $r_i \equiv r_j \pmod{p}$, then $p \mid (r_i - r_j) < (p-1)/2$, a contradiction unless $i = j$. Hence all elements of $\mathcal{R}$ are positive and incongruent modulo p. Since $r_j \in \mathcal{S}$ are all the elements of $\mathcal{R}$ such that $r_j > p/2$, then we may set $s_j \in \mathcal{R}$ such that $s_j < p/2$ for $j = 1, 2, \ldots, t$, where $s + t = (p-1)/2$. Thus, the values $p - r_j$ for $j = 1, 2, \ldots, s$ and s_j for $j = 1, 2, \ldots, t$ are all natural numbers less than $p/2$.

If $p - r_j \equiv s_i \pmod{p}$ for some $j = 1, 2, \ldots, s$ and some $i = 1, 2, \ldots, t$, then $p - cj \equiv ci \pmod{p}$, so $-cj \equiv ci \pmod{p}$. Since $\gcd(c, p) = 1$, then the latter congruence implies that $-j \equiv i \pmod{p}$. In other words, $p \mid (i + j)$. However, $1 < i + j \leq (s + t) = p - 1$, which is a contradiction. We have shown that the natural numbers $p - r_j$ for $j = 1, 2, \ldots, s$ and s_j for $j = 1, 2, \ldots, t$ together are just the natural numbers $1, 2, 3, \ldots, (p-1)/2$ in some order. Therefore,

$$\prod_{j=1}^{s}(p - r_j) \prod_{j=1}^{t} s_j \equiv \left(\frac{p-1}{2}\right)! \pmod{p}, \qquad (4.5)$$

so,

$$(-1)^s \prod_{j=1}^{s} r_j \prod_{j=1}^{t} s_j \equiv \left(\frac{p-1}{2}\right)! \pmod{p}, \qquad (4.6)$$

and since the r_j for $j = 1, 2, \ldots, s$ and s_j for $j = 1, 2, \ldots t$ are all the elements of $\mathcal{R}$, then

$$\prod_{j=1}^{s} r_j \prod_{j=1}^{t} s_j \equiv \prod_{j=1}^{(p-1)/2} jc \equiv \prod_{j=1}^{(p-1)/2} c \prod_{j=1}^{(p-1)/2} j \equiv c^{(p-1)/2}\left(\frac{p-1}{2}\right)! \pmod{p}. \qquad (4.7)$$

Now employing (4.5)–(4.7), we get,

$$(-1)^s c^{(p-1)/2}\left(\frac{p-1}{2}\right)! \equiv \left(\frac{p-1}{2}\right)! \pmod{p}.$$

Since $\gcd(p, \left(\frac{p-1}{2}\right)!) = 1$, we may cancel $\left(\frac{p-1}{2}\right)!$ from both sides to get,

$$(-1)^s c^{(p-1)/2} \equiv 1 \pmod{p},$$

so multiplying through by $(-1)^s$, we get

$$c^{(p-1)/2} \equiv (-1)^s \pmod{p}.$$

Now we invoke Euler's criterion given in Theorem 4.2 on page 179 to get,

$$\left(\frac{c}{p}\right) \equiv (-1)^s \pmod{p},$$

which yields the result since $p > 2$.   □

Example 4.6 *Let $c = 3$ and $p = 11$. To compute $\left(\frac{3}{11}\right)$ via Gauss' Lemma we need to compute the least residues of $3j$ modulo 11 for $j = 1, 2, 3, 4, 5$. They are 3, 6, 9, 1, and 4, of which 6 and 9 are greater than $11/2$. Hence, by Gauss' Lemma, $\left(\frac{3}{11}\right) = (-1)^2 = 1$.*

A more substantive application of Gauss' Lemma is the following answer to the query about the quadratic residuacity of 2 modulo p.

Theorem 4.6 | **The Quadratic Residuacity of 2 Modulo p**

For any odd prime p,

$$\left(\frac{2}{p}\right) \equiv (-1)^{(p^2-1)/8} \pmod{p}.$$

Proof. By Gauss' Lemma, we need to evaluate the number of values $2j$ for $j = 1, 2, \ldots, (p-1)/2$ that are greater than $p/2$. If $j \leq p/4$, then $2j < p/2$, so there are $\lfloor p/4 \rfloor$ of the $2j$ which are less than $p/2$. Thus, there are $s = (p-1)/2 - \lfloor p/4 \rfloor$ of them greater than $p/2$. By Gauss' Lemma this means that

$$\left(\frac{2}{p}\right) \equiv (-1)^{(p-1)/2 - \lfloor p/4 \rfloor} \pmod{p}.$$

Hence, it suffices to show that

$$s = \frac{p-1}{2} - \lfloor p/4 \rfloor \equiv \frac{p^2 - 1}{8} \pmod{2}.$$

We break this into cases based upon the congruence class of p modulo 8.

If $p = 8m + 1$, then $s = 4m - \lfloor 2m + 1/4 \rfloor = 4m - 2m = 2m$.

If $p = 8m - 1$, then $s = 4m - 1 - \lfloor 2m - 1/4 \rfloor = 4m - 1 - (2m - 1) = 2m$.

If $p = 8m + 3$, then $s = 4m + 1 - \lfloor 2m + 3/4 \rfloor = 4m + 1 - 2m = 2m + 1$.

If $p = 8m - 3$, then $s = 4m - 2 - \lfloor 2m - 1/2 \rfloor = 4m - 2 - (2m - 1) = 2m - 1$.

Hence, if $p \equiv \pm 1 \pmod{8}$, then s is even and if $p \equiv \pm 3 \pmod{8}$, then s is odd. It remains to show the same holds for $(p^2 - 1)/8$. If $p = 8m \pm 1$, then

$$s \equiv \frac{p^2 - 1}{8} = \frac{64m^2 \pm 16m + 1 - 1}{8} = 8m^2 \pm 2m \pmod{2},$$

and if $p = 8m \pm 3$, then

$$s \equiv \frac{p^2 - 1}{8} = \frac{64m^2 \pm 48m + 9 - 1}{8} = 8m^2 \pm 6m + 1 \pmod{2},$$

and the result follows.   □

Immediate from the above is the following formulation.

Corollary 4.2 *If p is an odd prime, then*

$$\left(\frac{2}{p}\right) = \begin{cases} 1 & \text{if } p \equiv \pm 1 \pmod 8, \\ -1 & \text{if } p \equiv \pm 3 \pmod 8. \end{cases}$$

Example 4.7 *By Theorem 4.6,*

$$\left(\frac{2}{11}\right) \equiv (-1)^{(11^2-1)/8} \equiv -1 \pmod{11}, \qquad \left(\frac{2}{13}\right) \equiv (-1)^{(13^2-1)/8} \equiv -1 \pmod{13},$$

$$\left(\frac{2}{7}\right) \equiv (-1)^{(7^2-1)/8} \equiv 1 \pmod 7, \text{ and } \left(\frac{2}{17}\right) \equiv (-1)^{(17^2-1)/8} \equiv -1 \pmod{17}.$$

To test understanding of the above, the reader may now go to Exercise 4.3 on page 187 to verify the quadratic residuacity of -2 modulo any odd prime p.

Example 4.8 *In this application, we employ Corollary 4.2 to prove that there are infinitely many primes of the form $8m + 7$. Assume, to the contrary, that there are only finitely many, say, $p_1, p_2, \ldots, p_s$, let $n = (4\prod_{j=1}^{s} p_j)^2 - 2$. Since $n/2 > 1$ and $n/2$ is odd, there exists some odd prime q dividing n, so*

$$\left(4\prod_{j=1}^{s} p_j\right)^2 \equiv 2 \pmod q,$$

from which we get that $(2/q) = 1$. Therefore, by Corollary 4.2, $q \equiv \pm 1 \pmod 8$. Since q cannot be one of the p_j for $j = 1, 2, \ldots, s$, then all odd primes dividing n are of the form $8t + 1$. Hence,

$$n = 2(8k + 1)$$

for some integer k, since products of primes of the form $8t + 1$ are also of that form. However, by Exercise 2.2 on page 83, there exists an $\ell \in \mathbb{N}$ such that

$$n = 16(8\ell + 1) - 2 = 2(64\ell + 7).$$

This implies that $8k + 1 = 64\ell + 7$, or by rewriting, $4(k - 8\ell) = 3$, which implies that $2 \mid 3$, a contradiction that proves the desired infinitude of primes of the form $8m + 7$.

We conclude this section with a result by Eisenstein — see Biography 4.3 on page 188 — that will be a technical result needed to prove the *Quadratic Reciprocity Law* in §4.2.

Lemma 4.1 $\boxed{\textbf{Eisenstein's Lemma}}$

Let $c \in \mathbb{Z}$ be odd, and $p > 2$ be a prime such that $p \nmid c$. Then $\left(\frac{c}{p}\right) = (-1)^M$, where $M = \displaystyle\sum_{j=1}^{(p-1)/2} \lfloor jc/p \rfloor$.

Proof. We use the same notation as in the proof of Gauss' Lemma. Since for each $j = 1, 2, \ldots, (p-1)/2$, there exist integers q_j, t_j such that

$$cj = q_j p + t_j \text{ with } 1 \le t_j < p,$$

then $q_j = \lfloor cj/p \rfloor$ since $cj/p = q_j + t_j/p < q_j + 1$. Hence, for each such j,

$$cj = \lfloor cj/p \rfloor p + t_j \text{ with } 1 \le t_j < p,$$

where $t_j = r_j$ if $t_j > p/2$, and $t_j = s_j$ if $t_j < p/2$.

Thus, we have

$$\sum_{j=1}^{(p-1)/2} jc = \sum_{j=1}^{(p-1)/2} p\lfloor jc/p \rfloor + \sum_{j=1}^{s} r_j + \sum_{j=1}^{t} s_j. \tag{4.8}$$

However, as shown in the proof of Gauss' Lemma, the values $p - r_j$ for $j = 1, 2, \ldots, s$ and s_j for $j = 1, 2, \ldots, t$ are just a rearrangement of the numbers $1, 2, \ldots, (p-1)/2$. Therefore,

$$\sum_{j=1}^{(p-1)/2} j = \sum_{j=1}^{s}(p - r_j) + \sum_{j=1}^{t} s_j = ps - \sum_{j=1}^{s} r_j + \sum_{j=1}^{t} s_j. \tag{4.9}$$

Subtracting (4.9) from (4.8), we get

$$(c - 1) \sum_{j=1}^{(p-1)/2} j = p \left(\sum_{j=1}^{(p-1)/2} \lfloor jc/p \rfloor - s \right) + 2 \sum_{j=1}^{s} r_j. \tag{4.10}$$

Now we reduce (4.10) modulo 2 to get $0 \equiv \left(\displaystyle\sum_{j=1}^{(p-1)/2} \lfloor jc/p \rfloor - s \right) \pmod 2$, since $c \equiv p \equiv 1 \pmod 2$, which means that $s \equiv \displaystyle\sum_{j=1}^{(p-1)/2} \lfloor jc/p \rfloor \pmod 2$. By Gauss' Lemma, we are now done. $\qquad\square$

Example 4.9 *We will find $\left(\frac{3}{7}\right)$ via Lemma 4.1 as a simple illustration:*

$$\sum_{j=1}^{3} \lfloor 3j/7 \rfloor = \lfloor 3/7 \rfloor + \lfloor 6/7 \rfloor + \lfloor 9/7 \rfloor = 0 + 0 + 1 = 1,$$

so $\left(\frac{3}{7}\right) = (-1)^1 = -1$.

Exercises

4.1. Find the value of the following Legendre Symbols.

(a) $\left(\frac{29}{31}\right)$ (b) $\left(\frac{51}{23}\right)$.

4.2. Find the value of the following Legendre Symbols.

(a) $\left(\frac{-15}{37}\right)$ (b) $\left(\frac{37}{41}\right)$.

4.3. Prove that if $p > 2$ is prime, then

$$\left(\frac{-2}{p}\right) = \begin{cases} 1 & \text{if } p \equiv 1, 3 \pmod 8, \\ -1 & \text{if } p \equiv 5, 7 \pmod 8. \end{cases}$$

4.4. Use the Legendre Symbol for the prime $p = 3$ to prove that the Diophantine equation $x^2 - 3y^2 = 17$ has no solutions.

4.5. Verify the Legendre Symbol identity, $\sum_{j=1}^{p-1} \left(\frac{j}{p}\right) = 0$, where $p > 2$ is prime.

4.6. Establish the Legendre Symbol identity, for any odd prime p,

$$\sum_{j=0}^{p-1} \left(\frac{(j-a)(j-b)}{p}\right) = \begin{cases} p-1 & \text{if } a \equiv b \pmod p, \\ -1 & \text{if } a \not\equiv b \pmod p. \end{cases}$$

4.7. Let $a \in \mathbb{Z}$, and $p \equiv q \equiv 3 \pmod 4$ distinct primes such that a is a quadratic residue modulo both p and q. Prove that the solutions of $x^2 \equiv a \pmod p$ and $x^2 \equiv a \pmod q$ are

$$x \equiv \pm a^{(p+1)/4} \pmod p \text{ and } x \equiv \pm a^{(q+1)/4} \pmod q,$$

respectively.

4.8. Let $f(x) = ax^2 + bx + c$ where $a, b, c \in \mathbb{Z}$, and set $\Delta = b^2 - 4ac$. Suppose that $p > 2$ is a prime such that $\left(\frac{\Delta}{p}\right) = 1$. Prove that

$$\sum_{x=0}^{p-1} \left(\frac{f(x)}{p}\right) = -\left(\frac{a}{p}\right).$$

4.9. Let $f(x) = ax^2 + bx + c$ where $a, b, c \in \mathbb{Z}$, and set $\Delta = b^2 - 4ac$. Suppose that $p > 2$ is a prime such that $p|\Delta$. Prove that

$$\sum_{x=0}^{p-1} \left(\frac{f(x)}{p}\right) = (p-1)\left(\frac{a}{p}\right).$$

Sums of the form $\sum \left(\frac{f(x)}{p}\right)$ *are called Jacobsthal sums.*

☆ 4.10. Assuming that p and $2p+1$ are both odd primes, prove that $(-1)^{(p-1)/2}2$ is a primitive root modulo $2p+1$.

(*Hint: Use Theorem 4.2 on page 179, the properties in Theorem 4.4 on page 182, and Theorem 4.6 on page 184.*)

4.11. Use the technique developed in Example 4.8 on page 185 to prove there are infinitely many primes of the form $8m+5$.

(*Hint: Assume there are only finitely many* $p_1, p_2, \ldots, p_s$, *let* $n = (\prod_{j=1}^{s} p_j)^2 + 4$, *and use part* (3) *of Theorem 4.4 on page 182.*)

4.12. Use the technique developed in Example 4.8 to prove there are infinitely many primes of the form $8m+3$.

(*Hint: Assume there are only finitely many* $p_1, p_2, \ldots, p_s$, *let* $n = (\prod_{j=1}^{s} p_j)^2 + 2$, *and use Exercise 4.3.*)

4.13. Use Exercise 4.3 to prove that the Diophantine equation $x^2 - 2y^2 = p$ has no solutions for any prime $p \equiv \pm 3 \pmod 8$.

4.14. Use Corollary 4.2 on page 185 to prove that the Diophantine equation $x^2 + 2y^2 = p$ has no solutions for any prime $p \equiv 5, 7 \pmod 8$.

4.15. If $p \equiv 1 \pmod 4$ is prime, prove that there are $(p-1)/4$ quadratic residues less than $p/2$.

4.16. If $p \equiv 1 \pmod 4$ is prime, prove that the sum of the quadratic residues of p, that are less than p is equal to $p(p-1)/4$.

4.2 The Quadratic Reciprocity Law

> *It is the tension between the scientist's laws and his own attempted breaches of them that powers the engines of science and makes it forge ahead.*
>
> From page 8 "Anomaly" of **Quiddities (1987)**
> **W.V.O. Quine (1908–2000), American mathematician and philosopher**

If we know the value of the Legendre Symbol (p/q) for distinct odd primes p and q, then do we then know the Legendre Symbol (q/p)? The answer was found by Gauss in 1796 and published in his masterpiece [16]. The actual formulation of the *quadratic reciprocity law* was first given by Legendre in 1785, but his attempts to prove it failed since his published proofs were all flawed.

Theorem 4.7 | **The Quadratic Reciprocity Law**

If $p \neq q$ are odd primes, then

$$\left(\frac{p}{q}\right)\left(\frac{q}{p}\right) = (-1)^{\frac{p-1}{2} \cdot \frac{q-1}{2}}.$$

Proof. First we establish the following result.

Claim 4.3

$$\frac{p-1}{2} \cdot \frac{q-1}{2} = \sum_{k=1}^{(p-1)/2} \lfloor kq/p \rfloor + \sum_{j=1}^{(q-1)/2} \lfloor jp/q \rfloor.$$

Let

$$\mathcal{S} = \{(jp, kq) : 1 \leq j \leq (q-1)/2; 1 \leq k \leq (p-1)/2\}.$$

The cardinality of $\mathcal{S}$ is $\frac{p-1}{2} \cdot \frac{q-1}{2}$. Also, it is an easy check to verify that $jp \neq kq$ for any $1 \leq j \leq (q-1)/2$, or $1 \leq k \leq (p-1)/2$. Furthermore, set

$$\mathcal{S} = \mathcal{S}_1 \cup \mathcal{S}_2,$$

where

$$\mathcal{S}_1 = \{(jp, kq) \in \mathcal{S} : jp < kq\},$$

and

$$\mathcal{S}_2 = \{(jp, kq) \in \mathcal{S} : jp > kq\}.$$

If $(jp, kq) \in \mathcal{S}_1$, then $j < kq/p$. Also, $kq/p \leq (p-1)q/(2p) < q/2$. Therefore, $\lfloor kq/p \rfloor < q/2$, from which it follows that

$$\lfloor kq/p \rfloor \leq (q-1)/2.$$

Hence, the cardinality of S_1 is $\sum_{k=1}^{(p-1)/2}\lfloor kq/p\rfloor$. Similarly, the cardinality of S_2 is $\sum_{j=1}^{(q-1)/2}\lfloor jp/q\rfloor$. This establishes Claim 4.3.

Now set $M=\sum_{k=1}^{(p-1)/2}\lfloor kq/p\rfloor$, and $N=\sum_{j=1}^{(q-1)/2}\lfloor jp/q\rfloor$. If we let $q=c$ in Eisenstein's Lemma 4.1 on page 186, then

$$\left(\frac{q}{p}\right)=(-1)^M.$$

Similarly,

$$\left(\frac{p}{q}\right)=(-1)^N.$$

Hence,

$$\left(\frac{q}{p}\right)\left(\frac{p}{q}\right)=(-1)^{M+N}.$$

The result now follows from Claim 4.3.   □

The following is an equivalent formulation of the recipocity law.

Corollary 4.3 *For $p\neq q$ odd primes,*

$$(q/p)=\begin{cases}-(p/q) & \text{if }p\equiv q\equiv 3\pmod 4\\ (p/q) & \text{otherwise.}\end{cases}$$

Example 4.10 *Let $p=7$ and $q=111$. Then by the quadratic reciprocity law,*

$$\left(\frac{p}{q}\right)\left(\frac{q}{p}\right)=\left(\frac{7}{111}\right)\left(\frac{111}{7}\right)=(-1)^{3\cdot 55}=-1=(-1)^{\frac{p-1}{2}\cdot\frac{q-1}{2}},$$

so

$$\left(\frac{111}{7}\right)=\left(\frac{-1}{7}\right)=-1.$$

Hence, $x^2\equiv 111\pmod 7$ has no solutions $x\in\mathbb{Z}$, namely 111 is not a quadratic residue modulo 7. However,

$$\left(\frac{7}{111}\right)=-\left(\frac{111}{7}\right)=-\left(\frac{-1}{7}\right)=1.$$

Hence, $x^2\equiv 7\pmod{111}$ has solutions $x\in\mathbb{Z}$, that is 7 is a quadratic residue modulo 111.

A more substantial illustration of the applicability of the quadratic reciprocity law is the following.

Example 4.11 *Suppose that we wish to determine the quadratic residuacity of* 3 *modulo any prime* $p > 3$. *By the quadratic reciprocity law,*

$$(3/p) = \begin{cases} (p/3) & \text{if } p \equiv 1 \pmod{4} \\ -(p/3) & \text{if } p \equiv 3 \pmod{4}. \end{cases}$$

Then given that

$$(p/3) = \begin{cases} 1 & \text{if } p \equiv 1 \pmod{3} \\ -1 & \text{if } p \equiv 2 \pmod{3}, \end{cases}$$

it follows that $(3/p) = 1$ *if and only if either*

$$p \equiv 1 \pmod{4} \text{ and } p \equiv 1 \pmod{3}, \tag{4.11}$$

or

$$p \equiv 3 \pmod{4} \text{ and } p \equiv 2 \pmod{3}. \tag{4.12}$$

Congruences (4.11)–(4.12) *translate into the following resolution of the quadratic residuacity of* 3 *modulo* p.

$$(3/p) = \begin{cases} 1 & \text{if } p \equiv \pm 1 \pmod{12} \\ -1 & \text{if } p \equiv \pm 5 \pmod{12}. \end{cases} \tag{4.13}$$

Another substantial application of the quadratic reciprocity law is the following primality test that adds to the methods we studied in §1.8 and §2.7.

Theorem 4.8 | **Pepin's Test**

The Fermat number $\mathcal{F}_n = 2^{2^n} + 1$, $n \in \mathbb{N}$, *is prime if and only if*

$$3^{(\mathcal{F}_n - 1)/2} \equiv -1 \pmod{\mathcal{F}_n}. \tag{4.14}$$

Proof. If $\mathcal{F}_n$ is prime, then by the quadratic reciprocity law,

$$\left(\frac{3}{\mathcal{F}_n}\right) = \left(\frac{\mathcal{F}_n}{3}\right) = \left(\frac{2}{3}\right) = -1, \tag{4.15}$$

where the first equality holds since $\mathcal{F}_n \equiv 1 \pmod{4}$ and $\mathcal{F}_n \equiv 2 \pmod{3}$, whereas the last equality comes from Corollary 4.2 on page 185.

By Euler's criterion given in Theorem 4.2 on page 179,

$$\left(\frac{3}{\mathcal{F}_n}\right) \equiv 3^{(\mathcal{F}_n - 1)/2} \pmod{\mathcal{F}_n}. \tag{4.16}$$

Putting congruences (4.15)–(4.16) together, we get,

$$3^{(\mathcal{F}_n - 1)/2} \equiv -1 \pmod{\mathcal{F}_n}, \tag{4.17}$$

which proves the necessity of (4.14).

Now assume that (4.14) holds. Raising both sides to the power 2, we get

$$3^{\mathfrak{F}_n-1} \equiv 1 \pmod{\mathfrak{F}_n}.$$

Hence, if $p \mid \mathfrak{F}_n$ is prime, then

$$3^{\mathfrak{F}_n-1} \equiv 1 \pmod{p}.$$

Therefore, $\mathrm{ord}_p(3) \mid (\mathfrak{F}_n - 1) = 2^{2^n}$. However, by (4.17), $\mathrm{ord}_p(3)$ does not divide 2^{2^n-1}. Hence, $\mathrm{ord}_p(3) = 2^{2^n} = \mathfrak{F}_n - 1$. However, by Fermat's Little Theorem, $\mathrm{ord}_p(3) = \mathfrak{F}_n - 1 \le p - 1$, so since $p \mid \mathfrak{F}_n$, then $p = \mathfrak{F}_n$. This proves the sufficiency of (4.14).   □

Note that Theorem 4.8 tells us that 3 is a primitive root of any Fermat prime. The following illustrates Pepin's test.

Example 4.12 *For* $\mathfrak{F}_3 = 257$, $3^{128} \equiv -1 \,(\mathrm{mod}\ 17)$, *so* $\mathfrak{F}_3$ *is prime. Also, for* $\mathfrak{F}_4 = 2^{2^4} + 1 = 65537$, $3^{2^{3}} \equiv -1 \,(\mathrm{mod}\ \mathfrak{F}_4)$, *so* $\mathfrak{F}_4$ *is prime. But for* $\mathfrak{F}_5 = 2^{32} + 1 = 4294967297$, $3^{2^{31}} \equiv 10324303 \not\equiv -1 \,(\mathrm{mod}\ \mathfrak{F}_5)$ *so* $\mathfrak{F}_5$ *is not prime, and indeed it is conjectured that* $\mathfrak{F}_n$, *for* $n \ge 5$, *are composite.*

We now present a generalization of the Legendre symbol.

Definition 4.3 | **The Jacobi Symbol**

Let $n > 1$ *be an odd natural number with* $n = \prod_{j=1}^{k} p_j^{e_j}$ *where* $e_j \in \mathbb{N}$ *and the* p_j *are distinct primes. Then the* Jacobi symbol *of* a *with respect to* n *is given by*

$$\left(\frac{a}{n}\right) = \prod_{j=1}^{k} \left(\frac{a}{p_j}\right)^{e_j},$$

for any $a \in \mathbb{Z}$, *where the symbols on the right are Legendre symbols.*

Biography 4.4 Carl Gustav Jacob Jacobi (1804–1851) *was born in Potsdam in Prussia on December 10, 1804, to a wealthy German banking family. In August of 1825, Jacobi obtained his doctorate from the University of Berlin in an area involving partial fractions. The next year he became a lecturer at the University of Königsberg and was appointed as a professor there in 1831. Jacobi's first major work was his application of* (his first love) *elliptic functions to number theory. Moreover, Jacobi and his good friend Dirichlet both generated their own brands of analytic number theory. As well, Jacobi was interested in the history of mathematics and was a prime mover in the publication of the collected works of Euler — a task, incredibly, not completed to this day. Outside of number theory, he made contributions to analysis, geometry, and mechanics. Although many of his colleagues felt that he might work himself to death, he died of smallpox on February 18, 1851.*

Example 4.13 *From the definition of Jacobi's symbol, we get,*

$$\left(\frac{2}{33}\right) = \left(\frac{2}{3}\right)\left(\frac{2}{11}\right) = (-1)(-1) = 1.$$

However, note that this does not mean that 2 is a quadratic residue modulo 33. In other words, if $(a/n) = 1$, when n is not prime, then we cannot conclude that $x^2 \equiv a \pmod{n}$ is solvable. The converse is true, namely if $x^2 \equiv a \pmod{n}$ has solutions, then $(a/n) = 1$, because then for any prime divisor p of n, we also have that $x^2 \equiv a \pmod{p}$. In the example here, 2 is a quadratic nonresidue of both 3 and 11, a fact that we know from Corollary 4.2 on page 185. If $(a/p) = 1$ for all primes dividing n, then $x^2 \equiv a \pmod{n}$ will have a solution. What we are saying here deserves to be summarized since we have already proved this in various stages.

Theorem 4.9 | **Jacobi and Quadratic Congruences**

Let $n > 1$ be an integer, $b \in \mathbb{Z}$ with $\gcd(b,n) = 1$, and

$$n = 2^{a_0} \prod_{j=1}^{m} p_j^{a_j}$$

the canonical prime factorization of n where $a_0 \geq 0$, and $a_j \in \mathbb{N}$ for the distinct odd primes p_j, $j = 1, 2, \ldots, m$. Then

$$x^2 \equiv b \pmod{n}$$

is solvable if and only if

$(a/p_j) = 1$ *for all* $j = 1, 2, \ldots, m$ *and* $a \equiv 1 \pmod{\gcd(8, 2^{a_0})}$.

Proof. We merely employ the Chinese Remainder Theorem in conjunction with what we proved in Exercises 3.23, 3.26 on page 159, and in Theorem 3.10 on page 155. We leave it to the reader to review these facts and patch them together for the desired result. $\square$

Example 4.14 *Suppose that we wish to find all solutions of the congruence $x^2 \equiv 35 \pmod{1829}$. Since $1829 = 31 \cdot 59$, then we must first look at the Legendre symbols $\left(\frac{35}{31}\right)$ and $\left(\frac{35}{59}\right)$. Indeed, we have that $\left(\frac{35}{29}\right) = \left(\frac{35}{59}\right) = 1$, so our congruence has solutions. In this case, we may employ Exercise 4.7 on page 187, and the Chinese Remainder Theorem to piece them together in this case. By that exercise,*

$$x \equiv \pm 35^{(31+1)/4} \equiv 35^8 \equiv \pm 2 \pmod{31}$$

and

$$x \equiv \pm 35^{(59+1)/4} \equiv 35^{25} \equiv \pm 25 \pmod{59}$$

are the solutions modulo 31 and 59, respectively. Now we use the Chinese Remainder Theorem and find that, for instance, $x = 2 + 31 \cdot 35 = 25 + 59 \cdot 18 = 1087$ is one of the solutions modulo $n = 1829$. The reader may verify in a similar fashion that the other three incongruent solutions are $x \in \{556, 742, 1273\}$.

Now we explore more fundamental properties of the Jacobi Symbol.

Theorem 4.10 | **Properties of the Jacobi Symbol**

Let $n \in \mathbb{N}$ be odd, and $a, b \in \mathbb{Z}$. Then

(1) $\left(\dfrac{ab}{n}\right) = \left(\dfrac{a}{n}\right)\left(\dfrac{b}{n}\right)$.

(2) $\left(\dfrac{a}{n}\right) = \left(\dfrac{b}{n}\right)$ *if* $a \equiv b \,(\mathrm{mod}\ n)$.

(3) $\left(\dfrac{-1}{n}\right) = (-1)^{(n-1)/2}$.

(4) $\left(\dfrac{2}{n}\right) = (-1)^{(n^2-1)/8}$.

Proof. Let $n = \prod_{j=1}^{k} p_j^{a_j}$ be the canonical prime factorization of n.

Proof of (1): We know from Theorem 4.4 on page 182 that $\left(\dfrac{ab}{p_j}\right) = \left(\dfrac{a}{p_j}\right)\left(\dfrac{b}{p_j}\right)$ for all $j = 1, 2, \ldots, k$. Hence,

$$\left(\frac{ab}{n}\right) = \prod_{j=1}^{k}\left(\frac{ab}{p_j}\right)^{a_j} = \prod_{j=1}^{k}\left(\frac{a}{p_j}\right)^{a_j}\left(\frac{b}{p_j}\right)^{a_j} = \left(\frac{a}{n}\right)\left(\frac{b}{n}\right).$$

Proof of (2): From Theorem 4.4 we get that $\left(\dfrac{a}{p_j}\right) = \left(\dfrac{b}{p_j}\right)$ for all $j = 1, 2, \ldots, k$. Hence,

$$\left(\frac{a}{n}\right) = \prod_{j=1}^{k}\left(\frac{a}{p_j}\right)^{a_j} = \prod_{j=1}^{k}\left(\frac{b}{p_j}\right)^{a_j} = \left(\frac{b}{n}\right).$$

Proof of (3): Theorem 4.4 says that $\left(\dfrac{-1}{p_j}\right) = (-1)^{(p_j-1)/2}$ for all $j = 1, 2, \ldots, k$. Therefore,

$$\left(\frac{-1}{n}\right) = \prod_{j=1}^{k}\left(\frac{-1}{p_j}\right)^{a_j} = \prod_{j=1}^{k}\left((-1)^{(p_j-1)/2}\right)^{a_j} = (-1)^{\sum_{j=1}^{k} a_j(p_j-1)/2}. \quad (4.18)$$

To complete this part we need to establish,

Claim 4.4 $n \equiv 1 + \sum_{j=1}^{k} a_j(p_j - 1)\,(\mathrm{mod}\ 4)$.

We use induction on k. If $k = 1$, then

$$n = p_1^{a_1} = [1 + (p_1 - 1)]^{a_1},$$

and by the Binomial Theorem this equals,

$$1 + a_1(p_1 - 1) + \sum_{j=2}^{k} \binom{a_j}{j} (p_j - 1)^j \equiv 1 + a_1(p_1 - 1) \pmod{4},$$

since $(p_j - 1)^j \equiv 0 \pmod 4$ for any $j > 1$. This is the induction step. Assume that the claim holds for all values less than or equal to $k = t \geq 2$. Then

$$n = \prod_{j=1}^{t+1} p_j^{a_j} = \left(\prod_{j=1}^{t} p_j^{a_j} \right) p_{t+1}^{a_{t+1}} \equiv$$

$$\left[1 + \sum_{j=1}^{t} a_j(p_j - 1) \right] \cdot [1 + a_{t+1}(p_{t+1} - 1)] \pmod{4},$$

by the induction hypothesis for $k = t$ and $k = 1$. Thus, we have that the above is congruent to

$$1 + a_{t+1}(p_{t+1} - 1) + \sum_{j=1}^{t} a_j(p_j - 1) \equiv 1 + \sum_{j=1}^{t+1} a_j(p_j - 1) \pmod{4},$$

since for any j, $(p_j - 1)(p_{t+1} - 1) \equiv 0 \pmod 4$, and this secures Claim 4.4.

Using Claim 4.4 and subtracting 1, then dividing through by 2 we get

$$(n - 1)/2 \equiv \sum_{j=1}^{k} a_j(p_j - 1)/2 \pmod{2},$$

which is the same exponent as -1 in (4.18).

Proof of (4): Theorem 4.6 on page 184 tells us that $\left(\frac{2}{p_j} \right) = (-1)^{(p_j^2 - 1)/8}$ for all $j = 1, 2, \ldots, k$. Hence,

$$\left(\frac{2}{n} \right) = \prod_{j=1}^{k} \left(\frac{2}{p_j} \right)^{a_j} = \prod_{j=1}^{k} \left((-1)^{(p_j^2 - 1)/8} \right)^{a_j} = (-1)^{\sum_{j=1}^{k} a_j(p_j^2 - 1)/8}. \qquad (4.19)$$

Using the Binomial Theorem as in the proof of part (3), we get that

$$(1 + p_j^2 - 1)^{a_j} \equiv 1 + a_j(p_j^2 - 1) \pmod{64},$$

since $p_j^2 \equiv 1 \pmod 8$. Therefore, by an induction argument as in the proof of part (3),

$$n^2 \equiv 1 + \sum_{j=1}^{k} a_j(p_j^2 - 1) \pmod{64}.$$

By subtracting 1 and dividing through by 8, we get

$$(n^2 - 1)/8 \equiv \sum_{j=1}^{k} a_j(p_j^2 - 1)/8 \pmod{8},$$

which the exponent of -1 in (4.19). □

We are now in a position to verify that quadratic reciprocity extends to the Jacobi symbol.

Theorem 4.11 | **The Reciprocity Law for the Jacobi Symbol**

If $m, n \in \mathbb{N}$ are odd and relatively prime, then

$$\left(\frac{m}{n}\right)\left(\frac{n}{m}\right) = (-1)^{\frac{m-1}{2} \cdot \frac{n-1}{2}}.$$

Proof. Let $m = \prod_{i=1}^{k} p_i^{a_i}$, and $n = \prod_{j=1}^{\ell} q_j^{b_j}$, where $p_i \neq q_j$ for any j, k. Thus,

$$\left(\frac{m}{n}\right)\left(\frac{n}{m}\right) = \prod_{j=1}^{\ell}\left(\frac{m}{p_j}\right)^{b_j} \prod_{i=1}^{k}\left(\frac{n}{q_i}\right)^{a_i} =$$

$$\prod_{j=1}^{\ell}\prod_{i=1}^{k}\left(\frac{q_i}{p_j}\right)^{b_j a_i} \prod_{i=1}^{k}\prod_{j=1}^{\ell}\left(\frac{p_j}{q_i}\right)^{a_i b_j} = \prod_{i=1}^{k}\prod_{j=1}^{\ell}\left[\left(\frac{p_i}{q_j}\right)\left(\frac{q_j}{p_i}\right)\right]^{a_i b_j},$$

and by Theorem 4.7 on page 189, this equals

$$\prod_{i=1}^{k}\prod_{j=1}^{\ell}(-1)^{a_i\left(\frac{p_i-1}{2}\right)b_j\left(\frac{q_j-1}{2}\right)} = (-1)^U,$$

where

$$U = \sum_{i=1}^{k}\sum_{j=1}^{\ell} a_i\left(\frac{p_i-1}{2}\right) b_j\left(\frac{q_j-1}{2}\right) = \sum_{i=1}^{k} a_i\left(\frac{p_j-1}{2}\right)\sum_{j=1}^{\ell} b_j\left(\frac{q_k-1}{2}\right).$$

However, as shown in Claim 4.3 on page 189,

$$\sum_{i=1}^{k} a_i\left(\frac{p_i-1}{2}\right) \equiv \frac{m-1}{2} \pmod 2,$$

and

$$\sum_{j=1}^{\ell} b_j\left(\frac{q_j-1}{2}\right) \equiv \frac{n-1}{2} \pmod 2,$$

so the result follows. □

Example 4.15 *If $m = 130449 = 3 \cdot 11 \cdot 59 \cdot 67$ and $n = 7735 = 5 \cdot 7 \cdot 13 \cdot 17$, then $\gcd(m, n) = 1$, and*

$$\left(\frac{130449}{7735}\right)\left(\frac{7735}{130449}\right) = (-1)^{\frac{130449-1}{2} \cdot \frac{7735-1}{2}} = (-1)^{65224 \cdot 3867} = 1.$$

$$\left(\frac{130449}{7735}\right) = \left(\frac{130449}{5}\right)\left(\frac{130449}{7}\right)\left(\frac{130449}{13}\right)\left(\frac{130449}{17}\right) =$$

$$\left(\frac{4}{5}\right)\left(\frac{4}{7}\right)\left(\frac{7}{13}\right)\left(\frac{8}{17}\right) = 1\cdot1\cdot(-1)\cdot1 = -1,$$

and

$$\left(\frac{7735}{130449}\right) = \left(\frac{7735}{3}\right)\left(\frac{7735}{11}\right)\left(\frac{7735}{59}\right)\left(\frac{7735}{67}\right) =$$

$$\left(\frac{1}{3}\right)\left(\frac{2}{11}\right)\left(\frac{6}{59}\right)\left(\frac{30}{67}\right) = 1\cdot(-1)\cdot(-1)\cdot(-1) = -1.$$

Example 4.15 shows that there are numerous ways of calculating the Jacobi symbol and the context determines which way to proceed. However, there is an efficient means of making this calculation that is straightforward to implement on a computer. Although it will appear more complicated than what we did in Example 4.15, we present it here since it is computationally efficient.

First, we define a sequence of integers related to the values in the Jacobi symbol that we wish to calculate. Let $1 < n < m$ be relatively prime odd integers, and set $S_0 = m$, $S_1 = n$. Then by repeated application of the division algorithm, for $j = 0, 1, 2, \ldots, \ell - 2$, we get

$$S_j = S_{j+1}q_{j+1} + 2^{\alpha_{j+1}}S_{j+2}, \tag{4.20}$$

where $S_\ell = 1$.

Now we establish the result that will give us the algorithm. The above notation is in force.

Theorem 4.12 | **An Algorithm for Computation of Jacobi Symbols**

Let $1 < n < m$ be relatively prime odd integers. Then

$$\left(\frac{m}{n}\right) = (-1)^{T+U},$$

where

$$T = \frac{1}{8}\sum_{j=1}^{\ell-1}\alpha_j(S_j^2 - 1),$$

and

$$U = \frac{1}{2}\sum_{j=1}^{\ell-2}(S_j - 1)(S_{j+1} - 1).$$

Proof. By Theorem 4.10 on page 194,

$$\left(\frac{m}{n}\right) = \left(\frac{S_0}{S_1}\right) = \left(\frac{2^{\alpha_1}S_2}{S_1}\right) = \left(\frac{2}{S_1}\right)^{\alpha_1}\left(\frac{S_2}{S_1}\right) = (-1)^{\alpha_1(S_1^2-1)/8}\left(\frac{S_2}{S_1}\right).$$

Also, by Theorem 4.11 on page 196,

$$\left(\frac{S_2}{S_1}\right) = (-1)^{\frac{S_1-1}{2}\cdot\frac{S_2-1}{2}}\left(\frac{S_1}{S_2}\right).$$

Therefore,

$$\left(\frac{m}{n}\right) = (-1)^{\frac{S_1-1}{2}\cdot\frac{S_2-1}{2}+\alpha_1(S_1^2-1)/8}\left(\frac{S_1}{S_2}\right).$$

By induction, this process continues and since for $j = 2, 3, \ldots, \ell - 1$,

$$\left(\frac{S_{j-1}}{S_j}\right) = (-1)^{\frac{S_j-1}{2}\cdot\frac{S_{j+1}-1}{2}+\alpha_j(S_j^2-1)/8}\left(\frac{S_j}{S_{j+1}}\right),$$

we get the result.   □

Example 4.16 *Let us take the values in Example 4.15 on page 196 and apply Theorem 4.12 to them. Thus, $S_0 = m = 130449$, $S_1 = n = 7735$, and*

$$S_0 = 130449 = S_1 q_1 + 2^{\alpha_1} S_2 = 7735 \cdot 16 + 2^0 \cdot 6689,$$

$$S_1 = 7735 = S_2 q_2 + 2^{\alpha_2} S_3 = 6689 \cdot 1 + 2^1 \cdot 523,$$

$$S_2 = 6689 = S_3 q_3 + 2^{\alpha_3} S_4 = 523 \cdot 12 + 2^0 \cdot 413,$$

$$S_3 = 523 = S_4 q_4 + 2^{\alpha_4} S_5 = 413 \cdot 1 + 2^1 \cdot 55,$$

$$S_4 = 413 = S_5 q_5 + 2^{\alpha_5} S_6 = 55 \cdot 7 + 2^2 \cdot 7,$$

$$S_5 = 55 = S_6 q_6 + 2^{\alpha_6} S_7 = 7 \cdot 7 + 2^1 \cdot 3,$$

$$S_6 = 7 = S_7 q_7 + 2^{\alpha_7} S_8 = 3 \cdot 2 + 2^0 \cdot 1,$$

so $\ell = 8$. Therefore,

$$T = \frac{1}{8}\sum_{j=1}^{\ell-1}\alpha_j(S_j^2 - 1) = \frac{1}{8}[0 \cdot (7735^2 - 1) + 1 \cdot (6689^2 - 1) + 0 \cdot (523^2 - 1) +$$

$$1 \cdot (413^2 - 1) + 2 \cdot (55^2 - 1) + 1 \cdot (7^2 - 1) + 0 \cdot (3^2 - 1)] \equiv 1 \pmod{2},$$

and

$$U = \frac{1}{2}\sum_{j=1}^{\ell-2}(S_j - 1)(S_{j+1} - 1) = \frac{1}{2}[(7735 - 1)(6689 - 1) + (6689 - 1)(523 - 1) +$$

$$(523 - 1)(413 - 1) + (413 - 1)(55 - 1) + (55 - 1)(7 - 1)] \equiv 0 \pmod{2}.$$

Hence,

$$\left(\frac{m}{n}\right) = \left(\frac{130449}{7735}\right) = (-1)^{T+U} = (-1)^{1+0} = -1.$$

◆ **Complexity Issues**

If $a > b > 1$ are relatively prime integers, then it can be shown that the Jacobi symbol (a/b) can be evaluated in $O(\log_2^2(b))$ bit operations. Recall, from Remark 1.1 on page 22, that the computational complexity of evaluating $\gcd(a,b)$ when $a > b$ is $O(\log_2^3(a))$. Hence, the number of divisions required to reach the last equation, namely when $j = \ell - 2$, in the sequence (4.20) on page 197 is less than the number of divisions required to find the gcd using the Euclidean algorithm. Thus, the Euclidean-like algorithm described above for evaluating Jacobi symbols provides a direct method that does not require selecting alternative methods depending upon the context of the problem at hand, and it is efficient.

Exercises

4.17. Evaluate each of the following Jacobi symbols.

    (a) $\left(\frac{7}{29}\right)$ (b) $\left(\frac{1029}{1111}\right)$.

4.18. Evaluate each of the following Jacobi symbols.

    (a) $\left(\frac{11}{59}\right)$ (b) $\left(\frac{2053}{2221}\right)$.

4.19. Prove that 7 is a primitive root of any prime of the form $p = 16^n + 1$, $n \in \mathbb{N}$.

    (*Hint: Use the quadratic reciprocity law, and Euler's criterion Theorem 4.2 on page 179.*)

4.20. Prove that if $n \in \mathbb{N}$ is odd and squarefree, then there exists an $a \in \mathbb{Z}$ such that $\left(\frac{a}{n}\right) = -1$.

    (*Hint: Use the Chinese Remainder Theorem.*)

In Exercises 4.21–4.26, we will be referring to the following generalization of the Jacobi symbol. Let $a, n \in \mathbb{Z}$. Define $\left(\frac{a}{n}\right) = 0$ if $\gcd(a,n) > 1$, and let $\left(\frac{a}{n}\right)$ be the Jacobi symbol if n is odd and $\gcd(a,n) = 1$. If $a \equiv 0, 1 \pmod{4}$ and $n = 2^b n_1$, where n_1 is odd and $b \geq 0$, then $\left(\frac{a}{n}\right) = \left(\frac{a}{2}\right)^b \left(\frac{a}{n_1}\right)$ where $\left(\frac{a}{n_1}\right)$ is the Jacobi symbol and

$$\left(\frac{a}{2}\right) = \begin{cases} 0 & \text{if } 2 \mid a \\ 1 & \text{if } a \equiv 1 \pmod 8 \\ -1 & \text{if } a \equiv 5 \pmod 8. \end{cases}$$

This symbol is called the *Kronecker symbol* — see Biography 4.5 on the following page.

4.21. Evaluate each of the following Kronecker symbols.

    (a) $\left(\frac{5}{18}\right)$ (b) $\left(\frac{21}{22}\right)$
    (c) $\left(\frac{113}{224}\right)$ (d) $\left(\frac{225}{124}\right)$.

4.22. Evaluate each of the following Kronecker symbols.

(a) $\left(\frac{5}{12}\right)$ (b) $\left(\frac{17}{32}\right)$ (c) $\left(\frac{97}{200}\right)$ (d) $\left(\frac{101}{324}\right)$.

4.23. Prove for $m, n \in \mathbb{N}$, $a \in \mathbb{Z}$ with $\gcd(a, mn) = 1$, that $\left(\frac{a}{mn}\right) = \left(\frac{a}{m}\right)\left(\frac{a}{n}\right)$.

4.24. Let $n \in \mathbb{N}$, and $a = 2^s \ell$ where $\ell \in \mathbb{Z}$ is odd, $s \in \mathbb{N}$, and $\gcd(a, n) = 1$. Prove that

$$\left(\frac{a}{n}\right) = \left(\frac{2}{n}\right)^s (-1)^{\frac{\ell-1}{2} \cdot \frac{n-1}{2}} \left(\frac{n}{|\ell|}\right).$$

Biography 4.5 Leopold Kronecker (1823–1891) *was born on December 7,
1823 in Liegnitz, Prussia. His well-educated, wealthy parents ensured that
Leopold had a private tutor at an early age. Later, he entered the Leipzig
Gymnasium where one his teachers was Ernst Eduard Kummer who became
his lifelong friend. Kronecker entered the University of Berlin in 1841, and
under the direction of Dirichlet, completed his Ph.D. in 1845, when he was
twenty-two. His thesis, which extended ideas of Gauss, was called De Unitat-
ibus Complexibus or On Complex Units. The years 1845–1854 were spent in
business, essentially managing the banking business of an uncle. However, he
was able to return to Berlin in 1855, and although he did not hold a university post, he was financially independent. In 1860 he was elected to the Berlin
Academy, which provided him with the opportunity and the right, to lecture at
Berlin University, a post he held during 1861–1883.*

His brand of mathematics would be called constructive *in the sense that he felt
all mathematics should be confined to the* finite, *rejecting the notion that tran-
scendental numbers could exist, and doubting the validity of* non-constructive
*proofs. This was perhaps best epitomized by his comment: "God made the in-
tegers, and all the rest is the work of man." This attitude brought him into
conflict with the likes of Georg Cantor (1845–1918) who had developed a theory of transfinite numbers. In 1883, he accepted a position as Professor at
the University of Berlin, succeeding his friend Kummer, who retired that year.
Kronecker died from a bronchial illness in Berlin on December 29, 1891, at the
age of sixty-nine.*

4.25. Prove that if $n \in \mathbb{N}$, $a \in \mathbb{Z}$ with $\gcd(a, n) = 1$, and a is odd, then

$$\left(\frac{a}{n}\right) = \left(\frac{n}{|a|}\right).$$

4.26. Prove that if $m, n \in \mathbb{N}$, $a \in \mathbb{Z}$ with $\gcd(a, mn) = 1$ and $m \equiv n \pmod{|a|}$, then

$$\left(\frac{a}{m}\right) = \left(\frac{a}{n}\right).$$

4.3 Factoring

> *The problem of distinguishing prime numbers from composite numbers and of resolving the latter into their prime factors is known to be one of the most important and useful in arithmetic.*
>
> **C.F. Gauss — see Biography 1.7 on page 33**

In §3.5, we saw the importance of factoring in public-key cryptosystems such as RSA. Indeed it was somewhat prescient of Gauss, in the above quote, to see the importance of factoring methods over two centuries ago. Thus, it is worth our having a closer look at the issue to which we devote this section. Some of the following is adapted from [31] and [34].

We first look at the following basic building block that we mentioned on page 167 with reference to related problems. Now we formally define it.

◆ The Integer Factoring Problem — (IFP)

Given $n \in \mathbb{N}$, find primes p_j for $j = 1, 2, \ldots, r \in \mathbb{N}$ with $p_1 < p_2 < \cdots < p_n$ and $e_j \in \mathbb{N}$ for $j = 1, 2, \ldots, r$, such that

$$n = \prod_{j=1}^{r} p_j^{e_j}.$$

A simpler problem than the IFP is the notion of *splitting* of $n \in \mathbb{N}$, which means the finding of factors $r, s \in \mathbb{N}$ such that $1 < r \leq s$ such that $n = rs$. Of course, with an RSA modulus, splitting and the IFP are the same thing. Yet, in order to solve the IFP for any integer, one merely splits n, then splits r and s if they are both composite, and so on until we have a complete factorization.

Now we discuss some older methods that still have relevance for the methods of today.

◆ Trial Division

The oldest method of splitting n is *trial division*, by which we mean dividing n by all primes up to $\sqrt{n}$. For $n < 10^8$, say, this is not an unreasonable method. However, for larger integers, we need more sophisticated methods.

◆ The MSR Test and Factoring

When we discussed the MSR probabilistic primality test on pages 119–121, we saw that we got factors of n whenever n is a pseudoprime to base a but not a *strong* pseudoprime to base a, namely when $a^{n-1} \equiv 1 \pmod{n}$. However, it is rare that the latter occurs. Suppose, on the other hand, that for a given modulus $n \in \mathbb{N}$ there exists an exponent $u \in \mathbb{N}$ such that

$$x^u \equiv 1 \pmod{n} \text{ for all } x \in \mathbb{N} \text{ with } \gcd(x, n) = 1,$$

where u is called a *universal exponent*. Then it may be possible to factor n as follows.

◆ Universal Exponent Factorization Method

Let u be a universal exponent for $n \in \mathbb{N}$ and set $u = 2^b m$ where $b \geq 0$ and m is odd. Execute the following steps.

(1) Choose a random base a such that $1 < a < n - 1$. If $\gcd(a, n) > 1$, then we have a factor of n, and we may terminate the algorithm. Otherwise go to step (2).

(2) Let $x_0 \equiv a^m \pmod{n}$. If $x_0 \equiv 1 \pmod{n}$, then go to step (1). Otherwise, compute $x_j \equiv x_{j-1}^2 \pmod{n}$ for all $j = 1, \ldots, b$. If

$$x_j \equiv -1 \pmod{n},$$

then go to step (1). If

$$x_j \equiv 1 \pmod{n}, \text{ but } x_{j-1} \not\equiv \pm 1 \pmod{n},$$

then $\gcd(x_{j-1} - 1, n)$ is a nontrivial factor of n, so we may terminate the algorithm.

The MSR test, which has similarities to the above, is not guaranteed to have a value such that $x_j \equiv 1 \pmod{n}$ as we *do* have in the universal exponent method (due to the existence of the exponent u).

In §3.5, we described public-key cryptographic methods, including RSA. When $n = pq$ is an RSA modulus, a universal exponent is sometimes taken to be $\mathrm{lcm}(p-1, q-1)$ instead of $\phi(n) = (p-1)(q-1)$. Yet these two values will be roughly the same since $\gcd(p-1, q-1)$ has an expectation of being small when p and q are chosen arbitrarily. Furthermore, recalling the discussion in §3.5 (see (3.14) on page 171) since $de - 1$ is a multiple of $\phi(n)$, then $de - 1$ is a universal exponent, and the above method can be used to factor n. Here's how.

Since $ed - 1 = 2^k s$ where $k \in \mathbb{N}$ and s is odd, and since $ed \equiv 1 \pmod{\phi(n)}$, then there exists an $a \in (\mathbb{Z}/n\mathbb{Z})^*$ such that $a^{2^k s} \equiv 1 \pmod{n}$. If $j \in \mathbb{N}$ is the least value such that $a^{2^j s} \equiv 1 \pmod{n}$, then $j \leq k$. If both

$$a^{2^{j-1}s} \not\equiv 1 \pmod{n} \text{ and } a^{2^{j-1}s} \not\equiv -1 \pmod{n}, \tag{4.21}$$

then $b = a^{2^{j-1}s}$ is a *nontrivial square root of* 1 *modulo* n. In other words,

$$n \mid (b+1)(b-1) \text{ with } n \nmid (b+1), \text{ and } n \nmid (b-1).$$

Therefore, we can factor n since

$$\gcd(b+1, n) = p \text{ or } \gcd(b+1, n) = q.$$

Hence, (4.21) is required to ensure that we can factor n. Indeed, the probability that (4.21) occurs can be made to approach 1 (for a sufficiently large number of trails testing the values of $a \in (\mathbb{Z}/n\mathbb{Z})^*$). Hence, knowledge of d can be converted into an algorithm for factoring n, with arbitrarily small probability of failure to do so.

This is the major value of our universal exponent test since actually finding u is difficult in practice.

There is a *least* universal exponent, and that value is given by Carmichael's function given in Definition 2.9 on page 94 — see Exercise 4.27 on page 208.

Example 4.17 *Let $n = 4189$ and suppose that we know $\lambda(n) = u = 4060$ is a universal exponent for n. Since $u = 2^2 \cdot 1015$, then we first choose $a = 2$ as a base and compute $2^{1015} \equiv 3480 \equiv x_0 \pmod{n}$. Then $3480^2 \equiv 1 \pmod{n}$. Since $x_0 \not\equiv \pm 1 \pmod{n}$, then $\gcd(x_0 - 1, n) = \gcd(3479, 4189) = 71$ is a factor of n. Indeed $n = 59 \cdot 71$.*

◆ **Fermat Factoring**

In 1643, Fermat discovered a factoring scheme based upon the following insight. If $n = rs$ is an odd natural number with $r < \sqrt{n}$, then

$$n = \left(\frac{s+r}{2}\right)^2 - \left(\frac{s-r}{2}\right)^2 = a^2 - b^2. \tag{4.22}$$

Therefore, in order to split n, we need only investigate the values,

$$x = a^2 - n \text{ for } a = \lfloor \sqrt{n} \rfloor + 1, \lfloor \sqrt{n} \rfloor + 2, \ldots, (n-1)/2,$$

until a perfect square is found. This is now called Fermat's *difference-of-squares factoring method*. It has been rediscovered many times and used as a basis for many modern factoring techniques since essentially we are looking at solutions of

$$x^2 \equiv y^2 \pmod{n} \text{ with } x \not\equiv \pm y \pmod{n}, \tag{4.23}$$

and

$$\gcd(x \pm y, n)$$

provides the nontrivial factors.

Although the order of magnitude of Fermat factoring can be shown to be $O(n^{1/2})$, Lehman has shown how to reduce the complexity to $O(n^{1/3})$ when combined with trial division. This is all contained in [24], complete with a computer program. There is also a method, from D.H. Lehmer, for speeding up the Fermat method when all factors are of the form $2k\ell + 1$ (see [25]).

◆ **Euler's Factoring Method**

This method applies only to integers of the form,

$$n = x^2 + ay^2 = z^2 + aw^2,$$

where $x \neq z$ and $y \neq w$. In other words, n can be written in two distinct ways in this special form for a given nonzero value of $a \in \mathbb{Z}$. Then

$$(xw)^2 \equiv (n - ay^2)w^2 \equiv -ay^2w^2 \equiv (z^2 - n)y^2 \equiv (zy)^2 \pmod{n},$$

from which we may have a factor of n, provided that $xw \not\equiv \pm zy \pmod{n}$. In this case, the (nontrivial) factors of n are given by $\gcd(xw \pm yz, n)$.

The Euler method essentially is predicated on the congruence (4.23), but unlike the Fermat method, not all integers have even one representation in the form $n = x^2 + ay^2$. In fact, the reader who is versed in some algebraic number theory will recognize these forms for n as norms from the quadratic field $\mathbb{Q}(\sqrt{-a})$. It can be shown that Euler's method requires at most $\lfloor \sqrt{n/a} \rfloor$ steps when $a > 0$.

Before describing the next method, we define what we mean by *sieving*. A *sieve* may be regarded as any process whereby we find numbers via searching up to a prescribed bound and eliminate candidates as we proceed until only the desired solution set remains. A (general) *quadratic* sieve is one in which about half of the possible numbers being sieved are removed from consideration, a technique used for hundreds of years as a scheme for eliminating impossible cases from consideration.

◆ **Legendre's Factoring Method**

This method is a precursor to what we know today as *continued fraction methods for factorization* that we will study in §5.4. Legendre reasoned in the following fashion. Instead of looking at congruences of the form (4.23), he looked at those of the form,

$$x^2 \equiv \pm py^2 \pmod{n} \text{ for primes } p, \tag{4.24}$$

since a solution to (4.24) implies that $\pm p$ is a quadratic residue of all prime factors of n. For instance, if the residue is 2, then all prime factors of n are congruent to $\pm 1 \pmod 8$ (see part (4) of Theorem 4.10 on page 194). Therefore, he would have halved the search for factors of n. Legendre applied this method for various values of p, thereby essentially constructing a quadratic sieve by getting many residues modulo n. This allowed him to eliminate potential prime divisors that sit in various linear sequences, as with the residue 2 example above. He realized that if he could achieve enough of these, he could eliminate primes up to $\sqrt{n}$, thereby effectively developing a test for primality!

Legendre was essentially building a sieve on the prime factors of n, which did not let him predict, for a given prime p, a different residue to yield a square. This meant that if he found a solution to

$$x^2 \equiv py^2 \pmod{n},$$

he could not predict a solution,

$$w^2 \equiv pz^2 \pmod{n},$$

distinct from the former. If he had been able to do this, he would have been able to combine them as

$$(xw)^2 \equiv (pzy)^2 \pmod{n}$$

and have a factor of n *provided that* $xw \not\equiv \pm pzy \pmod{n}$ since we are back to congruence (4.23).

Gauss invented a method that differed from Legendre's scheme only in the approach to finding small quadratic residues of n; but his approach makes it much more complicated (see [16, Articles 333 and 334, pages 403–406]).

In the 1920s, one individual expanded the idea, described above, of attempting to match the primes to create a square. We now look at his important influence.

◆ **Kraitchik's Factoring Method**

Maurice Kraitchik determined that it would suffice to find a *multiple* of n as a difference of squares in attempting to factor it. For this purpose, he chose a polynomial of the form, $kn = ax^2 \pm by^2$, for some integer k, which allowed him to gain control over finding two distinct residues at a given prime to form a square, which Legendre could not do. In other words, Kraitchik used quadratic polynomials to get the residues, then multiplied them to get squares (not a square times a small number). Kraitchik developed this method over a period of more than three decades, a method later exploited in the development of an algorithm that systematically extracted the best of the above ideas, which we will present in Chapter 5 when we have the full force of continued fractions at our disposal.

Biography 4.6 Maurice Borisovich Kraitchik (1882–1957) *obtained his Ph.D. from the University of Brussels in 1923. He worked as an engineer in Brussels and later as a Director at the mathematical sciences section of the Mathematical Institute for Advanced Studies there. From 1941–1946, he was associate professor at the New School for Social Research in New York. In 1946, he returned to Belgium, where he died on August 19, 1957. His work over thirty-five years on factoring methods stands tall today because he devised and used a variety of practical techniques that are found today in computer methods such as the Quadratic Sieve — see [34] for instance.*

We conclude this section with two algorithms due to one person.

In 1974, Pollard published a factorization scheme (see [40]) that utilizes Euler's generalization of Fermat's Little Theorem. He reasoned that if $(p-1) \mid n$ where p is prime, then $p \mid (t^n - 1)$ provided that $p \nmid t$, which follows from Euler's theorem, so p may be found by employing the Euclidean algorithm.

◆ **Pollard's p − 1 Algorithm**

Suppose that we wish to factor $n \in \mathbb{N}$, and that a smoothness bound B has been selected (see page 122). Then we execute the following.

(1) Choose a base $a \in \mathbb{N}$ where $2 \leq a < n$ and compute $g = \gcd(a, n)$. If $g > 1$, then we have a factor of n. Otherwise, go to step (2).

(2) For all primes $p \leq B$, compute $m = \left\lfloor \frac{\ln(n)}{\ln(p)} \right\rfloor$ and replace a by $a^{p^m} \pmod{n}$ using the repeated squaring method given on page 82. (Note that this iterative procedure ultimately gives $a^{\prod_{p \leq B} p^m}$ modulo n for the base a chosen in (1).)

(3) Compute $g = \gcd(a - 1, n)$. If $g > 1$, then we have a factor of n, and the algorithm is successful. Otherwise, the algorithm fails.

▼ Analysis

Let $\ell = \prod_{j=1}^{t} p_j^{a_j}$, where the $p_j^{a_j}$ are the prime powers with $p_j \leq B$. Since $p_j^{a_j} \leq n$, then $a_j \ln(p_j) \leq \ln(n)$, so $a_j \leq \left\lfloor \frac{\ln(n)}{\ln(p_j)} \right\rfloor$. Hence, $\ell \leq \prod_{j=1}^{t} p_j^{\lfloor \ln(n)/\ln(p_j) \rfloor}$. Now, if $p \mid n$ is a prime such that $p - 1$ is B-smooth, then $(p-1) \mid \ell$. Therefore, for any $a \in \mathbb{N}$ with $p \nmid a$, $a^\ell \equiv 1 \pmod{p}$, by Fermat's Little Theorem. Thus, if $g = \gcd(a^\ell - 1, n)$, then $p \mid g$. If $g = n$, then the algorithm fails. Otherwise, it succeeds.

Example 4.18 *Let $n = 330931$, and choose a smoothness bound $B = 13$, then select $a = 2$. We know that a is relatively prime to n so we proceed to step (2). The table shows the outcome of the calculations for step (2).*

p	2	3	5	7	11	13
m	18	11	7	6	5	4
a	167215	132930	87517	154071	330151	263624

Then we go to step (3) and check $\gcd(a-1, n) = \gcd(263623, 330931) = 5609$. *Thus, we have split* $n = 5609 \cdot 59$. *Indeed,* $n = 59 \cdot 71 \cdot 79$, *and we observe that both* $p = 71$ *and* $q = 79$ *are B-smooth since* $p - 1 = 2 \cdot 5 \cdot 7$ *and* $q - 1 = 2 \cdot 3 \cdot 13$.

The running time for Pollard's $p-1$ algorithm is $O(B \ln(n)/\ln(B))$ modular multiplications, assuming that $n \in \mathbb{N}$ and there exists a prime $p \mid n$ such that $p - 1$ is B-smooth. This is of course the drawback to this algorithm, namely, that it requires n to have a prime factor p such that $p - 1$ has only "small" prime factors.

Pollard also developed another method for factoring in 1975, called the *Monte Carlo factoring method*, also known as the *Pollard rho method*.

◆ Pollard Rho Method

Given $n \in \mathbb{N}$ composite, and p an (as yet unknown) prime divisor of it, perform the following steps.

(1) Choose an integral polynomial f with $\deg(f) \geq 2$ — usually $f(x) = x^2 + 1$ is chosen for simplicity.

(2) Choose a randomly generated integer $x = x_0$, the *seed*, and compute $x_1 = f(x_0)$, $x_2 = f(x_1), \ldots, x_{j+1} = f(x_j)$ for $j = 0, 1, \ldots B$, where the bound B is determined by step (3).

(3) Sieve through all differences $x_i - x_j$ modulo n until it is determined that

$$x_B \not\equiv x_j \pmod{n}$$

but $x_B \equiv x_j \pmod{p}$ for some natural number $B > j \geq 1$. Then

$$\gcd(x_B - x_j, n)$$

is a nontrivial divisor of n.

Now we illustrate the reason behind the name *Pollard rho method*. We take $n = 29$ as the modulus and $x_0 = 2$ as the seed, then we proceed through the Pollard rho method to achieve Diagram 4.1.

Diagram 4.1 Pollard's Rho Method Illustrated

We take $n = 29$ as the modulus and $x_0 = 2$ as the seed, then we proceed through the Pollard rho method to achieve the following diagram.

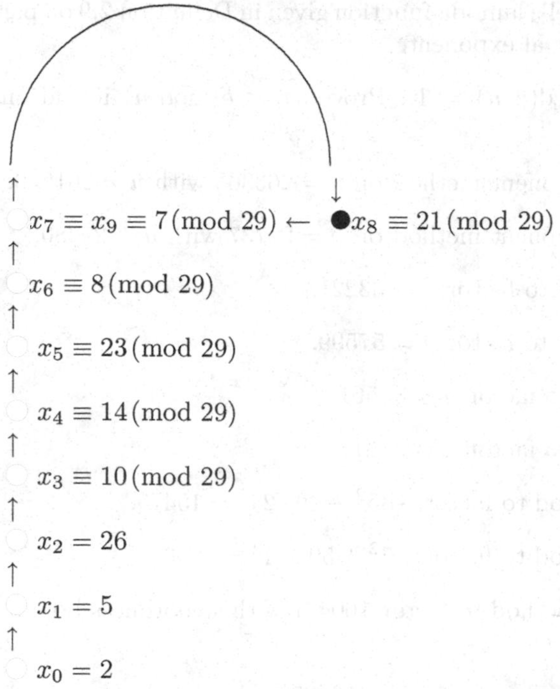

$$x_7 \equiv x_9 \equiv 7 \,(\text{mod } 29) \leftarrow \bullet x_8 \equiv 21 \,(\text{mod } 29)$$

$$x_6 \equiv 8 \,(\text{mod } 29)$$

$$x_5 \equiv 23 \,(\text{mod } 29)$$

$$x_4 \equiv 14 \,(\text{mod } 29)$$

$$x_3 \equiv 10 \,(\text{mod } 29)$$

$$x_2 = 26$$

$$x_1 = 5$$

$$x_0 = 2$$

Diagram 4.1 shows us that when we reach x_9, then we are in the period that takes us back and forth between the residue system of 7 and that of 21 modulo 29. This is the significance of the left pointing arrow from the position of x_8 back to the position of x_7, which is the same as the residue system of x_9. This completes the circuit. The shape of the symbol is reminiscent of the Greek symbol ρ, *rho*, pronounced *row*.

Example 4.19 If $n = 37351$, and $x_0 = 2$ is the seed with $f(x) = x^2 + 1$, then $x_1 = f(x_0) = 5$, $x_2 = f(x_1) = 26$, $x_3 = f(x_2) = 677$, $x_4 = f(x_3) = \overline{10118}$, $x_5 = f(x_4) = \overline{32185}$, and $x_6 = f(x_5) = \overline{18943}$, and $x_7 = f(x_6) = \overline{6193}$, where the bar notation denotes the fact that we have reduced the values to the least residue system modulo n. We find that all $\gcd(x_i - x_j, n) = 1$ for $i \neq j$ until

$$\gcd(x_7 - x_0, n) = \gcd(6191, 37351) = 41.$$

In fact, $37351 = 41 \cdot 911$.

Pollard's two methods above may be invoked when trial division fails to be useful. However, if the methods of Pollard fail to be useful, which they will for *large* prime factors, say, with the number of digits in the high teens, then we need more powerful machinery.

Exercises

4.27. Prove that Carmichael's lambda function given in Definition 2.9 on page 94 is the *minimal* universal exponent.

4.28. Let $a, n \in \mathbb{Z}$ with $\gcd(a, n) = 1$. Prove $ax \equiv b \pmod{n}$ if and only if $x \equiv a^{\lambda(n)-1} b \pmod{n}$.

4.29. Use the universal exponent method on $n = 263363$ with $u = 261960$.

4.30. Use the universal exponent method on $n = 29737$ with $u = 29380$.

4.31. Use Fermat's method to factor $n = 33221$.

4.32. Use Fermat's method to factor $n = 57599$.

4.33. Use Euler's method to factor $n = 35561$.

4.34. Use Euler's method to factor $n = 57611$.

4.35. Use Legendre's method to factor $-35^2 + 29 \cdot 24^2 = 15479$.

4.36. Use Legendre's method to factor $-5^2 + 59 \cdot 24^2 = 33959$.

4.37. Use Pollard's $p-1$ method to factor 160427 with smoothness bound $B = 13$.

4.38. Use Pollard's $p-1$ method to factor 453487 with smoothness bound $B = 13$.

4.39. Use Pollard's rho method to factor 60143 with seed $x_0 = 2$.

4.40. Use Pollard's rho method to factor 156809 with seed $x_0 = 2$.

Chapter 5

Simple Continued Fractions and Diophantine Approximation

> We might call Euclid's method the granddaddy of all algorithms, because it is the oldest nontrivial algorithm that has survived to the present day.
>
> **Donald Knuth (1938–),**
> **renowned computer scientist and Professor Emeritus of the Art of Computer Programming at Stanford University — see [22].**

5.1 Infinite Simple Continued Fractions

In §1.2, we introduced simple continued fractions, based on Euclid's algorithm, and proved, in Theorem 1.11 on page 24, that rational numbers are tantamount to finite continued fractions. We also introduced convergents and established, in Theorem 1.12 on page 25, certain of their properties. Also, in Exercise 1.33 on page 29, we established an identity based upon those representations of convergents. We continue now with a result that establishes more fundamental identities for them.

Theorem 5.1 | **Properties of Convergents**

Let $\alpha = \langle q_0; q_1, \ldots, q_\ell \rangle$ be a finite continued fraction expansion with $\ell \in \mathbb{N}$, and let the sequences $\{A_k\}$, $\{B_k\}$ be defined as in Theorem 1.12 for $k \geq -2$. Then the following hold:

(a) $A_k B_{k-1} - A_{k-1} B_k = (-1)^{k-1}$ $(k \geq 1)$,

(b) $A_k B_{k-2} - A_{k-2} B_k = (-1)^k q_k$ $(k \geq 1)$,

(c) $B_k \geq F_{k+1}$ for any $k \in \mathbb{N}$ where F_{k+1} is the $(k+1)^{st}$ Fibonacci number,

(d) $C_k - C_{k-1} = (-1)^{k-1}/(B_k B_{k-1})$ $(k \geq 1)$,

(e) $C_k - C_{k-2} = (-1)^k q_k/(B_k B_{k-2})$ $(k \geq 2)$.

Proof. Part (a) is Exercise 1.33. For parts (b)–(c), we use induction on k.
Proof of (b): If $k = 1$, then

$$A_k B_{k-2} - A_{k-2} B_k = A_1 \cdot 0 - 1 \cdot q_1 = (-1)^k q_k,$$

which is the induction step. Assume the induction hypothesis, for $k > 1$,

$$A_{k-1} B_{k-3} - A_{k-3} B_{k-1} = (-1)^{k-1} q_{k-1}.$$

Then

$$A_k B_{k-2} - A_{k-2} B_k = (q_k A_{k-1} + A_{k-2}) B_{k-2} - A_{k-2}(q_k B_{k-1} + B_{k-2}) =$$

$$q_k(A_{k-1} B_{k-2} - A_{k-2} B_{k-1}) = q_k(-1)^{k-2} = q_k(-1)^k,$$

by part (a).
Proof of (c): For $k = 1$, we have $B_1 = q_1 \geq F_2 = 1$. Assume that $B_k \geq F_{k+1}$ for all $k \leq n$. Then

$$B_{n+1} = q_n B_{n-1} + B_n \geq q_n F_n + F_{n+1} \geq F_n + F_{n+1} = F_{n+2}.$$

Proof of (d): We use part (a). Since

$$A_k B_{k-1} - A_{k-1} B_k = (-1)^{k-1},$$

dividing through by $B_k B_{k-1}$ we get

$$A_k/B_k - A_{k-1}/B_{k-1} = C_k - C_{k-1} = (-1)^{k-1}/(B_k B_{k-1}).$$

Proof of (e): We see that

$$C_k - C_{k-2} = A_k/B_k - A_{k-2}/B_{k-2} = \frac{A_k B_{k-2} - A_{k-2} B_k}{B_k B_{k-2}} = \frac{(-1)^k q_k}{B_k B_{k-2}},$$

by part (b), so we have proved the result. $\square$

Convergents also satisfy a certain ordering that we will need to establish facts about infinite continued fractions. Theorem 5.1 plays a role in the proof of the following.

Theorem 5.2 | Ordering of Convergents

If C_k is the kth convergent of the simple continued fraction expansion of $\langle q_0; q_1, \ldots, q_\ell \rangle$, then

$$C_1 > C_3 > C_5 > \cdots > C_{2k-1} > C_{2k} > C_{2k-2} > \cdots > C_4 > C_2 > C_0,$$

for any $k \in \mathbb{N}$.

Proof. By part (e) of Theorem 5.1,

$$C_k - C_{k-2} = (-1)^k q_k / (B_k B_{k-2}),$$

so when k is odd, $C_k < C_{k-2}$, and $C_k > C_{k-2}$ when k is even, namely

$$C_1 > C_3 > C_5 \cdots$$

and

$$C_0 < C_2 < C_4 \cdots.$$

By part (d) of Theorem 5.1, $C_k - C_{k-1} = (-1)^{k-1}/(B_k B_{k-1})$, so $C_{2j+1} > C_{2j}$, for any $j \in \mathbb{N}$. Hence,

$$C_{2j-1} > C_{2j+2k-1} > C_{2j+2k} > C_{2k}$$

for any $j, k \in \mathbb{N}$. □

In order to discuss infinite continued fractions, we need some facts about limits of sequences that we have presented in Theorem A.14 on page 307, and the discussion preceding it, to which we refer the reader in advance of the following material.

Theorem 5.3 | Limits of Convergents and Infinite Continued Fractions

If $q_0, q_1, q_2, \ldots$ is an infinte sequence of integers with $q_j > 0$ for $j > 0$, and if we set $C_k = \langle q_0; q_1, q_2, \ldots, q_k \rangle$, then

$$\lim_{k \to \infty} C_k = \alpha,$$

where α is called the infinite simple continued fraction

$$\langle q_0; q_1, q_2, \ldots \rangle.$$

Proof. By Theorem 5.2, the sequence C_{2j+1} for $j \geq 0$ is monotonically decreasing and bounded and the sequence C_{2j} for $j \geq 0$ is monotonically increasing and bounded. Thus, by part (a) of Theorem A.14, there exist $\alpha_1 \in \mathbb{R}$ such that $\lim_{j \to \infty} C_{2j+1} = \alpha_1$ and $\alpha_2 \in \mathbb{R}$ such that $\lim_{j \to \infty} C_{2j} = \alpha_2$. It remains to show that $\alpha_1 = \alpha_2$. By part (d) of Theorem 5.1,

$$C_{2j+1} - C_{2j} = \frac{1}{B_{2j+1} B_{2j}},$$

and by part (c) of Theorem 5.1,

$$\frac{1}{B_{2j+1}B_{2j}} \leq \frac{1}{(2j+1)(2j)}.$$

Hence,

$$\lim_{j\to\infty}(C_{2j+1} - C_{2j}) = 0,$$

so

$$\alpha_1 = \lim_{j\to\infty} C_{2j+1} = \lim_{j\to\infty} C_{2j} = \alpha_2.$$

Also, if $\alpha = \alpha_1 = \alpha_2$, then

$$\alpha = \lim_{j\to\infty} C_j = \lim_{j\to\infty} \langle q_0; q_1, q_2, \ldots, q_j \rangle = \langle q_0; q_1, q_2, \ldots, q_j, q_{j+1}, \ldots \rangle,$$

which is the infinite simple continued fraction

$$\alpha = q_0 + \cfrac{1}{q_1 + \cfrac{1}{q_2+}}$$

$$+\cfrac{1}{q_j + \cfrac{1}{q_{j+1}}}$$

as required.   □

We now demonstrate the analogue of Theorem 1.11 on page 24. We begin with one direction of this assertion.

Theorem 5.4 | **Infinite Simple Continued Fractions Are Irrational**

Let $q_0, q_1, q_2, \ldots, q_j$ be integers with $q_j > 0$ for $j > 0$. Then the infinite simple continued fraction $\langle q_0; q_1, q_2, \ldots, q_j, q_{j+1}, \ldots \rangle$ is an irrational number α, and no other infinite simple continued fraction represents α.

Proof. Let $\alpha = \langle q_0; q_1, q_2, \ldots, q_j, q_{j+1}, \ldots \rangle$ and let $C_j = A_j/B_j = \langle q_0; q_1, \ldots, q_j \rangle$ be the jth convergent of α. By Theorem 5.3 we have for any $j > 0$,

$$C_{2j} < \alpha < C_{2j+1}.$$

Thus,

$$0 < \alpha - C_{2j} < C_{2j+1} - C_{2j} = \frac{1}{B_{2j+1}B_{2j}},$$

where the equality comes from part (d) of Theorem 5.1. It follows that

$$0 < \alpha - C_{2j} < \alpha - \frac{A_{2j}}{B_{2j}} < \frac{1}{B_{2j+1}B_{2j}}.$$

By cross multiplying, we get

$$0 < \alpha B_{2j} - A_{2j} < \frac{1}{B_{2j+1}}. \tag{5.1}$$

Now we complete the proof by contradiction. Assume that $\alpha = a/b$ where $a, b \in \mathbb{Z}$ with $b \neq 0$. Then (5.1) becomes

$$0 < aB_{2j}/b - A_{2j} < \frac{1}{B_{2j+1}},$$

and multiplying through by b we get,

$$0 < aB_{2j} - A_{2j}b < \frac{b}{B_{2j+1}},$$

where $aB_{2j} - A_{2j}b \in \mathbb{Z}$ for all $j > 0$. However, $B_{2j+1} > 2j + 1$ for any $j \geq 0$ by part (c) of Theorem 5.1, so there must exist a value of $k \in \mathbb{N}$ such that $B_{2k+1} > b$. Hence,

$$0 < aB_{2k} - A_{2k}b < \frac{b}{B_{2k+1}} < 1,$$

which is a contradiction since there are no integers between 0 and 1. This establishes that α is irrational. This proves the existence. We now must prove uniqueness — that no other infinite simple continued fraction can represent α. Suppose that $\langle a_0; a_1, a_2, \ldots \rangle = \alpha$. Since

$$a_0 = C_0 < \alpha < C_1 = a_0 + 1/a_1 \leq a_0 + 1,$$

then $a_0 = \lfloor \alpha \rfloor$. However, $q_0 = \lfloor \alpha \rfloor$ by the same argument. Therefore, since

$$a_0 + \frac{1}{\langle a_1, a_2, \ldots \rangle} = q_0 + \frac{1}{\langle q_1, q_2, \ldots \rangle},$$

then

$$\langle a_1, a_2, \ldots \rangle = \langle q_1, q_2, \ldots \rangle,$$

and we repeat the above argument to get that $a_1 = q_1$, so by induction, we have uniqueness of representation of α by $\langle q_0; q_1, q_2, \ldots, q_j, q_{j+1}, \ldots \rangle$. □

Immediate from the above proof are the following two consequences.

Corollary 5.1 *Two distinct infinite simple continued fractions represent two distinct irrational numbers.*

Corollary 5.2 *If two infinite simple continued fractions $\langle a_0; a_1, a_2, \ldots \rangle$ and $\langle q_0; q_1, q_2, \ldots \rangle$ both represent the same irrational number, then $a_j = q_j$ for all $j \geq 0$.*

In order to prove the next result we need to reinterpret an earlier result as follows.

Theorem 5.5 | Infinite Continued Fraction Quotient Representation

Let $\alpha = \langle q_0; q_2, q_2, \ldots, \rangle$ be irrational and let α_k be a positive real number. Then

$$\alpha = \langle q_0; q_1, q_2, \ldots, q_{k-1}, \alpha_k \rangle = \frac{\alpha_k A_{k-1} + A_{k-2}}{\alpha_k B_{k-1} + B_{k-2}},$$

where A_j/B_j is the jth convergent of α.

Proof. This is established by an induction argument the same as that given in the proof of Theorem 1.12 on page 25, the only difference being that α_k is not an integer. $\square$

Now we demonstrate the converse of Theorem 5.4, that every irrational number has a unique infinite simple continued fraction expansion.

Theorem 5.6 | Irrationals Are Infinite Simple Continued Fractions

Let α_0 be an irrational number and recursively define for any $j \geq 0$,

$$q_j = \lfloor \alpha_j \rfloor \text{ and } \alpha_{j+1} = \frac{1}{\alpha_j - q_j}.$$

Then α_0 is the unique infinite simple continued fraction given by

$$\langle q_0; q_1, q_2, \ldots \rangle.$$

Proof. By definition, $q_j \in \mathbb{Z}$ for all $j \geq 0$. Also, given that α_0 is irrational, then $\alpha_0 \neq q_0$. Moreover, by induction α_j exists and is irrational for all $j \geq 0$ since

$$\alpha_j = q_j + \frac{1}{\alpha_{j+1}}. \tag{5.2}$$

Thus, $\alpha_j \neq q_j$ for all $j \geq 0$. By definition, $q_j < \alpha_j < q_j + 1$, so $0 < \alpha_j - q_j < 1$. Therefore, for any $j \geq 0$,

$$q_{j+1} = \lfloor \alpha_{j+1} \rfloor = \left\lfloor \frac{1}{\alpha_j - q_j} \right\rfloor \geq 1.$$

Now by applying (5.2) at each step, we get

$$\alpha_0 = q_0 + \frac{1}{\alpha_1} = q_0 + \cfrac{1}{q_1 + \cfrac{1}{\alpha_2}} = \ldots$$

$$q_0 + \cfrac{1}{q_1 + \cfrac{1}{q_2+}}$$

$$\ddots$$

$$+ \cfrac{1}{q_j + \cfrac{1}{\alpha_{j+1}}} = \langle q_0; q_1, q_2, \ldots, q_j, \alpha_{j+1} \rangle.$$

By Theorem 5.5,

$$\alpha_0 = \langle q_0; q_1, q_2, \ldots, q_j, \alpha_{j+1} \rangle = \frac{\alpha_{j+1} A_j + A_{j-1}}{\alpha_{j+1} B_j + B_{j-1}}.$$

Since $C_j = A_j / B_j$ is the jth convergent of $\langle q_0; q_1, q_2, \ldots \rangle$, then

$$\alpha_0 - C_j = \frac{\alpha_{j+1} A_j + A_{j-1}}{\alpha_{j+1} B_j + B_{j-1}} - \frac{A_j}{B_j} = \frac{-(A_j B_{j-1} - A_{j-1} B_j)}{(\alpha_{j+1} B_j + B_{j-1}) B_j},$$

and by part (a) of Theorem 5.1, this equals

$$\frac{-(-1)^{j-1}}{(\alpha_{j+1} B_j + B_{j-1}) B_j}.$$

However, $\alpha_{j+1} B_j + B_{j-1} > q_{j+1} A_j + B_{j-1} = B_{j+1}$, so

$$|\alpha - C_j| < \frac{1}{B_j B_{j+1}}.$$

Therefore, by part (c) of Theorem 5.1,

$$|\alpha - C_j| < \frac{1}{j(j+1)},$$

so $\langle q_0; q_1, q_2, \ldots \rangle = \lim_{j \to \infty} C_j = \alpha_0$, and we have the existence of the representation for α_0. By Corollary 5.2, the representation is unique. $\square$

Immediately from the proof of Theorem 5.6 is the following valuable result.

Corollary 5.3 *If $C_j = A_j / B_j$, for $j \in \mathbb{N}$, is the jth convergent of an irrational number α, then*

$$\left| \alpha - \frac{A_j}{B_j} \right| < \frac{1}{B_j^2}.$$

Theorem 5.6 tacitly provides an algorithm for finding the representation of an irrational number as an infinite simple continued fraction as illustrated below.

Example 5.1 *Let $\alpha_0 = (1 + \sqrt{29})/2$. Then $q_0 = \lfloor \alpha_0 \rfloor = 3$ and*

$$\alpha_1 = \frac{1}{\alpha_0 - q_0} = \frac{2}{\sqrt{29} - 5} = \frac{\sqrt{29} + 5}{2} = 5 + \frac{\sqrt{29} - 5}{2} = q_1 + \frac{1}{\alpha_2},$$

$$\alpha_2 = \frac{1}{\alpha_1 - q_1} = \frac{2}{\sqrt{29} - 5} = \frac{\sqrt{29} + 5}{2} = \alpha_1,$$

so we repeat the process. In other words, the infinite simple continued fraction expansion of α_0 is periodic, with period length 1, and is equal to $\langle 3, 5, 5, 5, \ldots \rangle$. In §5.2, we will study such periodic continued fractions. Also, see Exercise 5.1 on page 219.

The following shows how convergents give us rational approximations of irrationals.

Example 5.2 *Although we may use the above algorithm to find that*

$$\pi = \langle 3; 7, 15, 1, 292, 1, 1, 1, 2, 1, 3, 1, 14, 2, 1, 1, 2, 2, 2, 2, 1, 84, 2, 1, 1, 15, \ldots \rangle,$$

there is no known pattern in the sequence of partial quotients. The first convergent of π is $C_1 = 3 + 1/7 = 22/7$, which is the first rational approximation of it. Indeed, by Corollary 5.3 on the page before,

$$\left| \pi - \frac{22}{7} \right| < \frac{1}{7^2}.$$

We now show that the convergents of an irrational number are the best possible rational approximations in the sense that the jth convergent $C_j = A_j/B_j$ of the irrational α is closer to α than any other rational number with denominator less than B_j. First we need the following.

Lemma 5.1 *Let A_j/B_j for $j \in \mathbb{N}$ be the jth convergent of the infinite simple continued fraction expansion of the irrational number α. If $r, s \in \mathbb{Z}$ with $s > 0$ such that*

$$|s\alpha - r| < |B_j\alpha - A_j|, \tag{5.3}$$

then $s \geq B_{j+1}$.

Proof. We prove the result by contradiction, so assume that

$$s < B_{j+1} \tag{5.4}$$

and that (5.3) holds. Consider the system of equations,

$$A_j x + A_{j+1} y = r, \tag{5.5}$$

$$B_j x + B_{j+1} y = s. \tag{5.6}$$

Multiplying (5.5) by B_j and subtracting A_j times (5.6), we get

$$(A_{j+1}B_j - A_jB_{j+1})y = rB_j - sA_j,$$

but by part (a) of Theorem 5.1 on page 209, $A_{j+1}B_j - A_jB_{j+1} = (-1)^j$, so

$$y = (-1)^j(rB_j - sA_j,).$$

Now, multiplying (5.6) by A_{j+1} and subtracting this B_{j+1} times (5.5), we get, in a similar fashion,

$$x = (-1)^j(sA_{j+1} - rB_{j+1}). \tag{5.7}$$

We need the following to complete the result.

Claim 5.1 $x(B_j\alpha - A_j)$ *and* $y(B_{j+1}\alpha - A_{j+1})$ *have the same sign.*

We prove this by first showing that x and y have opposite sign then that $B_j\alpha - A_j$ and $B_{j+1}\alpha - A_{j+1}$ have opposite sign.

If $x = 0$, then by (5.7), $sA_{j+1} = rB_{j+1}$. However, by Exercise 1.33 on page 29, $\gcd(A_{j+1}, B_{j+1}) = 1$. Thus, $B_{j+1} \mid s$, so $s \geq B_{j+1}$ contradicting assumption (5.4). Thus, $x \neq 0$. If $y = 0$, then by (5.5)–(5.6), $s = B_jx$ and $r = A_jx$, so

$$|s\alpha - r| = |B_jx\alpha - xA_j| = |x||B_j\alpha - A_j| \geq |B_j\alpha - A_j|,$$

contradicting assumption (5.3). Thus, $y \neq 0$.

Now, if $y < 0$, then by (5.5), $B_jx = s - B_{j+1}y > 0$, which implies that $x > 0$, since $B_j > 0$. If $y > 0$, then by (5.6), $B_jx = s - B_{j+1}y \leq s - B_{j+1} < 0$, where the last inequality comes from assumption (5.4), so $x < 0$. Hence, x and y have opposite signs.

By Theorem 5.3 on page 211 we have for any $j > 0$, α lies between the convergents A_j/B_j and A_{j+1}/B_{j+1}, so $B_j\alpha - A_j$ and $B_{j+1}\alpha - A_{j+1}$ have opposite sign, given the ordering of convergents given in Theorem 5.2 on page 211. This completes the proof of Claim 5.1.

By Claim 5.1, using (5.5)–(5.6),

$$|s\alpha - r| = |(B_jx + B_{j+1}y)\alpha - (A_jx + A_{j+1}y)| = |x(B_j\alpha - A_j) + y(B_{j+1}\alpha - A_{j+1})| =$$

$$|x||B_j\alpha - A_j| + |y||B_{j+1}\alpha - A_{j+1}| \geq |x||B_j\alpha - A_j| \geq |B_j\alpha - A_j|,$$

contradicting assumption (5.3). This completes the proof. □

Theorem 5.7 | Law of Best Approximation |

Let α be an irrational number and let $C_j = A_j/B_j$ for $j \in \mathbb{N}$ be the convergents in the simple continued fraction expansion of α. Let $r, s \in \mathbb{Z}$ with $s > 0$. If $j \in \mathbb{N}$ with

$$|\alpha - r/s| < |\alpha - A_j/B_j|, \tag{5.8}$$

then $s > B_j$.

Proof. Suppose, to the contrary, that $s \leq B_j$, and that (5.8) holds. Therefore,

$$s|\alpha - r/s| \leq B_j|\alpha - r/s| < B_j|\alpha - A_j/B_j|,$$

or by rewriting, $|s\alpha - r| < |B_j\alpha - A_j|$, contracting Lemma 5.1 since $s < B_{j+1}$. $\square$

Immediate from Theorem 5.7 is the following result.

Corollary 5.4 *If s is a postive integer such that $s \leq B_j$, $j \in \mathbb{N}$, where A_j/B_j is the jth convergent of the irrational number α, then for any $r \in \mathbb{Z}$,*

$$\left| \alpha - \frac{A_j}{B_j} \right| \leq \left| \alpha - \frac{r}{s} \right|.$$

Example 5.3 *By Theorem 5.7 on the preceding page, the approximation $22/7$ of π in Example 5.2 is the best rational approximation of π for any rational number with denominator less than or equal to 7. Since $C_2 = 333/106$ is the second convergent of π, then this is the best rational approximation of π for any rational number with denominator less than or equal to 106. The reader may verify that each subsequent convergent $C_j = A_j/B_j$ is the best rational approximation of π for any rational number with denominator less than B_j.*

The concluding result of this section is the fact that any rational approximation of an irrational number α, which is sufficiently close to α, must be a convergent of the infinite simple continued fraction expansion of α.

Theorem 5.8 | **Rational Approximations**

Let α be an irrational number and let $r, s \in \mathbb{Z}$ with $\gcd(r, s) = 1$ and $s > 0$. If

$$|\alpha - r/s| < 1/(2s^2), \tag{5.9}$$

then r/s is a convergent in the infinite simple continued fraction expansion of α.

Proof. Assume that r/s is not a convergent of α. Since the integers B_j for $j \in \mathbb{N}$ form an increasing sequence, then here is some integer k such that

$$B_k \leq s < B_{k+1}. \tag{5.10}$$

By Lemma 5.1 on page 216,

$$|B_k\alpha - A_k| \leq |s\alpha - r| = s \left| \alpha - \frac{r}{s} \right| < \frac{1}{2s},$$

where the last inequality comes from (5.9), so by dividing through by B_k, we get $|\alpha - A_k/B_k| < 1/(2sB_k)$. Given that $r/s \neq A_k/B_k$, then the integer $|sA_k - rB_k|$ is positive. Hence,

$$\frac{1}{sB_k} \leq \left| \frac{sA_k - rB_k}{sB_k} \right| = \left| \frac{A_k}{B_k} - \frac{r}{s} \right| \leq \left| \frac{A_k}{B_k} - \alpha \right| + \left| \alpha - \frac{r}{s} \right| < \frac{1}{2sB_k} + \frac{1}{2s^2}.$$

Therefore, by subtracting $1/(2sB_k)$ from the left- and right-hand sides, we get $1/(sB_k) < 1/s^2$. By cross-multiplying and dividing through by s we get, $s < B_k$, a contradiction to (5.10). $\qquad\square$

Exercises

5.1. Let $\alpha = (1 + \sqrt{n^2 + 4})/2$, where $n \in \mathbb{N}$ is odd. Prove that

$$\alpha = \left\langle \frac{n+1}{2}; n, n, n, \ldots \right\rangle.$$

(See Example 5.1 on page 216.)

5.2. Let $n \in \mathbb{N}$. Prove that $\sqrt{n^2 + 1} = \langle n; 2n, 2n, 2n, \ldots \rangle$.

5.3. Let $n \in \mathbb{N}$. Prove that $\sqrt{n^2 + 2} = \langle n; n, 2n, n, 2n, \ldots \rangle$.

5.4. Let $n \in \mathbb{N}$. Prove that $\sqrt{n^2 + 2n} = \langle n; 1, 2n, 1, 2n, \ldots \rangle$.

5.5. Find the infinite simple continued fraction expansions of each of the following.
    (a) $(1 + \sqrt{53})/2$ (b) $\sqrt{26}$
    (c) $\sqrt{102}$ (d) $\sqrt{99}$

(*Hint: Use Exercises 5.1–5.4.*)

5.6. Find the infinite simple continued fraction expansions of each of the following.
    (a) $(1 + \sqrt{173})/2$ (b) $\sqrt{122}$
    (c) $\sqrt{443}$ (d) $\sqrt{168}$

(*See the hint to Exercise 5.5.*)

5.7. Let $\alpha \in \mathbb{R}$ have infinite continued fraction expansion $\langle q_0; q_1, \ldots \rangle$. Prove that if $q_1 > 1$, then $-\alpha = \langle -q_0 - 1; 1, q_1 - 1, q_2, q_3, \ldots \rangle$, and if $q_1 = 1$, then $-\alpha = \langle -q_0 - 1; q_2 + 1, q_3, q_4, \ldots \rangle$.

5.8. Suppose that $\alpha > 1$ is irrational and C_j is the jth convergent in the infinite simple continued fraction expansion of α. Prove that the $(j+1)$st convergent of $1/\alpha$ is $1/C_j$.

5.9. Suppose that α is irrational and C_j, for $j \in \mathbb{N}$, is the jth convergent in the infinite simple continued fraction expansion of α. Prove that for any $j \in \mathbb{N}$, at least one of $k \in \{j, j+1\}$ satisfies

$$\left| \alpha - \frac{A_k}{B_k} \right| < \frac{1}{2B_k^2}.$$

(*Hint: Use the fact established in the proof of Lemma 5.1 on page 216, that $\alpha - A_j/B_j$ and $\alpha - A_{j+1}/B_{j+1}$ have opposite sign.*)

5.10. Suppose that α is irrational and C_j, for $j \in \mathbb{N}$, is the jth convergent in the infinite simple continued fraction expansion of α. Prove that for any $j \in \mathbb{N}$, at least one of $k \in \{j, j+1, j+2\}$ satisfies $|\alpha - A_k/B_k| < 1/(\sqrt{5}B_k^2)$.

5.11. Let $\alpha > 0$ be irrational. Prove that there are infinitely many rationals r/s such that
$$\left| \alpha - \frac{r}{s} \right| < \frac{1}{\sqrt{5}s^2}.$$

(*Hint: Use Exercise 5.10.*)

This is a result due to Hurwitz.

Biography 5.1 Adolf Hurwitz (1859–1919) *was born in Hildesheim, Lower Saxony, Germany on March 26, 1859. He began his advanced education at the University of Munich in 1877. At Munich he attended lectures by Klein, and later, at the University of Berlin, he attended classes given by Kummer, Weierstrass, and Kronecker. In 1879, Hurwitz returned to Munich to continue work with Klein. Indeed, when Klein moved to the University of Leipzig in October of 1880, Hurwitz followed him and completed his doctorate, on elliptic modular functions, under Klein's supervision in 1881. Hurwitz, for a brief period, was a Privatdozent at the University of Göttingen. In 1884, he accepted an invitation from Lindemann to become an extraordinary Professor at Köningsberg, where he stayed for eight years. Two of his students there were Hilbert and Minkowski. Also, he got married there and the marrriage produced three children. In 1892, Hurwitz was appointed to the chair vacated by Frobenius at Eidgenössische Polytechnikum, Zürich, where he remained for the rest of his life.*

Hurwitz's work included the study of the genus of Riemann surfaces as well as automorphic groups of algebraic Riemann surfaces with genus bigger than 1, which he showed to be finite. He also worked on complex function theory, roots of Bessel functions, Fourier series, and in algebraic number theory. In the latter, he published a paper on the theory of quaternions for which he provided a theory of factorization. In 1896, he applied these ideas to solving the problem of representing integers as sums of four squares, some of which were published posthumously.

Hurwitz's health deteriorated so far that his kidneys became diseased and one was removed in 1905. He died in Zürich on November 18, 1919.

5.12. Let $\alpha = (1 + \sqrt{5})/2$ and let F_j denote the jth Fibonacci number for any $j \in \mathbb{N}$. Prove that if $n \in \mathbb{N}$ is odd, then $F_{n+2}/F_{n+1} - \alpha < 1/(\sqrt{5}F_{n+1}^2)$.

(*Hint: Use Theorem 1.3 on page 4.*)

5.13. If n is even in Exercise 5.12, prove that $\alpha - F_{n+2}/F_{n+1} > 1/(\sqrt{5}F_{n+1}^2)$. Conclude from these two exercises that
$$\left| \frac{F_{n+2}}{F_{n+1}} - \alpha \right| < \frac{1}{\sqrt{5}F_{n+1}^2} \text{ if and only if } n \text{ is odd.}$$

5.2   Periodic Simple Continued Fractions

Sure beauty's empires, like to greater states, have certain period sets, and hidden fates.

From **Sonnet (1646)**
John Suckling (1609–1642), English poet and dramatist

In Example 5.1 on page 216, we saw that

$$(1 + \sqrt{29})/2 = \langle 3; 5, 5, 5, \ldots \rangle$$

as an infinite simple continued fraction expansion. Also, in Exercises 5.1–5.4 on page 219, we witnessed other infinite continued fractions that followed a periodic pattern. Since the partial quotients repeat *ad infinitum*, it would be useful to have a more compact notation for such objects, which have a special name.

Definition 5.1 | **Periodic Simple Continued Fractions**

An infinite simple continued fraction $\alpha = \langle q_0; q_1, q_2, \ldots \rangle$ *is called* periodic *if there exists a nonnegative integer k and a positive integer ℓ such that $q_n = q_{n+\ell}$ for all integers $n \geq k$. We use the notation*

$$\alpha = \langle q_0; q_1, \ldots, q_{k-1}, \overline{q_k, q_{k+1}, \ldots, q_{\ell+k-1}} \rangle,$$

as a convenient abbreviation. The smallest such natural number $\ell = \ell(\alpha)$ is called the period length *of α, and $q_0, q_1, \ldots, q_{k-1}$ is called the* pre-period *of α. If k is the* least *nonnegative integer such that $q_n = q_{n+\ell}$ for all $n \geq k$, then $q_k, q_{k+1}, \ldots, q_{k+\ell-1}$ is called the* fundamental period *of α. If $k = 0$ is the least such value, then α is said to be* purely periodic, *namely*

$$\alpha = \langle \overline{q_0; q_1, \ldots, q_{\ell-1}} \rangle.$$

Thus, for instance, the example given before the definition becomes

$$(1 + \sqrt{29})/2 = \langle 3; \overline{5} \rangle$$

having period length 1. We also have the following illustrations.

Example 5.4 *By Exercise 5.2, $\sqrt{n^2 + 1} = \langle n; \overline{2n} \rangle$ for any $n \in \mathbb{N}$. Here $\ell = k = 1$.*

Example 5.5 *From Exercise 5.3, $\sqrt{n^2 + 2} = \langle n; \overline{n, 2n} \rangle$ for any $n \in \mathbb{N}$. Here $\ell = 2$ and $k = 1$.*

Example 5.6 *By Exercise 5.4, $\sqrt{n^2 + 2n} = \langle n; \overline{1, 2n} \rangle$ for any $n \in \mathbb{N}$. Here $\ell = 2$ and $k = 1$.*

Example 5.7 *We calculate that $\alpha = (1 + \sqrt{3})/2 = \langle \overline{2; 3} \rangle$, where $k = 0$ and $\ell = 2$, so α is purely periodic of period length 2.*

The examples given here suggest a pattern that we now formalize.

Definition 5.2 | Quadratic Irrationals

$\alpha \in \mathbb{R}$ is called a quadratic irrational *if it is an irrational number, which is the root of $f(x) = ax^2 + bx + c$ where $a, b, c \in \mathbb{Z}$, and $a \neq 0$.*

Remark 5.1 *What Definition 5.2 says is that there are integers a, b, c with $a \neq 0$ such that*

$$a\alpha^2 + b\alpha + c = 0.$$

For instance, the example which we cited at the outset of this section, $\alpha = (1 + \sqrt{29})/2$ is a root of

$$f(x) = x^2 - x - 7,$$

with the other root being $(1 - \sqrt{29})/2$. Also, in this example we see that the coefficient of x is $-1 = \alpha + \alpha' = -Tr(\alpha)$, where $Tr(\alpha)$ is called the trace *of α and the constant term is $\alpha\alpha' = N(\alpha)$, called the* norm *of α. This is a general fact, as is the form of $\alpha = (P + \sqrt{D})/Q$ for integers P, Q, D with $D > 0$ not a perfect square and $Q \neq 0$.*

Theorem 5.9 | The Form of a Quadratic Irrational

If $\alpha \in \mathbb{R}$, then α is a quadratic irrational if and only if there exist $P, Q, D \in \mathbb{Z}$ such that $Q \neq 0$, $D > 0$ is not a perfect square, and

$$\alpha = \frac{P + \sqrt{D}}{Q}. \tag{5.11}$$

Moreover, if α is a quadratic irrational, then $Q \mid (D - P^2)$, and both α and its conjugate,

$$\alpha' = (P - \sqrt{D})/Q,$$

are the roots of

$$f(x) = x^2 - Tr(\alpha)x + N(\alpha). \tag{5.12}$$

Proof. If α is a quadratic irrational, then there are integers a, b, c with $a \neq 0$ such that $a\alpha^2 + b\alpha + c = 0$. Therefore, by the quadratic formula,

$$\alpha = \frac{-b \pm \sqrt{b^2 - 4ac}}{2a}.$$

Since $\alpha \in \mathbb{R}$, and $\alpha \notin \mathbb{Q}$, then $b^2 - 4ac > 0$ and $b^2 - 4ac$ is not a perfect square. Let $P = \pm b$, $D = b^2 - 4ac$, and $Q = \pm 2a$, then α is of the form given in (5.11). Also, $P^2 - D = 4ac$ is divisible by Q.

Conversely, if α is of the form given in (5.11), then α is irrational since $D > 0$ is not a perfect square. To see this, if $(P + \sqrt{D})/Q = A/B \in \mathbb{Q}$, then $\sqrt{D} = QA/B - P = E/F \in \mathbb{Q}$, so $D = E^2/F^2$. Also,

$$Q^2\alpha^2 - 2PQ\alpha + (P^2 - Q) = 0,$$

so α is a quadratic irrational.

For the last assertion, if $\alpha = (P + \sqrt{D})/Q$, then

$$\alpha^2 - Tr(\alpha)\alpha + N(\alpha) = \frac{P^2 + D + 2\sqrt{D}}{Q^2} - \left(\frac{2P}{Q}\right)\left(\frac{P + \sqrt{D}}{Q}\right) + \frac{P^2 - D}{Q^2} =$$

$$\frac{P^2 + D + 2\sqrt{D} - 2P^2 - 2\sqrt{D} + P^2 - D}{Q^2} = 0,$$

and similarly α' is a root of (5.12), as required. $\qquad\square$

The following provides us with an algorithm for finding the infinite simple continued fraction expansion of quadratic irrationals.

Theorem 5.10 | **Algorithm for Quadratic Irrationals**

Let $\alpha = (P_0 + \sqrt{D})/Q_0$ *be a quadratic irrational, where* $D > 0$ *is not a perfect square* Q_0 *is a nonzero integer,* $P_0 \in \mathbb{Z}$ *and* $Q_0 \mid (D - P_0^2)$. *Recursively define for any* $j \geq 0$,

$$\alpha_j = (P_j + \sqrt{D})/Q_j,$$

$$q_j = \lfloor \alpha_j \rfloor,$$

$$P_{j+1} = q_j Q_j - P_j,$$

$$Q_{j+1} = (D - P_{j+1}^2)/Q_j.$$

Then

$$\alpha = \langle q_0; q_1, q_2, \ldots \rangle.$$

Proof. We use induction on j. Since α is a quadratic irrational, then the induction step is provided for showing that all P_j, Q_j are integers with $Q_j \neq 0$ and $Q_j \mid (P_j^2 - D)$. Now we assume the latter as the induction hypothesis, and prove the result for $j + 1$. We have $P_{j+1} = q_j Q_j - P_j \in \mathbb{Z}$ by the induction hypothesis. Moreover,

$$Q_{j+1} = \frac{D - P_{j+1}^2}{Q_j} = \frac{D - (q_j Q_j - P_j)^2}{Q_j} = \frac{D - P_j^2}{Q_j} + 2q_j P_j - q_j^2 Q_j,$$

which is an integer since $Q_j \mid (D - P_j^2)$ by the induction hypothesis. Moreover, since D is not a perfect square, then $D \neq P_{j+1}^2$, so $Q_{j+1} = (D - P_{j+1}^2)/Q_j \neq 0$.

Also, since $Q_j = (D - P_{j+1}^2)/Q_{j+1}$, then $Q_{j+1} \mid (D - P_{j+1}^2)$. This completes our induction.

To complete the proof, we need to exhibit the q_j as the partial quotients in the simple continued fraction expansion of α. To this end, we employ Theorem 5.6 on page 214, which allows us to make the desired conclusion if we can demonstrate that $\alpha_{j+1} = 1/(\alpha_j - q_j)$. We have,

$$\alpha_j - q_j = \frac{P_j + \sqrt{D}}{Q_j} - q_j = \frac{\sqrt{D} - (q_j Q_j - P_j)}{Q_j} = \frac{\sqrt{D} - P_{j+1}}{Q_j} =$$

$$\frac{(\sqrt{D} - P_{j+1})(\sqrt{D} + P_{J+1})}{Q_j(\sqrt{D} + P_{J+1})} = \frac{D - P_{j+1}^2}{Q_j(\sqrt{D} + P_{j+1})},$$

and since $(D - P_{j+1}^2) = Q_j Q_{j+1}$, then the latter equals

$$\frac{Q_j Q_{j+1}}{Q_j(\sqrt{D} + P_{j+1})} = \frac{Q_{j+1}}{\sqrt{D} + P_{j+1}} = \frac{1}{\alpha_{j+1}}.$$

Thus, we have completed the proof that

$$\alpha = \langle q_0; q_1, q_2, \ldots \rangle,$$

as required. $\qquad\qquad\qquad\qquad\qquad\qquad\qquad\qquad\qquad\qquad\qquad\qquad\square$

Example 5.8 *Let $\alpha = (4 + \sqrt{22})/6 = \alpha_0$. Then $P_0 = 4$, $Q_0 = 6$, $q_0 = \lfloor \alpha_0 \rfloor = \lfloor (4 + \sqrt{22})/6 \rfloor = 1$, and*

$$P_1 = q_0 Q_0 - P_0 = 1 \cdot 6 - 4 = 2, \quad Q_1 = \frac{D - P_1^2}{Q_0} = \frac{22 - 4}{6} = 3,$$

$$\alpha_1 = \frac{2 + \sqrt{22}}{3}, \qquad q_1 = \lfloor \alpha_1 \rfloor = 2,$$

$$P_2 = q_1 Q_1 - P_1 = 2 \cdot 3 - 2 = 4, \quad Q_2 = \frac{D - P_2^2}{Q_1} = \frac{22 - 16}{3} = 2,$$

$$\alpha_2 = \frac{4 + \sqrt{22}}{2}, \qquad q_2 = \lfloor \alpha_2 \rfloor = 4,$$

$$P_3 = q_2 Q_2 - P_2 = 4 \cdot 2 - 4 = 4, \quad Q_3 = \frac{D - P_3^2}{Q_2} = \frac{22 - 16}{2} = 3,$$

$$\alpha_3 = \frac{4 + \sqrt{22}}{3}, \qquad q_3 = \lfloor \alpha_3 \rfloor = 2,$$

$$P_4 = q_3 Q_3 - P_3 = 2 \cdot 3 - 4 = 2, \quad Q_4 = \frac{D - P_4^2}{Q_3} = \frac{22 - 4}{3} = 6,$$

$$\alpha_4 = \frac{2 + \sqrt{22}}{6}, \qquad q_4 = \lfloor \alpha_4 \rfloor = 1,$$

$$P_5 = q_4 Q_4 - P_4 = 1 \cdot 6 - 2 = 4, \quad Q_5 = \frac{D - P_4^2}{Q_4} = \frac{22 - 16}{6} = 1,$$

$$\alpha_5 = \frac{4 + \sqrt{22}}{1}, \quad q_5 = \lfloor \alpha_5 \rfloor = 8,$$

$$P_6 = q_5 Q_5 - P_5 = 8 \cdot 1 - 4 = 4, \quad Q_6 = \frac{D - P_6^2}{Q_5} = \frac{22 - 16}{1} = 6,$$

$$\alpha_6 = \frac{4 + \sqrt{22}}{6}, \quad q_6 = \lfloor \alpha_6 \rfloor = 1,$$

and the process repeats since $P_6 = P_0$ and $Q_6 = Q_0$. Thus, we have the purely periodic infinite simple continued fraction expansion with period length $\ell = 6$, given by

$$\alpha = \langle \overline{1; 2; 4; 2; 1; 8} \rangle.$$

Before we characterize quadratic irrationals in terms of continued fractions, we need the following technical result.

Lemma 5.2 *If α is a quadratic irrational, $a, b, c, d \in \mathbb{Z}$, and if $(a\alpha + b)/(c\alpha + d)$ is not rational, then $(a\alpha + b)/(c\alpha + d)$ is a quadratic irrational.*

Proof. By Theorem 5.9 on page 222, there exist $P, D, Q \in \mathbb{Z}$ with $D > 0$ not a perfect square, and $Q \neq 0$, such that

$$\alpha = \frac{P + \sqrt{D}}{Q}.$$

Therefore,

$$\frac{a\alpha + b}{c\alpha + d} = \frac{a\frac{P+\sqrt{D}}{Q} + b}{c\frac{P+\sqrt{D}}{Q} + d} = \frac{(aP + bQ) + a\sqrt{D}}{(cP + dQ) + c\sqrt{D}} =$$

$$\frac{[(aP + bQ) + a\sqrt{D}][(cP + dQ) - c\sqrt{D}]}{[(cP + dQ) + c\sqrt{D}][(cP + dQ) - c\sqrt{D}]} =$$

$$\frac{[(aP + bQ)(cP + dQ) - acD] + [a(cP + dQ) - c(aP + bQ)]\sqrt{D}}{(cP + dQ)^2 - c^2 D}.$$

Thus, by Theorem 5.9, $(a\alpha + b)/(c\alpha + d) \notin \mathbb{Q}$ is a quadratic irrational. $\square$

The following characterization is due to Lagrange — see Biography 2.7 on page 114.

Theorem 5.11 | **Quadratic Irrationals Are Periodic**

Let $\alpha \in \mathbb{R}$. Then α has a periodic infinite simple continued fraction expansion if and only if α is a quadratic irrational.

Proof. First suppose that α has a periodic infinite simple continued fraction expansion given by

$$\alpha = \langle q_0; q_1, \ldots, q_{k-1}, \overline{q_k, \ldots, q_{k+\ell-1}} \rangle,$$

and set

$$\beta = \langle \overline{q_k; q_{k+1}, \ldots, q_{k+\ell-1}} \rangle.$$

Then $\beta = \langle q_k; q_{k+1}, \ldots, q_{k+\ell-1}, \beta \rangle$, so, by Theorem 5.5 on page 214,

$$\beta = \frac{A_{\ell-1}\beta + A_{\ell-2}}{B_{\ell-1}\beta + B_{\ell-2}}, \tag{5.12}$$

where A_j/B_j $(j = \ell - 1, \ell - 2)$ are convergents of β. Cross-multiplying in Equation (5.12), we get

$$B_{\ell-1}\beta^2 + (B_{\ell-2} - A_{\ell-1})\beta - A_{\ell-2} = 0,$$

so β is a quadratic irrational. However, by Theorem 5.5 again,

$$\alpha = \langle q_0; q_1, \ldots, q_{k-1}, \beta \rangle = \frac{A_{k-1}\beta + A_{k-2}}{B_{k-1}\beta + B_{k-2}},$$

where A_j/B_j for $j = k - 1, k - 2$ are convergents of $\langle q_0; q_1, \ldots, q_{k-1} \rangle$. Since α has an infinite simple continued fraction expansion, $\alpha \notin \mathbb{Q}$. Thus, by Lemma 5.2, α is a quadratic irrational since β is one.

Conversely if α is a quadratic irrational, then by Theorem 5.9,

$$\alpha_o = \frac{P_0 + \sqrt{D}}{Q_0}, \tag{5.13}$$

and by Theorem 5.10,

$$\alpha = \langle q_0; q_1, \ldots \rangle, \tag{5.14}$$

where

$$\alpha_j = (P_j + \sqrt{D})/Q_j, \tag{5.15}$$

$$q_j = \lfloor \alpha_j \rfloor, \tag{5.16}$$

$$P_{j+1} = q_j Q_j - P_j, \tag{5.17}$$

$$Q_{j+1} = (D - P_{j+1}^2)/Q_j. \tag{5.18}$$

Since $\alpha = \langle q_0; q_1, \ldots, \alpha_j \rangle$, then by Theorem 5.5,

$$\alpha = \frac{A_{j-1}\alpha_j + A_{j-2}}{B_{j-1}\alpha_j + B_{j-2}}.$$

Taking conjugates of both sides and employing the basic facts in Exercises 5.14–5.15 on page 230, we get,

$$\alpha' = \frac{A_{j-1}\alpha_j' + A_{j-2}}{B_{j-1}\alpha_j' + B_{j-2}}.$$

Rewriting this equation so that we solve for α'_j, we have that

$$\alpha'_j = -\frac{B_{j-2}}{B_{j-1}}\left(\frac{\alpha' - \frac{A_{j-2}}{B_{j-2}}}{\alpha' - \frac{A_{j-1}}{B_{j-1}}}\right) = -\frac{B_{j-2}}{B_{j-1}}\left(\frac{\alpha' - C_{j-2}}{\alpha' - C_{j-1}}\right).$$

However, by Theorem 5.3 on page 211,

$$\lim_{j\to\infty}\left(\frac{\alpha' - C_{j-2}}{\alpha' - C_{j-1}}\right) = \left(\frac{\alpha' - \lim_{j\to\infty}C_{j-2}}{\alpha' - \lim_{j\to\infty}C_{j-1}}\right) = \left(\frac{\alpha' - \alpha}{\alpha' - \alpha}\right) = 1,$$

since $\alpha \neq \alpha'$. Thus, for sufficiently large j, say $j > M \in \mathbb{N}$ where M is fixed, $\alpha' < 0$ and $(\alpha' - C_{j-2})/(\alpha' - C_{j-1}) > 0$, but we know that $\alpha_k > 0$ for any $k \in \mathbb{N}$, so $\alpha_k - \alpha'_k > 0$ for any $k > M$. Therefore, for any $k > M$,

$$\alpha_k - \alpha'_k = \frac{P_k + \sqrt{D}}{Q_k} - \frac{P_k - \sqrt{D}}{Q_k} = \frac{2\sqrt{D}}{Q_k} > 0,$$

so $Q_k > 0$ for any $k > M$. Also, using (5.18),

$$Q_k \leq Q_k Q_{k+1} = D - P_{k+1}^2 \leq D,$$

and

$$P_{k+1}^2 < P_{k+1}^2 + Q_k Q_{k+1} = D,$$

so $|P_{k+1}| < \sqrt{D}$ and $0 < Q_k \leq D$ for $k > M$. Since $D \in \mathbb{N}$ is fixed, then Q_k and P_{k+1} can take on only finitely many possible values for $k > M$. Therefore, (P_k, Q_k) can assume only finitely many values for $k > M$. Hence, there exist $i, j \in \mathbb{Z}$ such that $P_i = P_j$ and $Q_i = Q_j$ with $i < j$. It follows from (5.13)–(5.18) that,

$$q_i = q_j, \quad q_{i+1} = q_{j+1}, \ldots,$$

so

$$\alpha = \langle q_0; q_1, \ldots, q_{k-1}, \overline{q_k, \ldots, q_{k+\ell-1}}\rangle,$$

and the proof is secured. $\square$

Using the notation of Theorem 5.10 on page 223, implicit in the proof of Theorem 5.11 is the following.

Corollary 5.5 *If* $\alpha_0 = \alpha = \langle q_0; q_1, \ldots, q_{k-1}, \overline{q_k, \ldots, q_{k+\ell-1}}\rangle$ *is a quadratic irrational, then for all* $j \geq 0$, $\alpha_{\ell j+n} = \alpha_n$ *for all* $n \geq k$.

In Example 5.8 on page 224, we saw a quadratic irrational that is purely periodic. These quadratic irrationals may also be classified. First we need a new notion.

Definition 5.3 | **Purely Periodicity and Reduction** |

A *quadratic irrational* α *is called* reduced *if both* $\alpha > 1$ *and* $-1 < \alpha' < 0$ *where* α' *is the conjugate of* α.

Theorem 5.12 | **Pure Periodicity Equals Reduction** |

The simple continued fraction expansion of a quadratic irrational α is purely periodic if and only if α is reduced.

Proof. Assume that α is purely periodic, namely $\alpha = \langle \overline{q_0; q_1, \ldots, q_{\ell-1}} \rangle$. Since $q_0 > 0$, then $\alpha > 1$. Also, since $\alpha = \langle q_0; q_1, q_2, \ldots, q_{\ell-1}, \alpha \rangle$, then by Theorem 5.5,

$$\alpha = \frac{\alpha A_{\ell-1} + A_{\ell-2}}{\alpha B_{\ell-1} + B_{\ell-2}},$$

where A_j/B_j are the jth convergents of α for $j = \ell - 1, \ell - 2$. By Rewriting, we get,

$$B_{\ell-1}\alpha^2 + (B_{\ell-2} - A_{\ell-1})\alpha - A_{\ell-2} = 0. \qquad (5.19)$$

Let β be the quadratic irrational $\beta = \langle \overline{q_{\ell-1}, q_{\ell-2}, \ldots, q_1, q_0} \rangle$. Then

$$\beta = \langle q_{\ell-1}, q_{\ell-2}, \ldots, q_1, q_0, \beta \rangle,$$

so by Theorem 5.5 again,

$$\beta = \frac{\beta A'_{\ell-1} + A'_{\ell-2}}{\beta B'_{\ell-1} + B'_{\ell-2}}, \qquad (5.20)$$

where A'_j/B'_j are the jth convergents of β for $j = \ell - 1, \ell - 2$. Now we need the following.

Claim 5.2 *Since $q_0 > 0$, then*

$$A_{\ell-1}/A_{\ell-2} = \langle q_{\ell-1}; q_{\ell-2}, \ldots, q_1, q_0 \rangle$$

and

$$B_{\ell-1}/B_{\ell-2} = \langle q_{\ell-1}; q_{\ell-2}, \ldots, q_2, q_1 \rangle.$$

We use induction on ℓ. If $\ell = 1$, then $A_0/A_{-1} = q_0/1 = q_0 = \langle q_0 \rangle$. Assume that $A_{\ell-1}/A_{\ell-2} = \langle q_{\ell-1}; q_{\ell-2}, \ldots, q_1, q_0 \rangle$. Then

$$\frac{A_\ell}{A_{\ell-1}} = \frac{q_\ell A_{\ell-1} + A_{\ell-2}}{A_{\ell-1}} = q_\ell + \frac{A_{\ell-2}}{A_{\ell-1}} = q_\ell + \frac{1}{\langle q_{\ell-1}; q_{\ell-2}, \ldots, q_1, q_0 \rangle} =$$

$$\langle q_\ell; q_{\ell-1}, q_{\ell-2}, \ldots, q_1, q_0 \rangle,$$

and a similar argument holds for $B_{\ell-1}/B_{\ell-2}$. This establishes the claim.

By Claim 5.2,

$$\frac{A_{\ell-1}}{A_{\ell-2}} = \langle q_{\ell-1}; q_{\ell-2}, \ldots, q_1, q_0 \rangle = \frac{A'_{\ell-1}}{B'_{\ell-1}}$$

and

$$\frac{B_{\ell-1}}{B_{\ell-2}} = \langle q_{\ell-1}; q_{\ell-2}, \ldots, q_2, q_1 \rangle = \frac{A'_{\ell-2}}{B'_{\ell-2}}.$$

However, by Excise 1.33 on page 29, $\gcd(A_{\ell-1}, A_{\ell-2}) = \gcd(A'_{\ell-1}, B'_{\ell-1}) = \gcd(B_{\ell-1}, B_{\ell-2}) = \gcd(A'_{\ell-2}, B'_{\ell-2}) = 1$, so

$$A_{\ell-1} = A'_{\ell-1}, \quad A_{\ell-2} = B'_{\ell-1}, \quad B_{\ell-1} = A'_{\ell-2}, \quad B_{\ell-2} = B'_{\ell-2}.$$

Putting these values into (5.20), we get,

$$\beta = \frac{\beta A_{\ell-1} + B_{\ell-1}}{\beta A_{\ell-2} + B_{\ell-2}}.$$

Rewriting the latter yields,

$$A_{\ell-2}\beta^2 + (B_{\ell-2} - A_{\ell-1})\beta - B_{\ell-1} = 0,$$

and multiplying this through by $-1/\beta^2$, gives

$$B_{\ell-1}(-1/\beta)^2 + (B_{\ell-2} - A_{\ell-1})(-1/\beta) - A_{\ell-2} = 0. \tag{5.21}$$

Hence, by Equations (5.19) and (5.21), we know that the two roots of

$$f(x) = B_{\ell-1}x^2 + (B_{\ell-2} - A_{\ell-1})x - A_{\ell-2}$$

are α and $-1/\beta$. Thus, by Theorem 5.9 on page 222,

$$\alpha' = -1/\beta, \text{ where } \beta = \langle \overline{q_{\ell-1}, q_{\ell-2}, \ldots, q_1, q_0} \rangle. \tag{5.22}$$

Thus, $\beta > 1$, so $-1 < \alpha' = -1/\beta < 0$, which means that α is a reduced quadratic irrational.

Conversely, assume that α is a reduced quadratic irrational. By Theorem 5.10 on page 223 and its proof, we have that the partial quotients of α are given, for $j \geq 0$, by

$$q_j = \lfloor \alpha_j \rfloor, \text{ where } \alpha_{j+1} = 1/(\alpha_j - q_j).$$

Taking conjugates and using Exercises 5.14–5.15 on page 230, we get

$$1/\alpha'_{j+1} = \alpha'_j - q_j. \tag{5.23}$$

Claim 5.3 $-1 < \alpha'_j < 0$ *for* $j \geq 0$.

Since $\alpha = \alpha_0$ is reduced, then $-1 < \alpha'_j < 0$, which is the induction step. Assume that $-1 < \alpha'_j < 0$. Since $\alpha \in \mathbb{N}$, given that $\alpha > 1$, then $q_j \in \mathbb{N}$ for $j \geq 0$. Therefore, (5.23) tells us that $1/\alpha'_{j+1} < -1$. Thus, $-1 < \alpha'_{j+1} < 0$, which secures the claim.

By (5.23), $\alpha'_j = q_j + 1/\alpha'_{j+1}$, and by Claim 5.3,

$$-1 < q_j + 1/\alpha'_{j+1} < 0.$$

By subtracting through by $1/\alpha'_{j+1}$, this becomes,

$$-1 - 1/\alpha'_{j+1} < q_j < -1/\alpha'_{j+1},$$

which says that
$$q_j = \lfloor -1/\alpha'_{j+1} \rfloor.$$

Now by Theorem 5.11 on page 225, there are integers i, k such that $\alpha_i = \alpha_k$ where $0 < i < k$. Thus, $\alpha'_i = \alpha'_k$ and

$$q_{i-1} = \lfloor -1/\alpha'_i \rfloor = \lfloor -1/\alpha'_k \rfloor = q_{k-1},$$

$$\alpha_{i-1} = q_{i-1} + 1/\alpha_i = q_{k-1} + 1/\alpha_k = \alpha_{k-1}.$$

We have shown that $\alpha_i = \alpha_k$ imples that $\alpha_{i-1} = \alpha_{k-1}$. Performing this argument i times gives us that $\alpha_0 = \alpha_{k-i}$, so

$$\alpha = \alpha_0 = \langle \overline{q_0; q_1, \ldots, q_{k-i-1}} \rangle,$$

which secures the proof. $\square$

Contained within the proof of Theorem 5.12 is the following.

Corollary 5.6 *Let α be a reduced quadratic irrational with continued fraction expansion $\alpha = \langle \overline{q_0; q_1, \ldots q_{\ell-1}} \rangle$. Then*

$$-1/\alpha' = \langle \overline{q_{\ell-1}; q_{\ell-2}, \ldots, q_0} \rangle.$$

Proof. This comes from (5.22). $\square$

Example 5.9 *Example 5.8 on page 224 gave us an instance of purely periodic continued fraction expansions, namely*

$$\alpha = \langle \overline{1; 2; 4; 2; 1; 8} \rangle.$$

Thus, by Corollary 5.6,

$$\frac{1}{\alpha'} = \langle \overline{8; 1; 2; 4; 2; 1} \rangle.$$

In §5.3, we will look more closely at the continued fraction expansions of quadratic irrationals, especially *surds,* namely those of the form $\sqrt{D}$ for $D \in \mathbb{N}$ not a perfect square, and the relationship with solutions of Pell's equations $x^2 - Dy^2 = \pm 1$.

Exercises

5.14. Prove that if $\alpha_j = (P_j + \sqrt{D})/Q_j$ are quadratic irrationals for $j = 1, 2$, then $(\alpha_1 \pm \alpha_2)' = \alpha'_1 \pm \alpha'_2$.

5.15. Prove that if $\alpha_j = (P_j + \sqrt{D})/Q_j$ are quadratic irrationals for $j = 1, 2$, then $(\alpha_1 \alpha_2)' = \alpha'_1 \alpha'_2$, and $(\alpha_1/\alpha_2)' = \alpha'_1/\alpha'_2$.

(*Note that Exercises 5.14–5.15 hold for the case where $\sqrt{D}$ has a coefficient bigger than 1 since $\alpha = (P + R\sqrt{D})/Q = (P + \sqrt{R^2 D})/Q$.*)

5.16. If $a, b \in \mathbb{Z}$ with $0 < b < a$ and a/b is not a perfect square, prove that there exists an $\ell \in \mathbb{N}$ such that

$$\sqrt{\frac{a}{b}} = \langle q_0; \overline{q_1, q_2, \ldots, q_{\ell-1}, 2q_0} \rangle.$$

(*Hint: Apply Theorem 5.12 on page 228 to* $\lfloor \sqrt{ab}/b \rfloor + \sqrt{ab}/b$.)

The first proof of this result was given by Lagrange in 1770 — see Biography 2.7 on page 114.

5.17. With reference to Exercise 5.16, prove that if $\ell > 1$, then $q_j = q_{\ell-j}$ for all $j = 1, 2, \ldots, \ell - 1$.

(*Hint: Apply Corollary 5.6 on the preceding page to* $\lfloor \sqrt{ab}/b \rfloor + \sqrt{ab}/b$.)

The first proof of this result was given by Gauss in 1828 — see Biography 1.7 on page 33. What the result says is that $q_1, q_2, \ldots, q_\ell$ is a *palindrome*, namely it reads the same forward or backward.

Use Exercises 5.16–5.17 to find the solutions to Exercises 5.18–5.21.

5.18. Find the simple continued fraction expansions of the following.

(a) $\sqrt{65}$. (b) $\sqrt{89}$.

5.19. Find the simple continued fraction expansions of the following.

(a) $\sqrt{17}$. (b) $\sqrt{29}$.

5.20. Find the quadratic irrational from the given continued fraction expansion given in the following.

(a) $\langle 2; \overline{12, 4} \rangle$. (b) $\langle 1; \overline{1, 5, 2, 2, 5, 1, 2} \rangle$.

5.21. Find the quadratic irrational from the given continued fraction expansion given in the following.

(a) $\langle 3; \overline{9, 6} \rangle$. (b) $\langle 3; \overline{1, 14, 1, 6} \rangle$.

5.22. Establish each of the following using (a) to prove (b).

(a) Prove that a quadratic irrational $\alpha = (P + \sqrt{D})/Q$ is reduced if and only if $0 < P < \sqrt{D}$ and $\sqrt{D} - P < Q < \sqrt{D} + P < 2\sqrt{D}$.

(b) Prove that if α is a reduced quadratic irrational, then so is $-1/\alpha'$.

5.23. With reference to Exercise 5.16, let $D \in \mathbb{N}$ not a perfect square with

$$\sqrt{D} = \langle q_0; \overline{q_1, q_2, \ldots, q_{\ell-1}, 2q_0} \rangle,$$

where $\ell \in \mathbb{N}$ is the period of $\sqrt{D}$. In the notation of Theorem 5.10, prove that $Q_j = 1$ for any $j \in \mathbb{N}$ if and only if $j \equiv 0 \pmod{\ell}$.

5.3   Pell's Equation and Surds

> *What is it that breathes fire into the equations and makes a universe for them*
> *to describe... Why does the universe go to all the bother of existing?*
>   From **A Brief History of Time** (1988)
>   **Stephen Hawking (1942–), English theoretical physicist**

In this section we will be concentrating on the Diophantine equation

$$x^2 - Dy^2 = n \qquad (5.24)$$

for integers D and n where D is not a perfect square. This equation is called a
norm-form equation, since

$$N(x + y\sqrt{D}) = x^2 - Dy^2 = n$$

is called the norm of $x + y\sqrt{D}$ — see also Remark 5.1 on page 222.
When $n = 1$, (5.24) is known as *Pell's
equation* after John Pell, who actually had little to do with its solution.
The misattribution is often said to
have been made by Euler who mistakenly attributed work of Brouncker on
the equation to Pell. The Pell equation has a long and distinguished history going back to Archimedes and
Diophantus — see Biography 1.15 on
page 48.   Also, the Hindu mathematician, Brahmagupta first studied
this equation for the case where D
is a prime — see Biography 1.11 on
page 43.   Although there were attempts to find the infinitely many solutions of the positive Pell equation,
posed in a letter by Fermat in 1657, it
took until 1768, when Lagrange published a proof. He based his proof on
the theory set out in a paper published by Euler in 1767 — see Biographies 1.17 on page 56 and 2.7 on
page 114. Lagrange's method for finding the solutions was based on the
employment of the simple continued
fraction expansion of $\sqrt{D}$, as had earlier incomplete attempts. In order to
present this elegant resolution of the

Biography 5.2 John Pell (1611–1685)
*was born on March 1, 1611 in South-
wick, Sussex, England.   He entered
Trinity College, Cambridge in 1624, re-
ceiving his Bachelor's degree in 1628
and his Master's degree in 1630. After graduating from Cambridge, he became a schoolmaster, and taught mathematics from 1638 to 1643. In 1638,
he published the* Idea of Mathematics,
*which was essentially a list of tasks that
Pell felt were necessary for the progress
of mathematics in England.   He also
worked on algebra, number theory, and
astronomy. His first posting as a Professor of Mathematics was at the Gymnasium Illustre in Amsterdam, where
he stayed until 1646, when he was appointed to a similar position at the University of Breda. In 1661, he finally followed his clergyman father's wish and
became chaplain to the Bishop of London. However, his life deteriorated, he
descended into abject poverty, and died
as such on December 12, 1685 in West-
minster, London.*

problem, we first need the following result that builds upon the results we have developed thus far in this chapter.

Theorem 5.13 | **Convergents as Solutions of Norm-Form Equations**

Let $D, n \in \mathbb{Z}$ where $D > 0$ is not a perfect square and $|n| < \sqrt{D}$. If there exist $x, y \in \mathbb{N}$ such that Equation (5.24) is satisfied, then x/y is a convergent in the simple continued fraction expansion of $\sqrt{D}$.

Proof. First we consider the case where

$$\sqrt{D} > n > 0. \tag{5.25}$$

We have the norm-form,

$$N(x + y\sqrt{D}) = (x + y\sqrt{D})(x - y\sqrt{D}) = n > 0,$$

so we must have that $x - y\sqrt{D} > 0$, namely

$$x > y\sqrt{D}. \tag{5.26}$$

Therefore

$$\frac{x}{y} - \sqrt{D} = \frac{x - y\sqrt{D}}{y} = \frac{x^2 - Dy^2}{y(x + y\sqrt{D})} < \frac{n}{y(2y\sqrt{D})} < \frac{\sqrt{D}}{2y^2\sqrt{D}} = \frac{1}{2y^2},$$

where the penultimate inequality follows from (5.26) and the last inequality follows from (5.25). By setting $\alpha = \sqrt{D}$, $r = x$, and $s = y$ in Theorem 5.8 on page 218, we may conclude that x/y is a convergent in the simple continued fraction expansion of $\sqrt{D}$.

Now we handle the case where $n < 0$. We may rewrite (5.24) as

$$y^2 - \frac{x^2}{D} = \frac{-n}{D},$$

and we may now apply the above argument in the same fashion to conclude that y/x is a convergent in the simple continued fraction expansion of $1/\sqrt{D}$. By Exercise 5.8 on page 219, x/y is such a convergent for $\sqrt{D}$. ☐

For the next result, which is also a precursor for the solution of Pell's equations, the reader should be familiar with the notation in Theorem 5.10 on page 223. The following looks at convergence of surds in relation to quadratic Diophantine equations.

Note that a *surd* is an irrational of the form $\sqrt{D}$. The original meaning of surd was *mute*, or *voiceless*. Today, in the area of phonetics, *surd* still means an unvoiced (as opposed to a voiced) consonant. In the literature, it has been said that al-Khwârizmî (see Biography 5.3 on page 235) referred to rationals and irrationals as *sounded* and *unsounded*, respectively, in his works. Also, Fibonacci adopted *surd* to mean a number that has no rational root — see Biography 1.1 on page 3.

Theorem 5.14 | Convergents of Surds and Norm-Form Equations |

    Let D be a positive integer that is not a perfect square, and let A_j/B_j be the jth convergent of $\sqrt{D}$. Then

$$A_j^2 - DB_j^2 = (-1)^{j-1}Q_{j+1},$$

for any $j \in \mathbb{N}$.

Proof. By Theorem 5.5 on page 214,

$$\sqrt{D} = \frac{\alpha_{j+1}A_j + A_{j-1}}{\alpha_{j+1}B_j + B_{j-1}}.$$

However, by Theorem 5.9 on page 222,

$$\alpha_{j+1} = \frac{P_{j+1} + \sqrt{D}}{Q_{j+1}},$$

so

$$\sqrt{D} = \frac{A_j(P_{j+1} + \sqrt{D}) + Q_{j+1}A_{j-1}}{B_j(P_{j+1} + \sqrt{D}) + Q_{j+1}B_{j-1}}.$$

By rewriting the latter to solve for coefficients of $\sqrt{D}$, we get,

$$A_jP_{j+1} + Q_{j+1}A_{j-1} - DB_j = (B_jP_{j+1} + Q_{j+1}B_{j-1} - A_j)\sqrt{D}.$$

Since $\sqrt{D}$ is irrational, then we have the following two equations

$$A_jP_{j+1} + Q_{j+1}A_{j-1} - DB_j = 0 \tag{5.27}$$

and

$$B_jP_{j+1} + Q_{j+1}B_{j-1} - A_j = 0. \tag{5.28}$$

Multiplying (5.27) by B_j and subtracting A_j times (5.28), we get,

$$A_j^2 - DB_j^2 = Q_{j+1}(A_jB_{j-1} - B_jA_{j-1}) = Q_{j+1}(-1)^{j-1},$$

where the last equality comes from part (a) of Theorem 5.1 on page 210. $\square$

    Now we are in a position to solve Pell's equations, $x^2 - Dy^2 = \pm 1$. Note that in the modern day, the norm-form equation $x^2 - Dy^2 = n$ for any integer n is often called *Pell's equation*. Earlier comments were on the historical significance of the term. However, it does not harm to reference such equations as such. Again, the reader should be familiar with the notation in Theorem 5.10 on page 223. Also, in what follows, when we speak of *positive solutions* (x, y) to Pell's equation, we mean solutions $x + y\sqrt{D}$ such that $x > 0$ and $y > 0$.

Theorem 5.15 | Solutions of Pell's Equations |

    Let D be a positive integer that is not a perfect square, and let A_j/B_j be the jth convergent in the simple continued fraction expansion of $\sqrt{D}$, that has period length $\ell \in \mathbb{N}$. Then

(a) The positive solutions of $x^2 - Dy^2 = 1$ are

$$(x, y) = \begin{cases} (A_{j\ell-1}, B_{j\ell-1}) \text{ for } j \in \mathbb{N} & \text{if } \ell \text{ is even} \\ (A_{2j\ell-1}, B_{2j\ell-1}) \text{ for } j \in \mathbb{N} & \text{if } \ell \text{ is odd.} \end{cases}$$

(b) The positive solutions of $x^2 - Dy^2 = -1$ are nonexistent when ℓ is even, and if ℓ is odd, they are given by,

$$(x, y) = (A_{(2j-1)\ell-1}, B_{(2j-1)\ell-1}) \text{ for } j \in \mathbb{N}.$$

Proof. By Theorem 5.14,

$$A_j^2 - DB_j^2 = (-1)^{j-1}Q_{j+1},$$

for any $j \in \mathbb{N}$. By the definition of ℓ, we have that $Q_{j\ell} = Q_0 = 1$ for any $j \in \mathbb{N}$. Hence,

$$A_{j\ell-1}^2 - DB_{j\ell-1}^2 = (-1)^{j\ell}Q_{j\ell} = (-1)^{j\ell}.$$

Thus, when ℓ is even $(A_{j\ell-1}, B_{j\ell-1})$ are solutions to $x^2 - Dy^2 = 1$ for any $j \in \mathbb{N}$. When ℓ is odd, $(A_{2j\ell-1}, B_{2j\ell-1})$ are solutions to $x^2 - Dy^2 = 1$ for all $j \in \mathbb{N}$, since $A_{2j\ell-1}^2 - DB_{2j\ell-1}^2 = (-1)^{2j\ell}$. Also, when ℓ is odd, $(A_{(2j-1)\ell-1}, B_{(2j-1)\ell-1})$ are solutions to $x^2 - Dy^2 = -1$ for any $j \in \mathbb{N}$.

It remains to show that there are no more solutions than those given above, which will hold if we can show that $Q_j = 1$ if and only if $\ell \mid j$ and that $Q_j \neq -1$ for all $j \in \mathbb{N}$. The former is the content of Exercise 5.23 on page 231. For the latter, if $Q_j = -1$, then $\alpha_j = (P_j + \sqrt{D})/Q_j = -P_j - \sqrt{D}$. However, $\alpha_j > 1$ is purely periodic, so by Theorem 5.12 on page 228, $1 < \alpha_j = -P_j - \sqrt{D}$ and $-1 < \alpha_j' = -P_j + \sqrt{D} < 0$. Hence, $\sqrt{D} < P_j < -\sqrt{D} - 1$, a contradiction. Therefore, $Q_j \neq -1$ for all nonnegative integers j, given that $Q_0 = 1$. □

Biography 5.3 Abu Ja'Far Muhammed ibn Mûsâ al-Khwârizmî (c. 780–850) *was both an astronomer and a mathematician. In around 825 A.D. he completed a book on arithmetic, which was later translated into Latin in the twelfth century under the title* Algorithmi de numero Indorum. *This book is one of the best-known means by which the Hindu-Arabic number system was introduced to Europe after being introduced into the Arab world (also see Biography 1.1 on page 3). This may account for the widespread, although mistaken, belief that our numerals are Arabic in origin. Not long after Latin translations of his book began appearing in Europe, readers began to attribute the new numerals to al-Khwârizmî and began contracting his name, concerning the use of these numerals, to* algorism, *and ultimately to* algorithm. *Also, al-Khwârizmî wrote a book on algebra,* Hisab al-jabr wa'lmuqābala. *The word algebra is derived from* al-jabr *or* restoration. *In the Spanish work* Don Quixote, *which came much later, the term* algebrist *is used for a bone-setter or restorer. Al-Khwârizmî lived during the caliphate of al-Mamun (809–833 A.D.) who had a vision in which he was visited by Aristotle. After this encounter he was driven to have the Greek classics translated into Arabic. Among them were Ptolemy's* Almagest *and the complete volumes of Euclid's* Elements.

Example 5.10 *Let $D = 425$. Then $\ell = 7$ for $\sqrt{D} = \langle 20; \overline{1,1,1,1,1,1,40} \rangle$. Also, $(x,y) = (A_{2\ell-1}, B_{2\ell-1}) = (A_{13}, B_{13}) = (143649, 6968)$ is the smallest positive solution of $x^2 - Dy^2 = 1$, and $(x,y) = (A_{\ell-1}, B_{\ell-1}) = (A_6, B_6) = (268, 13)$ is the smallest positive solution of $x^2 - Dy^2 = -1$.*

Remark 5.2 *Example 5.10 illustrates a fact that we may isolate as a consequence of Theorem 5.15, namely that if there is a solution (x,y) to a Pell equation, with x the least positive such value, then there is a corresponding $y > 0$ and (x,y) is the least positive solution, also called the fundamental solution.*

Corollary 5.7 *If $D > 0$ is not a perfect square and $\sqrt{D}$ has continued fraction expansion of period length ℓ, then the fundamental solution of $x^2 - Dy^2 = 1$ is given by*

$$(x_1, y_1) = \begin{cases} (A_{\ell-1}, B_{\ell-1}) & \text{if } \ell \text{ is even} \\ (A_{2\ell-1}, B_{2\ell-1}) & \text{if } \ell \text{ is odd.} \end{cases}$$

Moreover, if ℓ is odd, then the fundamental solution of $x^2 - Dy^2 = -1$ is given by

$$(x_1, y_1) = (A_{\ell-1}, B_{\ell-1}).$$

Corollary 5.7 may be employed to generate all the positive solutions of a Pell equation as follows.

Theorem 5.16 | **Generation of Solutions to Pell's Equations**

Suppose that D is a positive integer that is not a perfect square and (x_1, y_1) is the least positive solution to $x^2 - Dy^2 = \pm1$. Then all positive solutions of $x^2 - Dy^2 = 1$ are given by (x_j, y_j) for $j \in \mathbb{N}$, where

$$x_j + y_j\sqrt{D} = \left(x_1 + y_1\sqrt{D}\right)^j.$$

Moreover, if ℓ is odd, then all positive solutions of $x^2 - Dy^2 = -1$ are given by (x_j, y_j) for $j \in \mathbb{N}$, where

$$x_j + y_j\sqrt{D} = \left(x_1 + y_1\sqrt{D}\right)^{2j-1}.$$

Proof. We only prove the result for $x^2 - Dy^2 = 1$, since the other case follows in a similar fashion. We have that,

$$N(x_j + y_j\sqrt{D}) = x_j^2 - y_j^2 D = (x_1 + y_1\sqrt{D})^j(x_1 - y_1\sqrt{D})^j = (x_1^2 - y_1^2 D)^j = 1,$$

so (x_j, y_j) are positive solutions of $x^2 - Dy^2 = 1$. Suppose that there is a positive solution $r + s\sqrt{D}$ that is not of the form $x_j + y_j\sqrt{D}$. Since $x_j + y_j\sqrt{D} > x_{j-1} + y_{j-1}\sqrt{D} > 1$ for any $j > 1$, then the solutions become arbitrarily large as j increases. Hence, there must be a value of $j \in \mathbb{N}$ such that $(x_1 + y_1\sqrt{D})^j < r + s\sqrt{D} < (x_1 + y_1\sqrt{D})^{j+1}$. Therefore,

$$(x_1 + y_1\sqrt{D})^j < r + s\sqrt{D} < (x_1 + y_1\sqrt{D})^j(x_1 + y_1\sqrt{D}),$$

so multiplying through by the conjugate $x_j - y_j\sqrt{D}$, we get,

$$1 < (x_j - y_j\sqrt{D})(r + s\sqrt{D}) < x_1 + y_1\sqrt{D}.$$

Now let $u = rx_j - sy_jD$ and $v = sx_j - ry_j$. Then

$$u^2 - v^2D = (x_j^2 - Dy_j^2)(r^s - Ds^2) = 1.$$

We have produced a solution of $x^2 - Dy^2 = 1$ with

$$1 < u + v\sqrt{D} < x_1 + y_1\sqrt{D}. \tag{5.29}$$

Moreover, since $u + v\sqrt{D} > 1$ and $(u + v\sqrt{D})(u - v\sqrt{D}) = 1$, then $0 < u - v\sqrt{D} < 1$. Hence,

$$2u = (u + v\sqrt{D}) + (u - v\sqrt{D}) > 1 + 0 > 0,$$

and

$$2v\sqrt{D} = (u + v\sqrt{D}) - (u - v\sqrt{D}) > 1 - 1 = 0,$$

which shows that $u + v\sqrt{D}$ is a positive solution of $x^2 - Dy^2 = 1$. However, since $x_1 + y_1\sqrt{D}$ is the fundamental solution, then $x_1 < u$ and $y_1 < v$, which implies that

$$x_1 + y_1\sqrt{D} < u + v\sqrt{D},$$

but this contradicts (5.29), so the proof is complete. □

Example 5.11 *If $D = 19$, then we calculate that $\sqrt{D} = \langle 4; \overline{2,1,3,1,2,8} \rangle$ of period length $\ell = 6$, and*

$$x_1 + y_1\sqrt{D} = A_5 + B_5\sqrt{D} = A_{\ell-1} + B_{\ell-1}\sqrt{D} = 170 + 39\sqrt{19}$$

is the fundamental solution of $x^2 - Dy^2 = 1$. Moreover, all positive solutions of the latter are given by $x_j + y_j\sqrt{19} = (170 + 39\sqrt{19})^j$. For instance, $x_2 + y_2\sqrt{D} = A_{11} + B_{11}\sqrt{D} = 57799 + 13260\sqrt{19} = A_{2\ell-1} + B_{2\ell-1}\sqrt{D}$.

Remark 5.3 *If we look at the general norm-form equation $x^2 - Dy^2 = n$ and assume that we have a solution $r + s\sqrt{D}$, then we can find infinitely many solutions of this equation from this one solution as follows. Let $x_j + y_j\sqrt{D} = (x_1 + y_1\sqrt{D})^j$ where $x_1 + y_1\sqrt{D}$ is the fundamental solution of $x^2 - Dy^2 = 1$. Then*

$$n = (r^2 - s^2D)(x_j^2 - y_j^2D) = (rx_j \pm sy_jD)^2 - D(ry_j \pm sx_j)^2.$$

Thus, $(rx_j \pm sy_jD) + (ry_j \pm sx_j)\sqrt{D}$ is a solution to $x^2 - Dy^2 = n$ for any $j \in \mathbb{N}$, thereby providing infinitely many positive solutions.

Exercises

5.24. Let $\alpha = \alpha_0 = \sqrt{D}$ where D is a positive integer that is not a perfect square, and let ℓ be the period length of the simple continued fraction expansion of α. In the notation of Theorem 5.10 on page 223, prove that $Q_j = Q_{\ell-j}$ for all $j = 0, 2, \ldots, \ell$, and that $P_j = P_{\ell-j+1}$ for all $j = 1, 2, \ldots, \ell$.

5.25. Let α be as in Exercise 5.24. Prove that if ℓ is even, then $Q_{\ell/2} \mid 2D$.

5.26. Let $p > 2$ be prime and let ℓ be the period length of the simple continued fraction expansion of $\sqrt{p}$. Prove that if ℓ is even, then $Q_{\ell/2} = 2$.

(*Hint: Use Theorem 5.10.*)

(*The value $Q_{\ell/2}$ is called the* central norm *a term arising from Theorem 5.14, and has value in finding solutions to quadratic Diophantine equations and modular equations.*)

5.27. Let D be a positive integer divisible by a prime $p \equiv 3 \,(\text{mod } 4)$. Prove that the Pell equation $x^2 - Dy^2 = -1$ has no solution.

5.28. Let $p > 2$ be prime and let ℓ be the period length of the simple continued fraction expansion of $\sqrt{p}$. Prove that the following are equivlant.

   (a) $x^2 - py^2 = -1$ has a solution.

   (b) $p \equiv 1 \,(\text{mod } 4)$.

   (c) ℓ is odd.

(*Hint: Use Exercises 5.26–5.27 in conjunction with Theorems 5.14–5.15.*)

5.29. Let $p > 2$ be a prime and let ℓ be the period length of the simple continued fraction expansion of $\sqrt{p}$. Prove that $x^2 - py^2 = \pm 2$ has a solution $x, y \in \mathbb{Z}$ if and only if ℓ is even and $Q_{\ell/2} = 2$.

☆ 5.30. Let p be an odd prime and let ℓ be the period length of the simple continued fraction expansion of $\sqrt{p}$. If $x_1 + y_1\sqrt{p}$ is the fundamental solution of $x^2 - py^2 = 1$, prove that $x_1 \equiv 1 \,(\text{mod } p)$ if and only if $\ell \equiv 0 \,(\text{mod } 4)$ and $Q_{\ell/2} = 2$. (*Hint: Use Theorems 5.14–5.16 and Exercises 5.26 and 5.29.*)

5.31. Let p be an odd prime and let $x_1 + y_1\sqrt{p}$ be the fundamental solution of $x^2 - py^2 = 1$. Prove that $x_1 \equiv 1 \,(\text{mod } p)$ if and only if $p \equiv 7 \,(\text{mod } 8)$.

(*Hint: Use Exercises 4.3, 5.26, 5.28, and 5.30 in conjunction with Corollary 4.2 on page 185 and Theorem 5.14 on page 234.*)

(*The content of this exercise is a result first proved by Lagrange. This result was completely generalized to the following result in [33].*

   Theorem: *Suppose that D is a positive integer that is not a perfect square and $x_1 + y_1\sqrt{D}$ is the fundamental solution of $x^2 - Dy^2 = 1$, with ℓ being even, where ℓ is the period length of the simple continued fraction expansion of $\sqrt{D}$. Then $x_1 \equiv (-1)^{\ell/2} \,(\text{mod } D)$ if and only if $Q_{\ell/2} = 2$.*)

The *Farey series* of order $n \in \mathbb{N}$, denoted $\mathcal{F}_n$, is the set of rational numbers r/s with $0 \le r \le s \le n$ such that $\gcd(r, s) = 1$ in ascending order. For example,

$$\mathcal{F}_5 = \{0/1,\ 1/5,\ 1/4,\ 1/3,\ 2/5,\ 1/2,\ 3/5,\ 2/3,\ 3/4,\ 4/5,\ 1/1\}.$$

Also, two successive terms in a Farey series are called *adjacent*.

(*These numbers are named after John Farey who published an article on them in a British journal in 1816.*)

Answer each of Exercises 5.32–5.36 on Farey series.

☆ 5.32. Suppose that $r/s \in \mathcal{F}_n$ and $r/s = \langle q_0; q_1, \ldots, q_\ell \rangle$, where ℓ is odd, and set $t = \lfloor \frac{n - B_{\ell-1}}{s} \rfloor$, $u = \lfloor \frac{n + B_{\ell-1}}{s} \rfloor$. Prove that

$$\gcd(ur - A_{\ell-1},\ us - B_{\ell-1}) = \gcd(tr + A_{\ell-1},\ ts + B_{\ell-1}) = 1,$$

and

$$0 < \frac{ur - A_{\ell-1}}{us - B_{\ell-1}} < r/s < \frac{tr + A_{\ell-1}}{ts + B_{\ell-1}} < 1,$$

where the sequences of A_j, B_j are defined in Theorem 1.12 on page 25. Furthermore, prove that $(ur - A_{\ell-1})/(us - B_{\ell-1})$ and r/s are adjacent in $\mathcal{F}_n$; and that r/s and $(tr + A_{\ell-1})/(ts + B_{\ell-1})$ are adjacent in $\mathcal{F}_n$.

5.33. Find the simple continued fraction expansions of the terms in $\mathcal{F}_{10}$. (There are thirty-three of them.)

☆ 5.34. Suppose that $\overline{r}/\overline{s} < r/s < r'/s'$ are three adjacent terms in the Farey series $\mathcal{F}_n$. Prove that

$$r' = \left\lfloor \frac{n + \overline{s}}{s} \right\rfloor r - \overline{r},$$

and

$$s' = \left\lfloor \frac{n + \overline{s}}{s} \right\rfloor s - \overline{s}.$$

Conclude that

$$r/s = \frac{\overline{r} + r'}{\overline{s} + s'}.$$

5.35. Prove that if $r/s < \overline{r}/\overline{s}$ are adjacent, then $\overline{r}s - r\overline{s} = 1$.

☆ 5.36. Let $\alpha \in \mathbb{R}$ with $\overline{r}/\overline{s} < \alpha < r/s$ where $\overline{r}/\overline{s}$, and r/s are adjacent. Prove that one of r/s, $\overline{r}/\overline{s}$, or $\frac{r+\overline{r}}{s+\overline{s}}$ is a solution of

$$|\alpha - h/k| < \frac{1}{\sqrt{5}k^2},$$

where $h, k \in \mathbb{Z}$ with $k \ne 0$. Conclude that if α is irrational, there are infinitely many such solutions. Also establish that no larger value can replace $\sqrt{5}$. This proves that $\sqrt{5}$ is the *best possible*.

(*See Exercise 5.11 on page 220.*)

5.4 Continued Fractions and Factoring

> *If we find the answer to that [why it is that we and the universe exist], it would be the ultimate triumph of human reason — for then we would know the mind of God.*
>
> From Chapter 11 of **A Brief History of Time (1988)**
> **Stephen Hawking (1942–), English theoretical physicist**

In this concluding section for this chapter, we outline a method of factoring using continued fractions. The idea was developed by Brillhart and Morrison in 1975, which they based on earlier ideas of D.H. Lehmer and R.E. Powers posed in the 1930s — see [36].

Suppose that we have $n \in \mathbb{N}$ that we want to factor. If we can find $x, y \in \mathbb{N}$ such that $n = x^2 - y^2$, then we may have a nontrivial factor $x \pm y$ of n. Recall from §4.3 that this is the kind of reasoning used by Fermat. We now look at using the methods developed thus far in this chapter to extend this idea in order to factor n.

◆ The Continued Fraction Factoring Method

Let $n \in \mathbb{N}$, which is not a perfect square, and let $C_j = A_j/B_j$ be the jth convergent in the simple continued fraction expansion of $\sqrt{n}$. Then by Theorem 5.14 on page 234,

$$A_{j-1}^2 - nB_{j-1}^2 = (-1)^j Q_j \quad (j \geq 1).$$

If j is even and $Q_j = m^2$ for some $m \in \mathbb{N}$, then $n|(A_{j-1}^2 - m^2)$. If $\gcd(A_{j-1} \pm m, n) > 1$, namely if $A_{j-1} \pm m \neq 0, 1$, then we have a nontrivial factor of n. We use the algorithm described in Theorem 5.10 on page 223 to determine the simple continued fraction expansion of $\sqrt{n}$.

Example 5.12 *Let $n = 1501$ and $\alpha = \sqrt{n} = (P_0 + \sqrt{n})/Q_0$ where $P_0 = 0$ and $Q_0 = 1$. Then the simple continued fraction expansion of α is given in the following table.*

j	0	1	2	3	4
P_j	0	38	19	21	32
Q_j	1	57	20	53	9
q_j	38	1	2	1	7
A_j	38	39	116	155	
B_j	1	1	3	4	

We stop at $j = 4$ since $Q_4 = 9 = 3^2$ and we only need the value of A_{j-1}. Here $A_{j-1} = A_3 = 155$, so $155^2 \equiv 3^2 \,(\mathrm{mod}\ 1501)$. Since

$$155 - 3 = 2^3 \cdot 19 \ and \ 155 + 3 = 2 \cdot 79,$$

then we have a nontrivial factor of 1501. In fact, $1501 = 19 \cdot 79$.

The reader may complete the table using Theorem 5.10. One discovers that $\ell(\alpha) = 60$, and $P_{30} = 38 = P_{31}$, illustrating the symmetry described in Exercise 5.24 on page 238.

Also, the reader will find that $Q_{16} = 6^2$, $A_{15} = 87179575$, $A_{15} - 6 = 9337^2$, and $A_{15} + 6 = 19 \cdot 79 \cdot 241^2$, so we have another means of factoring n at $j = 16$.

Example 5.12 has several square values of $Q_j > 1$ that yield nontrivial factorizations of n. However, there are cases where there are *none*.

Example 5.13 *Let* $n = 327$ *and* $\alpha = \sqrt{327}$. *Then the table describing the continued fraction expansion of* α *is given by the following.*

j	0	1	2
P_j	0	18	18
Q_j	1	3	1
q_j	18	12	36

Since $\alpha = \sqrt{327} = \langle 18; \overline{12, 36} \rangle$, then we can never find a value of $Q_j > 1$ that is a square.

There is still hope of finding a nontrivial factor of n in such cases as that illustrated in Example 5.13. However, we need to modify the continued fraction method somewhat. If the continued fraction method fails for $\sqrt{n}$, then we may use that same algorithm on $\sqrt{mn}$ where we may suitably choose m, usually as the product of the first few primes to avoid introducing squares under the surd. If we get $Q_j = h^2$ for even j in the continued fraction expansion of $\sqrt{mn}$, then

$$A_{j-1}^2 - B_{j-1}^2 mn = (-1)^j Q_j,$$

so $A_{j-1}^2 \equiv h^2 \pmod{mn}$ and we may have a factor of n, since we may get $\gcd(A_{j-1} \pm h, n) > 1$.

Example 5.14 *Let* $\alpha = \sqrt{2 \cdot 327} = \sqrt{654}$. *We need not go far in the tabular description of* α *to get the desired factorization in this case.*

j	0	1	2	3	4	5	6
P_j	0	25	4	18	12	22	23
Q_j	1	29	22	15	34	5	25
q_j	25	1	1	2	1	9	1
A_j	25	26	51	128	179	1739	

Here, $Q_j = Q_6 = 5^2$, $A_{j-1} = A_5 = 1739$, so

$$1739^2 \equiv 5^2 \pmod{654}.$$

Therefore,

$$1739 - 5 = 2 \cdot 3 \cdot 17^2 \text{ and } 1739 + 5 = 2^4 \cdot 109.$$

Indeed, since $654 = 2 \cdot 327 = 2 \cdot 3 \cdot 109$. *then we have the complete factorization* $327 = 3 \cdot 109$.

Example 5.14 shows that the continued fraction algorithm may be extended to cases where the original method fails for n itself. This modified approach may also be used as an alternative if we find that we have gone through numerous values of j in the tabular expansion of the continued fraction expansion as above, without finding any square Q_j. Also, even when we find a square Q_j, it does not guarantee that we have a nontrivial factor of n. For instance, if $n = 161$, then $Q_j = Q_4 = 4^2$ and $A_3 = 165$, but $A_3 - 4 = 161$, so no factor is found.

◆ Complexity Issues

It can be shown that the number of bit operations required to factor n by trial division is $O(\sqrt{n}(\log_2 n)^2)$. We also mentioned Fermat's method discussed on page 203. To get

$$n = x^2 - y^2,$$

we need to try $n + y^2$ for $y \in \mathbb{N}$ until we get a square. For instance, in the most interesting case, where $n = pq$ for $p > q$ both primes, we compute $n + y^2$ until we get $y = (p - q)/2$. Furthermore, we cannot get a square value before this. The reason is that by Exercise 5.40, there is *exactly* one representation of n as $n = x^2 - y^2$ where $x - y > 1$, namely for $x = (p + q)/2$ and $y = (p - q)/2$. The only other such representation for positive x and y is

$$n = [(pq + 1)/2]^2 - [(pq - 1)/2]^2.$$

Hence, Fermat's method requires $(p - q)/2$ iterations, namely $O(\sqrt{p}) < O(\sqrt{n})$ bit operations. Hence this is not much better than trial division for values of p and q that are far apart, and any good cryptologist can make them so. There are more sophisticated continued fraction algorithms for factoring, but these are beyond the scope of this book. See [49], for instance.

Exercises

5.37. Use the continued fraction method to factor 1517.

5.38. Use the continued fraction method to factor 2279.

5.39. Prove that if $a = x^2 - y^2$ and $b = u^2 - v^2$ for integers x, y, u, v, then

$$ab = (xu + yv)^2 - (yu + xv)^2,$$

and

$$ab = (xu + -yv)^2 - (yu - xv)^2.$$

☆ **5.40.** Let the number of distinct prime factors of $n \in \mathbb{N}$ be $d \in \mathbb{N}$, and assume that n can be represented as a difference of squares. Prove that the number of representations of n as $n = x^2 - y^2$ with $x, y \in \mathbb{N}$, and $\gcd(x, y) = 1$ is 2^{d-1}.

(*Hint: Use induction on d and Exercise 5.39.*)

Chapter 6

Additivity — Sums of Powers

This chapter is dedicated to additivity in number theory. In particular, the problem of representing integers as a sum of squares goes back to Diophantus, who ostensibly knew that every $n \in \mathbb{N}$ is a sum of at most four integer squares of positive numbers — see Biography 1.15 on page 48. Later, Fermat, then Euler, and more substantially Lagrange made contributions to the representation problem. We begin with the simplest of the additive representations.

6.1   Sums of Two Squares

We have encountered some special cases of sums of two squares such as with the Fibonacci numbers in Exercise 1.7 on page 11 and parts (e)–(f) of Exercise 1.13 on page 13. We now seek a more general accounting as to which $n \in \mathbb{N}$ can be represented as a sum of two squares of natural numbers. We begin with a result that allows us to focus on primes.

Lemma 6.1 | **Products of Sums of Two Squares**

If m, n are both sums of two squares, then mn is a sum of two squares.

Proof. If $m = a_1^2 + a_2^2$ and $n = b_1^2 + b_2^2$, then

$$mn = (a_1^2 + a_2^2)(b_1^2 + b_2^2) = (a_1 b_1 + a_2 b_2)^2 + (a_1 b_2 - a_2 b_1)^2,$$

by Remark 1.6 on page 46. $\square$

In view of Lemma 6.1, we may concentrate on prime representation as a sum of two squares since we may then multiply them together to get another sum of two squares. Representation of an integer as the sum of two squares was studied by Fibonacci in the thirteenth century and Bachet in the seventeenth century. The first to correctly formulate necessary and sufficient conditions for an integer to be a sum of two squares was Albert Girard (1595–1632), and so the following is often referenced as *Girard's Theorem*. Several years later on December 25, 1640, Fermat stated these conditions independently in a letter to Mersenne, and claimed he had an iron-clad proof, which he did not publish. The first *published* proof was given by Euler in 1754.

Theorem 6.1 $\boxed{\textbf{Primes as Sums of Two Squares}}$

An odd prime $p = a^2 + b^2$ for $a, b \in \mathbb{N}$ if and only if $p \equiv 1 \,(\mathrm{mod}\ 4)$. Moreover, when such a representation exists it is unique (ignoring the order of the summands).

Proof. If $p = a^2 + b^2$, for $a, b \in \mathbb{N}$, then a and b have opposite parity, say a is odd and b is even without loss of generality. Then $p \equiv 1 + 0 \equiv 1 \,(\mathrm{mod}\ 4)$.

Conversely, if $p \equiv 1 \,(\mathrm{mod}\ 4)$, then

$$-1 \equiv a^2 \quad (\mathrm{mod}\ p) \tag{6.1}$$

for some $a \in \mathbb{N}$ by part (3) of Theorem 4.4 on page 182. Since $\gcd(a, p) = 1$, then then there are unique $a, b \in \mathbb{N}$ such that $p = a^2 + b^2$ by Example 1.21 on page 45. $\Box$

Remark 6.1 *Note that the uniqueness of representation in Theorem 6.1 is "ignoring the order of the summands," meaning that we do not distinguish between $p = x^2 + y^2$ and $p = y^2 + x^2$. In fact, we can stipulate a canonical representation for odd primes such as $p = x^2 + y^2$ where $0 < x < y$. Then there is only one possible interpretation.*

Example 6.1 *The prime $p = 17$ is uniquely written as $p = 17 = 1^2 + 4^2$. We rely on this uniqueness of representation since trivial sums of squares would otherwise arise. For instance, if we allow the summands to be squares of negative integers and distinguish the order of the summands, then we get the following eight representations.*

$$17 = 1^2 + 4^2 = 4^2 + 1^2 = (-4)^2 + 1^2 = 1^2 + (-4)^2 =$$

$$(-1)^2 + 4^2 = 4^2 + (-1)^2 = (-1)^2 + (-4)^2 = (-4)^2 + (-1)^2.$$

However, there is essentially only one *representation, namely $17 = 1^2 + 4^2$, which is unique ignoring order and sign, sometimes referenced as "unique up to order and sign." Not only primes may be represented as a sum of two squares. For instance, since $2 = 1^2 + 1^2$ and $13 = 3^2 + 2^2$, then by applying Lemma 6.1 on the preceding page, $26 = 1^2 + 5^2$. Thus, we need to look further to characterize those positive integers which are sums of two squares.*

Theorem 6.2 | **Integers as Sums of Two Squares**

Let $N \in \mathbb{N}$ with $N = m^2 n$ where n is squarefree. Then N can be represented as a sum of two integer squares if and only if n is not divisible by any prime $p \equiv 3 \pmod{4}$.

Proof. If n has no prime divisor of the form $p \equiv 3 \pmod{4}$, then by Lemma 6.1 and Theorem 6.1, $n = x^2 + y^2$, so

$$N = m^2 n = (mx)^2 + (my)^2.$$

Now suppose that $N = m^2 n = x^2 + y^2$, and there exists a prime $p \equiv 3 \pmod{4}$ dividing n. Let $\gcd(x, y) = g$. Then $g \mid m$ since n is squarefree. Therefore,

$$M = \frac{N}{g^2} = \left(\frac{m}{g}\right)^2 n = \left(\frac{x}{g}\right)^2 + \left(\frac{y}{g}\right)^2 = r^2 + s^2,$$

where $\gcd(r, s) = 1$. Since $p \mid n$, then $p \mid M$, so

$$r^2 + s^2 \equiv 0 \pmod{p}.$$

Since $\gcd(r, s) = 1$, then $\gcd(p, r) = 1 = \gcd(p, s)$. Let $r_1 \in \mathbb{Z}$ such that $r_1 r \equiv 1 \pmod{p}$. Then,

$$(r_1 r)^2 + r_1^2 s^2 \equiv 0 \pmod{p}.$$

In other words, $1 + (r_1 s)^2 \equiv 0 \pmod{p}$, which implies that $(r_1 s)^2 \equiv -1 \pmod{p}$, and this contradicts Example 4.5 on page 182. $\square$

Immediate from the above is the following.

Corollary 6.1 *A natural number n can be represented as the sum of two integer squares if and only if every prime factor of the form $p \equiv -1 \pmod{4}$ appears to an even power in the canonical prime factorization of n.*

Example 6.2 *The prime 3 is not representable as the sum of two integer squares, but*

$$3^2 = 9 = 0^2 + 3^2.$$

Also, $3^3 \cdot 5$ cannot be represented as a sum of two integer squares but

$$5^2 = 25 = 0^2 + 5^2 = 3^2 + 4^2.$$

Notice that 0 is allowed as a summand for composite numbers, which was clearly impossible for prime numbers.

Example 6.2 suggests that for composite numbers, the number of distinct representations as a sum of two squares can be larger than one. Now the question arises as to how many there are in total for a given integer. To answer this query, we first need a result that will allow us to translate the problem to one involving quadratic residues.

Lemma 6.2 | Unique Sums of Two Squares |

Let -1 be a quadratic residue modulo the integer $n > 1$. Each solution

$$n = a^2 + b^2 \text{ for some } a, b \in \mathbb{N} \text{ with } \gcd(a, b) = 1, \qquad (6.2)$$

determines a unique $m \in \mathbb{N}$, modulo n, such that $a \equiv mb \pmod{n}$. Furthermore, $m^2 \equiv -1 \pmod{n}$ and different nonnegative solutions satisfying (6.2), determine different solutions modulo n.

Conversely, given an $m \in \mathbb{N}$ such that

$$m^2 \equiv -1 \pmod{n},$$

then there are unique $a, b \in \mathbb{N}$ with $\gcd(a, b) = 1$ such that

$$n = a^2 + b^2 \text{ with } a \equiv mb \pmod{n}.$$

Proof. If there is an $m \in \mathbb{N}$ such that $m^2 \equiv -1 \pmod{n}$, then there are unique $a, b \in \mathbb{N}$ with $\gcd(a, b) = 1$ such that $n = a^2 + b^2$ with $a \equiv mb \pmod{n}$ by Example 1.21 on page 45.

Conversely, if $n = a^2 + b^2$ for relatively prime natural numbers a and b, then set $m = ab^{-1}$, where b^{-1} is the unique inverse of b modulo n (see Definition 2.5 on page 80). Thus, $m^2 \equiv -1 \pmod{n}$ and $a \equiv mb \pmod{n}$. If there is another integer m' such that $a \equiv m'b \pmod{n}$, then

$$m' \equiv ab^{-1} \equiv m \pmod{n},$$

so m is uniquely determined modulo n. $\square$

Remark 6.2 *Lemma 6.2 may be reinterpreted as saying that there is a one-to-one correspondence between the ordered pairs of relatively prime integers $\{a, b\}$, with*

$$n = a^2 + b^2,$$

and the least positive residues $\overline{ab^{-1}} = \overline{m}$ modulo n, where

$$m^2 \equiv -1 \pmod{n},$$

and b^{-1} is the unique multiplicative inverse of b modulo n. In turn, these least positive residues are in one-to-one correspondence with the incongruent solutions of $x^2 \equiv -1 \pmod{n}$. Hence, the number of distinct representations of n as a primitive sum of two integer squares is exactly one-half the number of incongruent solutions of $x^2 \equiv -1 \pmod{n}$. For instance,

$$5 = 1^2 + 2^2 = a^2 + b^2,$$

and the ordered pair $(a, b) = (1, 2)$ corresponds, uniquely, to the least positive residue

$$\overline{ab^{-1}} = \overline{1 \cdot 2^{-1}} = \overline{3} = \overline{m}$$

modulo $n = 5$. *In turn, this corresponds, uniquely, to the solution* $x = 3$ *of* $x^2 \equiv -1 \,(\text{mod } 5)$. *Similarly, the other solution of* $x^2 \equiv -1 \,(\text{mod } 5)$, $x = 2$, *corresponds uniquely to the ordered pair* $(2, 1)$. *Thus,* $x^2 \equiv -1 \,(\text{mod } 5)$ *has the two incongruent solutions* $x = 2, 3$, *and* $n = 5$ *has the unique representation* $n = 1^2 + 2^2$ *as a sum of two squares, since we do not take the order of the factors into account.*

To formalize some terminology from the above, we have the following.

Definition 6.1 | Primitive Representations

A primitive representation of $n \in \mathbb{N}$ *as a sum of two integer squares is of the form*

$$n = a^2 + b^2$$

for $a, b \in \mathbb{Z}$ *with* $\gcd(a, b) = 1$. *If* $\gcd(a, b) > 1$, *then the representation is called imprimitive.*

We can summarize the above in the following.

Theorem 6.3 | Total Number of Primitive Representations

The number $r_2(n)$ *of primitive representations of* $n > 1$ *as a sum of two integer squares is given by*

$$r_2(n) = \begin{cases} 0 & \text{if } 4 \mid n \text{ or if there is a prime } p \equiv 3 \pmod{4} \text{ dividing } n; \\ 2^{d-1} & \text{if } 4 \nmid n, \text{ there is no prime } p \equiv 3 \pmod{4} \text{ dividing } n, \text{ and } \\ & d \text{ is the number of distinct odd prime divisors of } n. \end{cases}$$

Proof. By Theorem 6.2 on page 245, $r_2(n) = 0$ if there is a prime $p \equiv 3 \,(\text{mod } 4)$ dividing n. If $4 \mid n$, then $x^2 + y^2 \equiv 0 \,(\text{mod } 4)$. However, since $\gcd(x, y) = 1$, then both x and y must be odd, so $x^2 + y^2 \equiv 2 \,(\text{mod } 4)$, a contradiction. Suppose that

$$n = 2^{a_0} p_1^{a_1} p_2^{a_2} \cdots p_d^{a_d},$$

is the canonical prime factorization of n (where $a_0 \in \{0, 1\}$). Then by Theorem 3.10 on page 155,

$$x^2 \equiv -1 \pmod{p_j^{a_j}}$$

has exactly two incongruent solutions for each $j = 1, 2, \ldots, d$. Hence, $x^2 \equiv -1$ (mod n) has exactly 2^d incongruent solutions. Therefore, by Lemma 6.2, and Remark 6.2, 2^{d-1} is the number of primitive representations of n as a sum of two integer squares (up to order and sign). $\qquad \square$

From the above we get the following.

Corollary 6.2 *If* $n \in \mathbb{N}$ *where* $n > 1$ *is odd, then the following are equivalent.*

(a) n has a unique representation, up to order and sign, as a sum of two integer squares, *and* this representation is primitive.

(b) n is prime.

Proof. If n is prime, the the result follows from Theorem 6.3, so part (b) implies part (a). If part (a) holds, then by Theorem 6.3, $n = p^a$ for a prime $p > 2$ and some $a \in \mathbb{N}$. There is a primitive representation for $p = x^2 + y^2$. If $a > 1$ and a is odd, then

$$p^a = (p^{(a-1)/2}x)^2 + (p^{(a-1)/2}y)^2,$$

which is an imprimitive representation, contradicting (a). If $a > 0$ is even, then

$$p^a = (p^{a/2})^2 + 0^2,$$

again an imprimitive representation, contradicting (a). Hence, $a = 1$ and we have (b).   □

Remark 6.3 *The reader may solve Exercise 6.1 on page 251 to see that any $n \in \mathbb{N}$ is a sum of two integer squares if and only if it is a sum of two squares of rational numbers.*

We may now turn to the results on continued fractions that we developed in Chapter 5 and apply them to sums of two squares for some elegant results. We need the following refinement of the notion of pure periodicity given in Definition 5.1 on page 221.

Definition 6.2 | **Pure Symmetric Periods**

If $\alpha = \langle \overline{q_0; q_1, \ldots, q_{\ell-1}} \rangle$ is a reduced quadratic irrational, then α is said to have pure symmetric period if $q_0, q_1, \ldots, q_{\ell-1}$ is a palindrome, namely $q_j = q_{\ell-j-1}$ for $j = 0, 1, \ldots, \ell - 1$.

Theorem 6.4 | **Sums of Two Squares and Pure Symmetric Periods**

Let $\alpha = (P + \sqrt{D})/Q$ be a reduced quadratic irrational where $D > 1$ is not a perfect square. Then the following are equivalent.

(a) α has pure symmetric period.

(b) $N(\alpha) = \alpha\alpha' = -1$.

(c) $D = P^2 + Q^2$.

Proof. If α has pure symmetric period, then

$$\alpha = \langle \overline{q_0; q_1, q_2, \ldots, q_{\ell-1}} \rangle = \langle \overline{q_{\ell-1}; q_{\ell-2}, \ldots, q_1, q_0} \rangle = -1/\alpha',$$

where the last equality comes from Corollary 5.6 on page 230. In other words, $N(\alpha) = \alpha\alpha' = -1$, so part (a) implies part (b).

If we have that

$$N(\alpha) = (P^2 - D)/Q^2 = -1,$$

then

$$D = P^2 + Q^2,$$

so part (b) implies part (c). If $D = P^2 + Q^2$, then

$$N(\alpha) = (P^2 - D)/Q^2 = -1,$$

so $\alpha = -1/\alpha'$. Therefore, by Corollary 5.6, α has pure symmetric period. Thus, part (c) implies part (a), and the logical circle is complete. $\square$

Example 6.3 *Let*

$$\alpha = (P + \sqrt{D})/Q = (9 + \sqrt{145})/8.$$

Since

$$\alpha > 1, \ and \ -1 < \alpha' = (9 - \sqrt{145})/8 < 0,$$

then α is a reduced quadratic irrational. Also,

$$\alpha = \langle \overline{2; 1, 1, 1, 2} \rangle,$$

so α has pure symmetric period and

$$N(\alpha) = (81 - 145)/64 = -1,$$

and

$$145 = 1^2 + 12^2 = 8^2 + 9^2,$$

the only two (primitive) representations by Theorem 6.3 on page 247 since

$$r_2(D) = 2 = 2^{d-1}$$

given that $145 = 5 \cdot 29$.

Remark 6.4 *Theorem 6.4 on the facing page is one of the prettiest results to emerge from the connection between continued fractions and sums of two squares.*

Note that the representation given in Theorem 6.4 need not be primitive. For instance, if

$$\alpha = (5 + \sqrt{50})/5,$$

then α is a reduced quadratic irrational with pure symmetric period given by
$\alpha = \langle \overline{2} \rangle$ *and*
$$50 = 5^2 + 5^2.$$

We also have the primitive representation, $50 = 1^2 + 7^2$, which arises from the simple continued fraction expansion of $\sqrt{50}$ in the following fashion. Indeed, since $\sqrt{50}$ is not reduced, this points to the fact that there is more underlying this connection that does not involve reduced quadratic irrationals.

Let $\alpha = \sqrt{D}$ where $D > 1$ is not a perfect square. If the period length ℓ of the simple continued fraction expansion of α is odd, then by Exercise 5.24 on page 238,
$$Q_{(\ell-1)/2} = Q_{(\ell+1)/2},$$

and by Theorem 5.10 on page 223,

$$D = P^2_{(\ell+1)/2} + Q_{(\ell+1)/2} Q_{(\ell-1)/2} = P^2_{(\ell+1)/2} + Q^2_{(\ell+1)/2},$$

so we always have that D is a sum of two squares when ℓ is odd for $\sqrt{D}$.

Furthermore, we see by Theorem 5.15 on page 234, that when ℓ is odd, then

$$x^2 - Dy^2 = -1 \tag{6.3}$$

has a solution. For instance, $D = 17$ has $\ell = 1$ and

$$17 = 1^2 + 4^2 = Q_1^2 + P_1^2 = P^2_{(\ell+1)/2} + Q^2_{(\ell+1)/2},$$

the only representation as a sum of two squares since it is prime. So it would appear from (6.3) and part (b) of Theorem 6.4 that sums of two squares are related to -1 norms. However, if ℓ is even, namely (6.3) has no solution, it is still possible for D to be a sum of two squares. For instance, for $D = 205$, $\ell(\sqrt{D}) = 8$ and
$$205 = 3^2 + 14^2 = 6^2 + 13^2,$$

the only two primitive representations since $r_2(D) = 2$ given that $D = 5 \cdot 41$. Also, there are values of D for which there are sums of squares but all are imprimitive. For instance, $D = 45$ has $\ell(\sqrt{D}) = 6$ and

$$D = 45 = 3^2 + 6^2,$$

which is imprimitive and the only representation of D as a sum of two squares. The imprimitive representations are of little interest since they are merely multiples of primitive one. For example

$$5 = 1^2 + 2^2 \text{ and so } 45 = 5 \cdot 3^2 = 3^2 + 6^2.$$

We do not delve into the imprimitive representations in the same detail as the primitive ones for this reason.

Exercises

6.1. Prove that the natural number $n > 1$ is a sum of two integer squares if and only if it is the sum of two squares of rational numbers.

6.2. Prove that n is the *difference* of two integer squares if and only if $n \not\equiv 2$ (mod 4).

6.3. Prove that any odd prime is the difference of two successive squares. Conclude that this representation as a difference of squares is unique.

   (*Hint: See Exercise 5.40 on page 242.*)

6.4. Let $n = pq$ where p and q are odd primes. Prove that n may be written as a difference of squares in exactly two distinct ways and provide those representations.

6.5. Prove that if p is a prime such that

$$p = q^2 + r^2$$

   where q and r are primes, then one of q or r must be equal to 2.

6.6. Prove that $n \in \mathbb{N}$ is a sum of two squares if and only if n can be written in the form

$$n = 2^a b^2 c$$

   where $a \geq 0$, bc is odd, c is squarefree, and every prime divisor of c is of the form $4m + 1$.

6.7. Prove that a prime p is a sum of two squares if and only if there is an integer x such that

$$x^2 + 1 \equiv 0 \pmod{p}.$$

6.8. Prove that n has as many representations as a sum of two squares as does $2n$.

6.9. Express 221 as a sum of two squares using Theorem 6.4 on page 248.

6.10. Express 65 as a sum of two squares using Theorem 6.4.

6.11. Prove that every Fermat number

$$\mathcal{F}_n = 2^{2^n} + 1$$

   for $n \in \mathbb{N}$ is a sum of two squares.

6.12. Prove that every odd perfect number is the sum of squares.

   (*Hint: Use Exercise 2.41 on page 106.*)

6.2   Sums of Three Squares

> *Conduct is three-fourths of our life and its largest concern.*
> From Chapter 1 of **Literature and Dogma (1873)**
> **Matthew Arnold (1822–1888), English poet and essayist**

In §6.1, we classified all those natural numbers that can be represented as a sum of two integer squares. Now we turn to a classification of those that may be represented as a sum of three integer squares, allowing for 0^2, so those in the previous section all qualify.

The history of sums of three squares goes back to Diophantus, who posited that no natural number $n \equiv -1 \pmod 8$ is a sum of three integer squares — see Biography 1.15 on page 48. In 1638, Descartes proved this to be the case — see Biography 6.1 on the facing page. The following criterion was first proved, in a complicated manner, by Legendre in 1798, then more clearly, by Gauss in 1801 — see Biographies 4.1 on page 181 and 1.7 on page 33. However, the proof of the sufficiency of Condition (6.4) below is beyond the scope of this book, and so we prove only the necessity, referring the reader to another source for the sufficiency.

Theorem 6.5 | Sums of Three Squares |

A natural number n can be represented as the sum of three integer squares if and only if n is not of the form

$$4^a(8b + 7), \text{ where } a, b \text{ are nonnegative integers.} \tag{6.4}$$

Proof. To prove that (6.4) is necessary, assume that

$$n = 4^a(8b + 7) = x^2 + y^2 + z^2,$$

for nonnegative $x, y, z \in \mathbb{Z}$. Furthermore, we may assume that n is the *least* such representable value. If any of the values of x, y, z is odd, then a straightforward check shows that $n \equiv 1, 2 \pmod 4$ if one, respectively two of them are odd, and $n \equiv 3 \pmod 8$ in the case where all three are odd. Hence, since $n \equiv 0 \pmod 4$ or $n \equiv 7 \pmod 8$, then all of x, y, z must be even. Thus,

$$4^{a-1}(8b + 7) = \frac{n}{4} = \left(\frac{x}{2}\right)^2 + \left(\frac{y}{2}\right)^2 + \left(\frac{z}{2}\right)^2,$$

contradicting the minimality of n. We have shown that Condition (6.4) is necessary for n to be a sum of three integer squares.

For the sufficiency, the reader is referred to [4]. Although the latter is a simplified version of Gauss' proof, it is still complicated and the reader should have some knowledge of Minkowski's convex body result in the geometry of numbers area. For this reason, the proof is beyond the level of this introductory text.   □

Exercises

6.13. Show that the Diophantine equation $x^2 + y^2 + z^2 = x^3 + y^3 + z^3$ has infinitely many solutions.

6.14. Show that $x^2 + y^2 + z^2 = 3xyz$ has infinitely many solutions.

6.15. Are there solutions $a, b \in \mathbb{N}$ such that $a^2 + (a+1)^2 = b^4 + (b+1)^4$?

6.16. Show that the product of four consecutive natural numbers cannot be a square.

6.17. Solve the Diophantine equation $x^3 + y^3 = 6xy$.

6.18. Solve the Diophantine equation $z^2 = (x^2 - 1)(y^2 - 1) + 1981$.

6.19. Solve the Diophantine equation $x^3 + 8x^2 - 6x + 8 = y^3$ for x, y nonnegative integers.

6.20. Prove that every $n \in \mathbb{N}$ is the sum of at most three triangular numbers.

(*Hint: Use Exercise 2.45 on page 106 in conjunction with Theorem 6.5.*)

This result was first conjectured by Fermat — see Biography 1.10 on page 38.

6.21. Prove that every odd natural number is of the form $x^2 + y^2 + 2z^2$ where $x, y, z \in \mathbb{Z}$.

6.22. Prove that every natural number is either of the form $x^2 + y^2 + z^2$ or of the form $x^2 + y^2 + 2z^2$ where $x, y, z \in \mathbb{Z}$.

Biography 6.1 René Descartes (1596–1650) *was born on March 31, 1596 in La Haye, France. He is arguably the founder of modern philosophy, and most certainly one of the most prominent scholars in human history.*

He studied under the Jesuits at La Flèche College from 1606 to 1614. In 1616, he received a law degree from the University of Poiters. Then, after traveling through Europe for more than a decade, he settled in Holland in 1628. During his travels he met Mersenne, with whom he kept a life-long friendship and mathematical correspondence — see Biography 1.9 on page 36. Descartes saw mathematics as the only certain body of knowledge, so reasoned that all understanding must be based on mathematics.

In 1641, he published his magnum opus, A Discourse on Method. Meditations on First Philosophy. *His method of systematic doubt to reach the truth had a tremendous impact on philosophical thought thereafter. His famous Latin phrase* Cogito ergo sum, I think, therefore, I am, *epitomized his viewpoint that he may doubt, he cannot doubt that he exists. In 1644, he published his four-part* Principia Phiosophiea, *which attempted to situate the entire universe on a mathematical base. In 1649, Queen Christina of Sweden summoned Descartes to come to Stockholm, but after only a few months in the bitterly cold weather, Descartes, who had bouts of bad health throughout his life, died of pneumonia on February 26, 1650.*

6.3   Sums of Four Squares

> *Freedom is the freedom to say that two plus two make four. If that is granted,*
> *all else follows.*
>   From Part I, Chapter 7 of **Nineteen Eighty-Four (1949)**
>   **George Orwell (Eric Blair) (1903–1950), English novelist**

The primary goal of this section is to show that four integer squares suffice
to represent all natural numbers. To this end, we first show that it suffices to
prove this result for primes. This is one of Euler's results on the topic.

Lemma 6.3 [**Euler's First Contribution**]

If m and n are sums of four squares, then so is mn.

Proof. Let $m = m_1^2 + m_2^2 + m_3^2 + m_4^2$ and $n = n_1^2 + n_2^2 + n_3^2 + n_4^2$. We leave the
details of the following identity as a verification for the reader.

$$mn = (m_1^2 + m_2^2 + m_3^2 + m_4^2)(n_1^2 + n_2^2 + n_3^2 + n_4^2) = \qquad (6.5)$$

$$(m_1n_1 + m_2n_2 + m_3n_3 + m_4n_4)^2 + (m_1n_2 - m_2n_1 + m_3n_4 - m_4n_3)^2$$

$$+(m_1n_3 - m_2n_4 - m_3n_1 + m_4n_2)^2 + (m_1n_4 + m_2n_3 - m_3n_2 - m_4n_1)^2,$$

which yields the result. $\square$

Example 6.4 *Since* $7 = 2^2 + 1^2 + 1^2 + 1^2$ *and* $6 = 2^2 + 1^2 + 1^2 + 0^2$, *then by*
(6.5),
$$42 = 6 \cdot 7 = (2^2 + 1^2 + 1^2 + 0^2)(2^2 + 1^2 + 1^2 + 1^2) =$$
$$(2 \cdot 2 + 1 \cdot 1 + 1 \cdot 1 + 0 \cdot 1)^2 + (2 \cdot 1 - 1 \cdot 2 + 1 \cdot 1 - 0 \cdot 1)^2$$
$$+(2 \cdot 1 - 1 \cdot 1 - 1 \cdot 2 + 0 \cdot 1)^2 + (2 \cdot 1 + 1 \cdot 1 - 1 \cdot 1 - 0 \cdot 2)^2 =$$
$$6^2 + 1^2 + 1^2 + 2^2.$$

Euler also contributed the following crucial piece.

Lemma 6.4 [**Euler's Second Contribution**]

If $p > 2$ is prime, then

$$x^2 + y^2 + 1 \equiv 0 \pmod{p} \qquad (6.6)$$

has a solution in nonnegative integers x, y with $x \le (p-1)/2$ and $y \le (p-1)/2$.

Proof. The proof centers around the two sets

$$\mathcal{S}_1 = \left\{1 + j^2 : \text{ for } j = 0, 1, 2, \ldots, (p-1)/2\right\},$$

and

$$\mathcal{S}_2 = \left\{-j^2 : \text{ where } j = 0, 1, 2, \ldots, (p-1)/2\right\}.$$

If $1 + j^2 \equiv 1 + k^2 \pmod{p}$ for $j, k \in \{0, 1, 2, \ldots, (p-1)/2\}$, then either $j \equiv k$ $\pmod{p}$ or $j \equiv -k \pmod{p}$. However, $0 < j + k < p$, so if the latter congruence holds, $j = k = 0$. If the former congruence holds, then $j = k$. We have shown that no two elements of $\mathcal{S}_1$ are congruent modulo p. Similarly, no two elements of $\mathcal{S}_2$ are congruent modulo p. Hence, the sets $\mathcal{S}_1$ and $\mathcal{S}_2$ together contain $2[1 + (p-1)/2] = p + 1$ integers. By the Pigeonhole Principle some integer in $\mathcal{S}_1$ must be congruent to some integer in $\mathcal{S}_2$ since there are only p distinct nonnegative integers less than p. Hence, there exists nonnegative integers $x, y \leq (p-1)/2$ such that

$$1 + x^2 \equiv -y^2 \pmod{p},$$

which is the desired result. $\square$

Corollary 6.3 *For a prime $p > 2$, there exists an integer $k < p$ such that kp is the sum of four integer squares.*

Proof. By Lemma 6.4, there exist integers x, y with $0 \leq x < p/2$ and $0 \leq y < p/2$ such that

$$x^2 + y^2 + 1 + 0^2 = kp$$

for some integer k. Given the bounds on x and y

$$kp = x^2 + y^2 + 1^2 + 0^2 < \frac{p^2}{4} + \frac{p^2}{4} + 1 < p^2,$$

so $k < p$, which is the result. $\square$

We are now in a position to prove that every prime is a sum of four squares.

Theorem 6.6 | **Sums of Four Squares Suffice for Primes**

If p is a prime then there exist integers x, y, z, w such that

$$p = x^2 + y^2 + z^2 + w^2.$$

Proof. Since $2 = 1^2 + 1^2 + 0^2 + 0^2$, then we may assume that p is odd. Let $k \in \mathbb{Z}$ be the smallest positive integer such that $kp = x^2 + y^2 + z^2 + w^2$, where $k < p$ by Corollary 6.3. If $2 \mid k$, then an even number of x, y, z, w have the same parity. Thus, without loss of generality, we may assume that $x \equiv y \pmod 2$ and $w \equiv z \pmod 2$. Therefore,

$$\frac{kp}{2} = \left(\frac{x-y}{2}\right)^2 + \left(\frac{x+y}{2}\right)^2 + \left(\frac{z-w}{2}\right)^2 + \left(\frac{z+w}{2}\right)^2,$$

which contradicts the minimality of k since all of the summands are integers. We have shown that k is odd. It remains to show that $k = 1$.

Assume that $k > 1$. Then by choosing the remainders of x, y, z, w when divided by k, respectively, we may select integers a, b, c, d such that

$$a \equiv x \pmod{k} \quad b \equiv y \pmod{k} \quad c \equiv z \pmod{k} \quad d \equiv w \pmod{k},$$

where

$$|a| < \frac{k}{2} \qquad |b| < \frac{k}{2} \qquad |c| < \frac{k}{2} \qquad |d| < \frac{k}{2}. \tag{6.7}$$

(Note that if the remainder r, upon selecting say a, is larger than $k/2$, then we merely replace r by $r - k$, so that (6.7) always holds.)

Therefore,

$$a^2 + b^2 + c^2 + d^2 \equiv x^2 + y^2 + z^2 + w^2 \equiv 0 \pmod{k},$$

so

$$a^2 + b^2 + c^2 + d^2 = kt$$

for some nonnegative integer t. Therefore, by (6.7),

$$0 \leq kt = a^2 + b^2 + c^2 + d^2 < 4 \left(\frac{k}{2} \right)^2 = k^2.$$

If $t = 0$, then $a = b = c = d = 0$, so k divides each of x, y, z, w, from which we get that $k^2 \mid kp$ so $k \mid p$, a contradiction since p is prime and $1 < k < p$. Thus, $1 \leq t < k$. Hence, by (6.5),

$$k^2 tp = (kp)(kt) = (x^2 + y^2 + z^2 + w^2)(a^2 + b^2 + c^2 + d^2) =$$

$$(xa + yb + zc + wd)^2 + (xb - ya + zd - wc)^2$$

$$+(xc - yd - za + wb)^2 + (xd + yc - zb - wa)^2.$$

However,

$$r = xa + yb + zc + wd \equiv a^2 + b^2 + c^2 + d^2 \equiv 0 \pmod{k},$$

$$s = xb - ya + zd - wc \equiv ab - ba + d^2 - d^2 \equiv 0 \pmod{k},$$

$$t = xc - yd - za + wb \equiv ac - bd - ca + db \equiv 0 \pmod{k},$$

and

$$u = xd + yc - zb - wa \equiv ad + bc - cb - da \equiv 0 \pmod{k}.$$

Therefore,

$$tp = \left(\frac{r}{k} \right)^2 + \left(\frac{s}{k} \right)^2 + \left(\frac{t}{k} \right)^2 + \left(\frac{u}{k} \right)^2,$$

is a sum of integer squares, but $1 \leq t < k$, contradicting the minimality of k. Hence, $k = 1$, as required. $\square$

Lagrange was the first to provide a complete proof of the following, which he achieved in the early 1770s, and he acknowledged the contributions of Euler given in Lemmas 6.3–6.4. Shortly thereafter, Euler found a simpler proof, and this is essentially what has been given here.

Theorem 6.7 | Lagrange's Four-Squares Theorem |

Every natural number may be represented as the sum of four squares of nonnegative integers.

Proof. Since $1 = 1^2 + 0^2 + 0^2 + 0^2$, and any natural number $n > 1$ has a canonical prime factorization, then by Lemma 6.3 and Theorem 6.6 the result follows. $\square$

Now that the problem of squares has been solved, the next natural question to ask is for higher powers. For instance, how many cubes does it take to represent any natural number? The answer, provided by L.E. Dickson is nine. For instance,

$$23 = 2^3 + 2^3 + 1^3 + 1^3 + 1^3 + 1^3 + 1^3 + 1^3 + 1^3.$$

The general question as to how many k^{th} powers it takes to represent a given natural number was raised in a book, published in 1770, entitled *Meditationes Algebraicae* by Edward Waring — see Biography 6.2 on the following page. In this book, he stated that each positive integer is the sum of at most 9 cubes, a sum of at most 19 fourth powers, and posed other conjectures for higher powers. Thus, Waring's name is attached to the problem of determining the number $g(k)$ of kth powers that represent any natural number, where $g(k)$ depends only on k and not on the number being represented.

◆ **Waring's Problem**

For a given $n > 1$, what is the number $g(k) \in \mathbb{N}$ such that any $n \in \mathbb{N}$ can be represented in the form

$$n = r_1^k + r_2^k + \cdots + r_{g(k)}^k,$$

where the not necessarily distinct r_j are nonnegative integers for $j = 1, 2, \ldots, g(k)$?

For instance, we have shown that $g(2) = 4$, which is Lagrange's result, and it is known that $g(3) = 9$, which is Dickson's result. It is worth noting however, that only 23 given above, and $239 = 4^3 + 4^3 + 3^3 + 3^3 + 3^3 + 3^3 + 1^3 + 1^3 + 1^3$, require 9 cubes in their representation. All numbers larger than 239 require at most 8 cubes. Furthermore, in 1942, Linnik proved that only a finite number of $n \in \mathbb{N}$ require 8 cubes, so for some $M \in \mathbb{N}$ all $n > M$ need at most 7 cubes. Thus, if we let $G(k)$ be the least positive integer such that *all sufficiently large* integers can be represented as a sum of at most $G(k)$ k^{th} powers, then we know, at the time of this writing, that $4 \leq G(3) \leq 7$. Indeed, the only definitely known values are $G(2) = 4$ and $G(4) = 16$. For $g(k)$ the known values are

$$g(2) = 4, \; g(3) = 9, \; g(4) = 19, \; g(5) = 37, \; g(6) = 73, \; g(7) = 143, \; g(8) = 279.$$

Indeed, it has been established that the following formula holds for all but a finite number of values of k, and there is considerable evidence that it holds for all $k \in \mathbb{N}$, namely

$$g(k) = \left\lfloor \left(\frac{3}{2} \right)^k \right\rfloor + 2^k - 2.$$

In 1909, Hilbert proved that $g(k)$ exists for all $k \in \mathbb{N}$, but his proof was not constructive, so no explicit method of finding $g(k)$ is known.

Biography 6.2 Edward Waring (1734–1798) *was born in Old Heath, Shropshire, England. His early education was at Shrewsbury School. On March 24, 1753, he entered Magdalene College on a scholarship, and soon impressed his teachers with his mathematical ability. On April 24, 1754, he was elected a fellow of Magdalene College, from which he graduated in 1757 with his B.A. Despite his young age he was nominated for the Lucasian Chair of Mathematics at Cambridge in 1759, and after some attempts to block his nomination, he was confirmed on January 28, 1760 as Lucasian professor. His aforementioned work* Meditationes Algebraicae *was an important contribution. It contained not only number theory and geometry, but also contributions to what we now call Galois theory. It was in honour of this work that he was elected a Fellow of the Royal Society in 1763, and received its Copley Medal in 1784.*

While holding his chair in mathematics, he surprisingly turned to the study of medicine and graduated with his M.D. in 1767. He even practiced for a while in various London hospitals. Part of what allowed him to digress was that the Lucasian Chair, did not require that he do any lectures as part of his duties. This was due largely to his atrocious communication skills.

He was considered by his colleagues to be both vain and modest, with the former being predominant. He was also said, near the end of his life, to sink into a deep religious depression that bordered on insanity. He died on August 15, 1798, in Pontesbury, Shropshire, England.

Exercises

6.23. Use Lemma 6.3 on page 254 to write 35 as a sum of four squares.

6.24. Use Lemma 6.3 to write 55 as a sum of four squares.

6.25. Let $n \equiv 0 \pmod{8}$ be a natural number. Prove that n is the sum of the squares of eight odd integers.

6.26. Prove that if $n \in \mathbb{N}$ is odd, then n is the sum of four squares, two of which are consecutive. (*Hint: Use Theorem 6.5 on page 252.*)

6.27. Prove that there are infinitely many primes of the form $a^2 + b^2 + c^2 + 1$.

   (*Hint: Use Example 4.8 on page 185 together with Theorem 6.5 on page 252.*)

6.28. Prove that there are no integers x, y, z such that $x^2 + y^2 + z^2 + x + y + z = 1$.

   (*Hint: Use Theorem 6.5.*)

6.4 Sums of Cubes

The pen is mightier than the sword, but no match for a gun.
— Mike Love, on the Beach Boys' album "Surf's Up" in "Student Demonstration Time," added the phrase "but no match for a gun" to the proverb coined by Edward Bulwer-Lytton in 1839 for his play *Richelieu; or the conspiracy.*
Edward George Earl Bulwer-Lytton (1803–1873), English novelist, playwright, and politician

At the end of §6.3, we looked at Waring's problem involving the representation of integers as the sum of kth powers for $k > 1$. We have solved the problem in this chapter for $k = 2$. Now we look at $k = 3$.

Theorem 6.8 | **Representations as Sums of Two Cubes**

If n has a representation as a sum of two cubes, then, $n \not\equiv \pm 3, \pm 4 \,(\text{mod } 9)$. If $p > 2$ is prime, then p is the sum of two cubes of integers if and only if $p = 3y(y+1) + 1$ where $y \in \mathbb{N}$.

Proof. If $n = x^3 + y^3$, then since $x^3, y^3 \equiv 0, 1, 8 \,(\text{mod } 9)$, then

$$x^3 + y^3 \equiv 0, 1, 2, 7, 8 \quad (\text{mod } 9),$$

so $n \not\equiv \pm 3, \pm 4 \,(\text{mod } 9)$.

If p is prime, and $p = x^3 + y^3$, then

$$p = (x + y)(x^2 - xy + y^2).$$

Thus, $x + y = 1$ or $x^2 - xy + y^2 = 1$. If both $x, y \in \mathbb{N}$, then the former is impossible, so $x + y = p$ and the latter holds. However,

$$x^2 - xy + y^2 = (x - y)^2 + xy = 1,$$

so $x = y$ and $xy = 1$. Hence, $p = 2$. Now assume that $p > 2$ and $p = x^3 - y^3$ for $x, y \in \mathbb{N}$. Thus, $p = (x - y)(x^2 + xy + y^2)$, which implies that $x - y = 1$ and

$$p = x^2 + xy + y^2 = (y + 1)^2 + (y + 1)y + y^2 = 3y(y + 1) + 1.$$

Conversely, if $p = 3y(y + 1) + 1$, then $p = (y + 1)^3 - y^3$. □

Example 6.5 *If $p = 7$, then p is of the form in Theorem 6.8 with $y = 1$ and we have $p = 2^3 - 1^3$. Indeed, as the proof of Theorem 6.8 shows, $p = (y + 1)^3 - y^3$ for each such prime. However, 5 is not so representable.*

Theorem 6.9 | Sums of Two Cubes Not Equal to a Cube

Suppose that $n > 2$ is a natural number not divisible by the cube of any natural number bigger than 1. Then if

$$x^3 + y^3 = nz^3 \qquad (6.8)$$

has a solution in integers x, y, z with $\gcd(x, y) = 1$, and $z \neq 0$, it has infinitely many such solutions.

Proof. Let x, y, z with $\gcd(x, y) = 1$, and $z \neq 0$ be a solution of (6.8). Then $\gcd(x, z) = \gcd(y, z) = 1$ since any divisor $d > 1$ of x and z, say, must satisfy that $d^3 \mid y^3 = nz^3 - x^3$, and this would contradict the hypothesis that $\gcd(x, y) = 1$. Let

$$g = \gcd(x(x^3 + 2y^3), -y(2x^3 + y^3), z(x^3 - y^3)). \qquad (6.9)$$

Then

$$x(x^3 + 2y^3) = gd_1, \qquad (6.10)$$

$$-y(2x^3 + y^3) = gd_2, \qquad (6.11)$$

and

$$z(x^3 - y^3)) = gd_3 \qquad (6.12)$$

for some $d_j \in \mathbb{Z}$ for $j = 1, 2, 3$. It follows that

$$d_1^3 + d_2^3 = nd_3^3.$$

We cannot have that $x = y$ since the relative primality of the two values would then imply that $x = y = \pm 1$. Thus, by (6.8), $nz^3 = \pm 2$ an impossibility since the hypothesis states that $n > 2$. Therefore, $x \neq y$, so by (6.12), $d_3 \neq 0$.

Claim 6.1 $\gcd(d_1, d_2) = 1$.

If $g = \gcd(d_1, d_2)$, then

$$g^3 \mid (d_1^3 + d_2^3) = nd_3^3.$$

However, if $g > 1$ and $\gcd(g, d_3) = 1$, then $g^3 \mid n$ contradicting the hypothesis that n is not divisible by such a value. Therefore, $\gcd(g, d_3) > 1$, in which case $\gcd(d_1, d_2, d_3) > 1$, which is impossible by Claim 1.3 on page 26. Thus, $g = 1$ which is the claim.

Claim 6.2 $\gcd(g, x) = \gcd(g, y) = \gcd(g, z) = 1$.

From Claim 6.1 it follows that $\gcd(d_1, d_3) = 1$, since $d_1^3 + d_2^3 = nd_3^3$. Also, since $\gcd(x, y) = 1$, then $\gcd(x, y^4) = 1$, which we may use to prove that $\gcd(g, x) = 1$ for if $\gcd(gd_2, x) = g_1$, then by (6.11), $g_1 \mid y^4$, so $g_1 = 1$. In particular, this says that $\gcd(g, x) = 1$. Similarly, using (6.10), we may conclude that $\gcd(gd_1, y) = 1$, so $\gcd(g, y) = 1$.

To complete Claim 6.2, it remains to show that $\gcd(g, z) = 1$. Let $d = \gcd(g, z)$. Since $\gcd(x, z) = \gcd(y, z) = 1$, then $\gcd(xy^3, z) = 1$. If $d \mid g$ and $d \mid z$, then by (6.8), $d \mid (x^3 + y^3) \mid x(x^3 + y^3)$ and by (6.10), $d \mid x(x^3 + 2y^3)$, from which it follows that $d \mid xy^3$. However, $d \mid z$ and $\gcd(xy^3, z) = 1$ so $d = 1$ which secures Claim 6.2.

By Claim 6.2, and (6.10)–(6.12), g must divide each of

$$x^3 + 2y^3, 2x^3 + y^3, \text{ and } x^3 - y^3,$$

from which it follows that g divides $x^3 + 2y^3 + 2(x^3 - y^3) = 3x^3$, but since $\gcd(g, x) = 1$ by Claim 6.2, then $g \mid 3$, so $g \leq 3$.

If $x = 0$, then $y = \pm 1$ since $\gcd(x, y) = 1$, but this contradicts (6.8) since $n > 2$. Thus, $x \neq 0$. Similarly, $y \neq 0$. Moreover, we have already shown that $x \neq y$ so $|x - y| \geq 1$.

Claim 6.3 $|x^3 - y^3| \geq 4$.

If $xy < 0$, then

$$x^2 - xy + y^2 = (x + y)^2 - 3xy \geq 4, \qquad (6.13)$$

since $x + y \neq 0$ by (6.8), since $n > 2$, and $x = -y$ would imply that $z = 0$. Now since, $xy < 0$, then $|x^3 - y^3| \geq |x^3 + y^3|$, so by (6.13),

$$|x^3 - y^3| \geq |x^3 + y^3| = |x + y|((x + y)^2 - 3xy) \geq |x + y| \cdot 4 \geq 4.$$

If $xy > 0$, then

$$x^2 + xy + y^2 = (x - y)^2 + 3xy \geq 1 + 3xy \geq 4.$$

Thus,

$$|x^3 - y^3| = |x - y||(x - y)^2 + 3xy| \geq 4.$$

This completes Claim 6.3.

By Claim 6.3, since $g \leq 3$, then by (6.12), $|d_3| = |z||x^3 - y^3|/g > |z|$. What this demonstrates is that if (6.8) holds with the conditions in the hypothesis, then we can get another solution (d_1, d_2, d_3) with $\gcd(d_1, d_2) = 1$ and $|d_3| > |z|$. This proves that there are infinitely many solutions. $\qquad \square$

Theorem 6.10 | The Number of Sums of Two Cubes

For any $r \in \mathbb{N}$ there exists an $n \in \mathbb{N}$ that is representable as a sum of two cubes of integers in at least r different ways.

Proof. By Theorem 6.9, there are infinitely many solutions to $x^3 + y^3 = 7z^3$ with $\gcd(x, y) = 1$. Let (x_j, y_j, z_j) for $j = 1, 2, \ldots$ be such solutions with $0 < |z_j| < |z_{j+1}|$ for each $j \in \mathbb{N}$. Without loss of generality, we may assume that $z_j > 0$ for each j by selecting alternate signs for x_j and y_j if necessary.

Let

$$n = 7z_1^3 z_2^3 \cdots z_r^3, \quad u_j = \frac{z_1 z_2 \cdots z_r}{z_j} x_j, \quad v_j = \frac{z_1 z_2 \cdots z_r}{z_j} y_j.$$

We have that $u_j, v_j \in \mathbb{Z}$ and $n = u_j^3 + v_j^3$. To complete the proof, we need only show that $u_i \neq u_j$ for any natural numbers $i \neq j$. If $u_i = u_j$, then since all z_k are nonzero, $x_i/z_i = x_j/z_j$, but $\gcd(x_i, z_i) = \gcd(x_j, z_j) = 1$, so $x_i = x_j$ and $z_i = z_j$, which is a contradiction if $i \neq j$ since the solutions are distinct, so $i = j$. A similar argument shows that $u_i \neq v_j$. Hence, there are r different representations of n as a sum of two cubes of integers. $\qquad\square$

Example 6.6 *Suppose that $r = 3$ in Theorem 6.10. Then we employ the technique in the proof as taken from Theorem 6.9 to provide an integer n representable as a sum of two cubes in three ways.*

$$x^3 + y^3 = 7z^3$$

has solution $x_1 = 2$, $y_1 = -1$ and $z_1 = 1$. From this employing (6.9)–(6.12), we get $x_2 = 4$, $y_2 = 5$, and $z_2 = 3$. Then using (6.9)–(6.12) on these values we get $x_3 = 1256$, $y_3 = -1265$, and $z_3 = -183$. Now we take

$$n = 7z_1^3 z_2^3 z_3^3 = 7 \cdot 1^3 \cdot 3^3 \cdot (-183)^3 = -1158284043.$$

We have that

$$u_1 = z_2 z_3 x_1 = -1098, \quad v_1 = z_2 z_3 y_1 = 549,$$

$$u_2 = z_1 z_3 x_2 = -732, \quad v_2 = z_1 z_3 y_2 = -915,$$

and

$$u_3 = z_1 z_2 x_3 = 3768, \quad v_3 = z_1 z_2 y_3 = -3795.$$

In each case we get that

$$n = -1158284043 = u_j^3 + v_j^3,$$

for $j = 1, 2, 3$, which the reader may check.

This demonstrates the constructive algorithm given by employing (6.9)–(6.12) for finding the representations of n in r different ways, a valuable methodology.

Theorem 6.11 $\boxed{\text{**Sums of Three Rational Cubes Suffice**}}$

Any positive rational number is the sum of three rational cubes.

Proof. Given $r \in \mathbb{Q}$ we need to find positive $x, y, z \in \mathbb{Q}$ such that

$$r = x^3 + y^3 + z^3. \tag{6.14}$$

Claim 6.4 (6.14) *is equivalent to*

$$r = (x + y + z)^3 - 3(y + z)(z + x)(x + y).$$

We expand

$$(x + y + z)^3 = x^3 + y^3 + z^3 + 3xy^2 + 3xz^2 + 3x^2y + 6xyz + 3x^2z + 3yz^2 + 3y^2z,$$

and

$$3(y + z)(z + x)(x + y) = 3(xy^2 + xz^2 + x^2y + 2xyz + x^2z + yz^2 + y^2z),$$

from which the claim follows.
    In Claim 6.4, set

$$X = y + z, \quad Y = z + x, \quad \text{and } Z = x + y.$$

Then the representation in Claim 6.4 becomes

$$8r = (X + Y + Z)^3 - 24XYZ. \tag{6.15}$$

Now set $a = (X + Z)/Z$ and $b = Y/Z$. Then (6.15) becomes

$$8rZ^{-3} = (a + b)^3 - 24b(a - 1). \tag{6.16}$$

We now simplify by requiring that Z and b satisfy

$$r = 3Z^3 b, \tag{6.17}$$

so (6.16) becomes

$$(a + b)^3 = 24ab. \tag{6.18}$$

If we set

$$a = 24u^2/(u + 1)^3 \text{ and } b = 24u/(u + 1)^3, \tag{6.19}$$

then this is a solution to (6.18) for all $u \in \mathbb{Q}$. However, this was contingent upon (6.17) holding, which now becomes

$$r(u + 1)^3 = 72Z^3 u.$$

Let $v \in \mathbb{Q}$ and set $u = r/(72v^3)$. Then $Z = v(u + 1)$. It follows that a solution of (6.15) is

$$X = (a - 1)Z, \quad Y = bZ, \quad Z = v(u + 1), \tag{6.20}$$

where a and b are given by (6.19) with $u = rv^{-3}/72$. Therefore a solution of (6.14) is given by

$$2x = Y + Z - X, \quad 2y = Z + X - Y, \quad 2z = X + Y - Z. \tag{6.21}$$

To complete the proof, we need to verify that v may be selected such that x, y, z are all positive. If v is positive, then u and Z are positive. Also, by (6.20)–(6.21),

$$\frac{2x}{Z} = b + 1 - (a - 1) = 2 + b - a, \quad \frac{2y}{Z} = a - b, \text{ and } \frac{2z}{Z} = a + b - 2,$$

all of which are positive provided that $a > b$, and $a - b < 2 < a + b$, namely from (6.19),

$$u > 1, \text{ and } 12u(u - 1) < (u + 1)^3 < 12u(u + 1),$$

and the latter holds if $2 \geq u > 1$. Since we may select v so that

$$2 \geq u = r/(72v^3) > 1,$$

then we are done. $\square$

Example 6.7 *We have that*

$$434 = 3^3 + 4^3 + 7^3 \text{ and } 862 = 2^3 + 5^3 + 9^3$$

as illustrations of Theorem 6.11.

Remark 6.5 *It can be demonstrated that every positive rational number is the sum of three rational positive cubes in infinitely many ways. Moreover, it can be shown that for any integer $r > 2$, any positive rational number has infinitely many representations as the sum of the cubes of r positive rational numbers. However, some techniques in proving these results are not "fundamental," so Theorem 6.11 will suffice as a closing feature of this section.*

Exercises

6.29. Prove that if $n \equiv 0 \pmod 6$, then n is a sum of four cubes.

6.30. Use Exercise 6.29 to prove that any integer has infinitely many representations as a sum of five cubes.

6.31. Prove that there are infinitely many natural numbers x, y, z, w such that $x^3 = 2 + y^3 + z^3 + w^3$.

6.32. Prove that there are infinitely many natural numbers x, y, z, w such that $x^3 = 3 + y^3 + z^3 + w^3$.

6.33. Prove that there are infinitely many natural numbers n for which $n + x^3 + y^3 = z^3 + w^3$ has infinitely many solutions $x, y, z, w \in \mathbb{N}$.

6.34. Prove that there are infinitely many natural numbers n for which $n + x^3 = y^3 + z^3 + w^3$ has infinitely many solutions $x, y, z, w \in \mathbb{N}$.

   (*Hint: Select natural numbers of the form $n = u^3 - v^3$.*)

Chapter 7

Diophantine Equations

> *In language there are only differences.*
> From **Course in General Linguistics (1916)**
> **Ferdinand de Saussure (1857–1913), Swiss Linguistics scholar**

In this chapter we look at Diophantine equations, many instances of which we have already encountered. For instance, in Exercises 5.39–5.40 on page 242, we looked at differences of squares which is a special case of the topic in the following section.

7.1 Norm-Form Equations

Positive norm-form equations are those of the form

$$x^2 - Dy^2 = n, \tag{7.1}$$

where D is a positive integer that is not a perfect square, and $n \in \mathbb{Z}$ is nonzero. The case where $D = 1$, as noted above, was covered earlier, which is why we may assume D *not* to be a perfect square. Now we look at integers $D > 1$. We encountered norm-form equations earlier in §5.3 in Equation (5.24) on page 232.

Definition 7.1 | **Primitive Solutions**

If $x, y \in \mathbb{Z}$, $n \in \mathbb{N}$ such that $N(x + y\sqrt{D}) = x^2 - Dy^2 = n$, then $x + y\sqrt{D}$ is called a solution *of Equation (7.1). If $\gcd(x, y) = 1$, then $x + y\sqrt{D}$ is called a* primitive solution. *We use the notations (x, y), and $x + y\sqrt{D}$ interchangeably to denote solutions of Equation (7.1).*

Now suppose that $u + v\sqrt{D}$ is a solution of the Pell equation (see Theorem 5.15 on page 234 for all solutions of Pell's equations, both positive and negative, via continued fractions):

$$x^2 - Dy^2 = 1. \tag{7.2}$$

If $x_1 + y_1\sqrt{D}$ is a solution of Equation (7.1), then so is

$$(u + v\sqrt{D})(x_1 + y_1\sqrt{D}) = ux_1 + vy_1 D + (uy_1 + vx_1)\sqrt{D},$$

since $N(u + v\sqrt{D})N(x_1 + y_1\sqrt{D}) = N(x_1 + y_1\sqrt{D})$. Indeed, we saw in Remark 5.3 on page 237 how to generate infinitely many solutions of Equation (7.1) from one solution in this fashion. Solutions x_1, y_1 *associated* with solutions of Equation (7.1) in this fashion have a special consideration.

Definition 7.2 | Classes of Solutions |

If $\alpha_j = x_j + y_j\sqrt{D}$ for $j = 1,2$ are primitive solutions of Equation (7.1), then they are said to be in the same class provided that there is a solution $\beta = u + v\sqrt{D}$ of Equation (7.2) such that $\alpha_1\beta = \alpha_2$. If α_1 and $\alpha_1' = x_1 - y_1\sqrt{D}$ are in the same class, then the class is called ambiguous. *In a given class, let $\alpha_0 = x_0 + y_0\sqrt{D}$ be a primitive solution with least possible positive y_0. If the class is ambiguous, then we require that $x_0 \geq 0$. Also, $|x_0|$ is the least possible value for any x with $x + y\sqrt{D}$ in its class, and so α_0 is uniquely determined — see Claim 7.1 on the facing page. We call α_0 the* fundamental solution *in its class.*

It is always convenient to have a simple criterion for membership in a given set.

Proposition 7.1 | Equivalence of Primitive Solutions |

Two primitive solutions $x_j + y_j\sqrt{D}$ for $j = 1,2$ of Equation (7.1) are in the same class if and only if both

$$(x_1x_2 - y_1y_2 D)/n \in \mathbb{Z} \text{ and } (y_1x_2 - x_1y_2)/n \in \mathbb{Z}.$$

Proof. By Definition 7.2, there is a solution $u + v\sqrt{D}$ of Equation (7.2) such that

$$(u + v\sqrt{D})(x_1 + y_1\sqrt{D}) = ux_1 + vy_1 D + (uy_1 + vx_1)\sqrt{D} = x_2 + y_2\sqrt{D}.$$

Thus,

$$x_2 = ux_1 + vy_1 D, \tag{7.3}$$

and

$$y_2 = uy_1 + vx_1. \tag{7.4}$$

Multiplying Equation (7.3) by y_1, and subtracting x_1 times Equation (7.4), we get $y_1 x_2 - x_1 y_2 = -v(x_1^2 - y_1^2 D) = -vn$, so

$$(y_1 x_2 - x_1 y_2)/n \in \mathbb{Z}.$$

Similarly, multiplying Equation (7.3) by x_1 and subtracting $y_1 D$ times Equation (7.4), we get $x_1 x_2 - y_1 y_2 D = u(x_1^2 - y_1^2 D) = un$, so

$$(x_1 x_2 - y_1 y_2 D)/n \in \mathbb{Z}.$$

$\square$

Now we determine bounds on the coefficients of the fundamental solution.

Theorem 7.1 $\boxed{\text{Bounds on Fundamental Solutions I}}$

Let $n > 1$ in Equation (7.1), and let $\alpha_0 = x_0 + y_0 \sqrt{D}$ be the fundamental solution in its class. If $\beta_0 = u_0 + v_0 \sqrt{D}$ is the fundamental solution of Equation (7.2), then

$$0 < |x_0| \le \sqrt{(u_0 + 1)n/2}, \tag{7.5}$$

and

$$0 \le y_0 \le \frac{v_0 \sqrt{n}}{\sqrt{2(u_0 + 1)}}. \tag{7.6}$$

Proof. First we establish that α_0 is unique based upon the minimality of $|x_0|$.

Claim 7.1 *α_0 is uniquely determined in the sense that $|x_0|$ is the smallest possible in its class.*

Suppose that $\alpha_0 = x_0 + y_0 \sqrt{D}$, with $y_0 > 0$, is the smallest possible primitive solution in its class. Let $\alpha_1 = x_1 + y_1 \sqrt{D}$, with $y_1 = y_0$. Then by Proposition 7.1 on the preceding page,

$$(y_0 x_1 - x_0 y_0)/n \in \mathbb{Z},$$

so

$$y_0 x_1 \equiv x_0 y_0 \pmod{n}.$$

Therefore, since the solutions are primitive, we may divide through by y_0 since $x_0^2 - y_0^2 D = n$, to get

$$x_1 \equiv x_0 \pmod{n}.$$

If $x_1 = x_0 + nt$ for some $t \in \mathbb{Z}$, then

$$(x_0 + nt)^2 - y_0^2 D = n,$$

so

$$x_0^2 + 2ntx_0 + n^2 t^2 - y_0^2 D = n.$$

Hence,
$$2ntx_0 + n^2t^2 = 0,$$
so $x_0 = -nt/2$, and $x_1 = nt/2$. Therefore, $x_0 = -x_1$. By Definition 7.1 on page 265, the class is ambiguous and only $x_0 > 0$ is allowed. Thus, α_0 is unique. This secures Claim 7.1.

Now observe that a solution in the class of α_0 is
$$(x_0 + y_0\sqrt{D})(u_0 - v_0\sqrt{D}) = x_0u_0 - y_0v_0D + (y_0u_0 - x_0v_0)\sqrt{D}.$$

Next, we see that from Equations (7.1)–(7.2),
$$v_0^2 y_0^2 D^2 = (x_0^2 - n)(u_0^2 - 1) < x_0^2 u_0^2,$$

so
$$|x_0|u_0 - v_0y_0D > 0,$$
and by the minimality of $|x_0|$ from Claim 7.1,
$$|x_0|u_0 - v_0y_0D \geq |x_0|.$$

Thus, by squaring, using Equations (7.1)–(7.2), and rewriting, we get
$$x_0^2(u_0 - 1)^2 \geq (x_0^2 - n)(u_0^2 - 1),$$

and by dividing through by $u_0 - 1$, we get
$$x_0^2(u_0 - 1) \geq (x_0^2 - n)(u_0 + 1).$$

By rewriting with the summands involving x_0^2 on the right, we get
$$n(u_0 + 1) \geq x_0^2[(u_0 + 1) - (u_0 - 1)] = 2x_0^2.$$

Therefore,
$$0 < |x_0| \leq \sqrt{(u_0 + 1)n/2},$$

which is inequality (7.5).

From Equations (7.1)–(7.2), we get
$$x_0^2 - n = Dy_0^2 = y_0^2(u_0^2 - 1)/v_0^2,$$

so by inequality (7.5),
$$n(u_0 + 1)/2 - n \geq x_0^2 - n = y_0^2(u_0^2 - 1)/v_0^2,$$

and multiplying both sides by $v_0^2/(u_0^2 - 1)$, we get
$$nv_0^2(u_0 + 1)/[2(u_0^2 - 1)] - nv_0^2/(u_0^2 - 1) \geq y_0^2.$$

Since $nv_0^2(u_0 + 1)/[2(u_0^2 - 1)] - nv_0^2/(u_0^2 - 1) = nv_0^2/[2(u_0 + 1)]$, then

$$0 \leq y_0 \leq \frac{v_0\sqrt{n}}{\sqrt{2(u_0 + 1)}},$$

which is inequality (7.6), and so we have the entire result. $\square$

Now we determine such bounds when $n < 0$ in Equation (7.1).

Theorem 7.2 | **Bounds on Fundamental Solutions II** |

Let $n < -1$ in Equation (7.1), and let $\alpha_0 = x_0 + y_0\sqrt{D}$ be the fundamental solution in its class. If $\beta_0 = u_0 + v_0\sqrt{D}$ is the fundamental solution of Equation (7.2), then

$$0 \le |x_0| \le \sqrt{(u_0 - 1)|n|/2},\tag{7.7}$$

and

$$0 < y_0 \le \frac{v_0\sqrt{|n|}}{\sqrt{2(u_0 - 1)}}.\tag{7.8}$$

Proof. As in the proof of Theorem 7.1 on page 267,

$$x_0 u_0 - y_0 v_0 D + (y_0 u_0 - x_0 v_0)\sqrt{D}$$

is a solution in the class of α_0. Since

$$u_0^2 y_0^2 = (Dv_0^2 + 1)((x_0^2 - n)/D) > v_0^2 x_0^2,$$

then $y_0 u_0 - x_0 v_0 > 0$. Therefore, by the minimality of y_0, $y_0 u_0 - x_0 v_0 > y_0$, or by rewriting, multiplying by D, and squaring we get $Dx_0^2 v_0^2 \le (u_0 - 1)^2 y_0^2 D$. Using Equations (7.1)–(7.2), we get

$$x_0^2(u_0^2 - 1) \le (u_0 - 1)^2(x_0^2 - n).$$

By rewriting with the x_0^2 terms on the left, we get

$$x_0^2[(u_0^2 - 1) - (u_0 - 1)^2] \le -n(u_0 - 1)^2,$$

so $2x_0^2 \le -n(u_0 - 1)$, namely

$$0 \le |x_0| \le \sqrt{(u_0 - 1)|n|/2},$$

which is inequality (7.7). By the same reasoning as in the proof of Theorem 7.1, we can use this inequality on $x_0^2 - n = Dy_0^2$ to get

$$-n(u_0 - 1)/2 - n \ge y_0^2(u_0^2 - 1)/v_0^2,$$

from which it follows that

$$-nv_0^2(u_0 + 1)/[2(u_0^2 - 1)] \ge y_0^2,$$

namely

$$0 \le y_0 \le \frac{v_0\sqrt{|n|}}{\sqrt{2(u_0 - 1)}},$$

which is inequality (7.8). $\square$

Remark 7.1 *From the above, there are only finitely many classes of primitive solutions of Equation (7.1). The reason is that since n is fixed, the fundamental solutions of all the classes can be found after a finite number of iterations by using inequalities (7.5)–(7.8) in Theorems 7.1–7.2. Also, Theorems 7.1–7.2 tell us how to find all of the solutions in the finitely many classes of Equation (7.1) for any $n \in \mathbb{Z}$. If $u_0 + v_0\sqrt{D}$ is a fundamental solution of Equation (7.2) and if $x_0 + y_0\sqrt{D}$ runs over all the fundamental solutions of the classes of Equation (7.1), then*

$$\pm(u_0 + v_0\sqrt{D})^m(x_0 + y_0\sqrt{D}) \quad (m \in \mathbb{Z})$$

provide all its solutions. See Remark 5.3 on page 237.

Example 7.1 *If $D = 13$ and $n = -27$, then a fundamental solution for the equation $x^2 - 13y^2 = -27$ is $\alpha_0 = 5 + 2\sqrt{13}$. Since a fundamental solution of*

$$x^2 - 13y^2 = 1$$

is

$$\beta_0 = 649 + 180\sqrt{13} = u_0 + v_0\sqrt{D},$$

then all solutions in the class of α_0 are given by $\beta_0^m \alpha_0$ where $m \in \mathbb{Z}$. For instance, if $m = -1$, then

$$\beta_0^{-1}\alpha_0 = \beta'\alpha_0 = (649 - 180\sqrt{13})(5 + 2\sqrt{13}) = -1435 + 398\sqrt{13},$$

is a solution in the class of α_0. Also, if $m = 1$, then

$$\beta_0\alpha_0 = (649 + 180\sqrt{13})(5 + 2\sqrt{13}) = 7925 + 2198\sqrt{13},$$

is a solution in the class of α_0.

By Theorem 7.2, any fundamental solution $x_0 + y_0\sqrt{13}$ must satisfy

$$0 < y_0 \le 180\sqrt{\frac{27}{2(649 - 1)}} < 26.$$

A check shows that the only such natural numbers y_0 that satisfy $x^2 - 13y^2 = -27$ are $y_0 = 2$ and $y_0 = 6$. However, $y_0 = 6$ does not yield a primitive solution. Therefore, α_0 and

$$-\alpha_0' = -5 + 2\sqrt{13}$$

are the only possible fundamental solutions. An example of a solution in the class of $-\alpha_0'$ is for $m = 1$, namely

$$(649 + 180\sqrt{13})(-5 + 2\sqrt{13}) = 1435 + 398\sqrt{13}.$$

In Example 1.22 on page 46, and in Exercise 1.62 on page 48, we determined those primes representable in the form $x^2 + 2y^2$. In Theorem 6.1 on page 244, we found those primes representable as a sum of two squares. Also, Exercise 1.61 on page 48 answered the question about those primes representable in the form $p = x^2 + 3y^2$. Now we look at the analogous question for primes p with $\pm p = x^2 - cy^2$ for $c = 1, 2, 3$.

Proposition 7.2 (Primes Representable as Norm Forms)
Let $p > 2$ be prime.

(a) There exist $x, y \in \mathbb{N}$ such that $\pm p = x^2 - 2y^2$ if and only if $p \equiv \pm 1 \pmod{8}$.

(b) There exist $x, y \in \mathbb{N}$ such that $p = x^2 - 3y^2$ if and only if $p \equiv 1 \pmod{12}$.

(c) There exist $x, y \in \mathbb{N}$ with $-p = x^2 - 3y^2$ if and only if $p \equiv -1 \pmod{12}$.

Proof. If $\pm p = x^2 - 2y^2$ and $p \equiv \pm 3 \pmod{8}$, then since x must be odd, $-2y^2 \equiv 2, -4 \pmod{8}$, so $y^2 \equiv -1, 2 \pmod{4}$, a contradiction. This is one direction of part (a). For the converse of part (a) we proved this in Example 1.23 on page 47, via Corollary 4.2 on page 185 for the $+$ sign. The identity

$$p = 2y^2 - x^2 = (x + 2y)^2 - 2(x + y)^2$$

proves it for the $-$ sign. This completes part (a).

For part (b), if $p = x^2 - 3y^2$, then the Legendre symbol $\left(\frac{p}{3}\right) = 1$, so by Example 4.11 on page 191, $p \equiv 1, 7 \pmod{12}$. If $p \equiv 7 \pmod{12}$, then since one of x or y is even, we have either

$$p = x^2 - 3y^2 \equiv x^2 \equiv 7 \equiv 3 \pmod{4},$$

or

$$p = x^2 - 3y^2 \equiv -3y^2 \equiv 7 \pmod{4}, \text{ which implies that } y^2 \equiv 3 \pmod{4},$$

both of which are contradictions since -1 is not a quadratic residue modulo 4. Hence,

$$p \equiv 1 \pmod{12}.$$

Conversely, if $p \equiv 1 \pmod{12}$, then by Example 4.11, there exists a $w \in \mathbb{Z}$ such that

$$w^2 \equiv 3 \pmod{p}.$$

By Thue's Theorem on page 44, there exist natural numbers $x, y < \sqrt{p}$ such that

$$wy \equiv \pm x \pmod{p}.$$

Therefore, since $3 - p < 3y^2 - x^2 < 3p - 1$,

$$x^2 - 3y^2 = -p, \text{ or } -2p.$$

However, since $p \equiv 1 \pmod{12}$, then if $x^2 - 3y^2 = -p$, we have

$$x^2 \equiv -1 \pmod 3,$$

a contradiction. Thus, $x^2 - 3y^2 = -2p$, so

$$\left(\frac{x+3y}{2}\right)^2 - 3\left(\frac{x+y}{2}\right)^2 = p.$$

For part (c), assume that $p \equiv -1 \pmod{12}$. We proceed exactly as in part (2) to get that

$$x^2 - 3y^2 = -p, \text{ or } -2p.$$

In this case, if $x^2 - 3y^2 = -2p$, then $x^2 \equiv 2 \pmod 3$, a contradiction. Hence,

$$x^2 - 3y^2 = -p,$$

as required.   □

Example 7.2 *If $p = 1553 \equiv 1 \pmod 8$, then*

$$p = 41^2 - 2\cdot 8^2 \text{ and } -p = 25^2 - 2\cdot 33^2.$$

Example 7.3 *If $p = 1031 \equiv -1 \pmod 8$, then*

$$p = 37^2 - 2\cdot 13^2 \text{ and } -p = 11^2 - 2\cdot 24^2.$$

Example 7.4 *If $p = 2137 \equiv 1 \pmod{12}$, then $p = 50^2 - 3\cdot 11^2$.*

Example 7.5 *If $p = 1223 \equiv -1 \pmod{12}$, then $-p = 10^2 - 3\cdot 21^2$.*

Example 7.6 *If $p = 9601 \equiv 1 \pmod{12}$, then $p = 98^2 - 3\cdot 1^2$. Also, $p = 9601 \equiv 1 \pmod 8$ and $p = 99^2 - 2\cdot 10^2$.*

We close this section with some comments on Diophantine equations of the form $x^2 - Dy^2 = n$ where $n \in \mathbb{Z}$, $D < 0$, and $-D$ is not a perfect square, called *negative norm-form equations*. For instance, the case where $n = p^a$ for $a \in \mathbb{N}$ and p a prime is called the *generalized Ramanujan-Nagell equation*. The original equation that stimulated the interest is $x^2 - Dy^2 = 2^a$, called the *Ramanujan-Nagell equation*. Ramanujan knew of solutions to the equations for $a = 3,4,5,7,15$ and in 1948, the Norwegian mathematician Nagell proved that these are all of the solutions. (See Biography 7.1 on the facing page.) For a detailed history and proofs of the above facts plus more, see [29]. To solve negative norm-form equations requires some algebraic number theoretic techniques that we leave for a second course in number theory.

Biography 7.1 Srinivasa Ramanujan (1887–1920) *was born in the Tanjore district of Madras, India. He graduated from high school in 1904 and won a scholarship to the University of Madras. He enrolled in the fine arts curriculum, but his interest in mathematics caused him to neglect other subjects to the point where he lost that scholarship. During this time he had notebooks that he filled with mathematical discoveries, many of which were original and some of which were rediscoveries. He did not finish his university degree and this caused problems in his finding a job. However, in 1912 he was hired as an accounts clerk that allowed him to support himself and his wife.*
He continued his mathematical investigations, publishing his first paper in India in 1910. However he sought help from outside India, and contacted some British mathematicians. One of them was G.H. Hardy (1877–1947) who saw the raw talent and invited Ramanujan to England in 1914. Hardy tutored Ramanujan and they collaborated for five years. There they wrote a series of papers together, and Hardy became his friend and mentor. During this time Ramanujan contributed to number theory, including elliptic functions, infinite series, and continued fractions. His work caused him to become one of the youngest members ever appointed as a Fellow of the Royal Society. In 1917 he became quite ill and returned to India in 1919. He died there the following year from tuberculosis.

Exercises

7.1. Find the fundamental solution of $x^2 - 19y^2 = -2$ and the fundamental solution of $x^2 - 19y^2 = 1$. Then use Remark 7.1 on page 270 to find two more solutions of the former norm-form equation.

7.2. Find the fundamental solution of $x^2 - 29y^2 = -5$ and the fundamental solution of $x^2 - 29y^2 = -1$. Then use Remark 7.1 to find two more solutions of the former norm-form equation.

7.3. Using Theorem 5.15 on page 234, find the fundamental solution of $x^2 - 97y^2 = -1$.

7.4. Using Exercise 7.3, find the fundamental solution of $x^2 - 97y^2 = 1$.

7.5. Represent $p = 3391$ in the form $x^2 - 2y^2$.

7.6. Represent $p = 6733$ in the form $p = x^2 - 3y^2$.

7.7. Represent $-p$ in the form $x^2 - 3y^2$ where $p = 9539$.

7.8. Represent $-p$ in the form $x^2 - 2y^2$ where $p = 7703$.

7.2 The Equation $ax^2 + by^2 + cz^2 = 0$

> *Form follows function.*
> From **The Tall Office Building Artistically Considered (1896)**
> **Louis Henri Sullivan (1856–1924), American architect**

One of the most elegant results in the theory of quadratic Diophantine equations is Legendre's following result — see Biography 4.1 on page 181.

Theorem 7.3 | Legendre's Theorem

Let $a, b, c \in \mathbb{Z}$ be nonzero, squarefree, and not all of the same sign. Then

$$f(x, y, z) = ax^2 + by^2 + cz^2 = 0 \tag{7.9}$$

has a solution $x, y, z \in \mathbb{Z}$ with $(x, y, z) \neq (0, 0, 0)$ if and only if $-ab$, $-bc$, and $-ca$ are quadratic residues of $|c|$, $|a|$, and $|b|$, respectively.

Proof. Suppose that Equation (7.9) has a nontrivial solution (x, y, x) with $\gcd(x, y, z) = 1$. Suppose there is a prime p dividing $\gcd(x, c) = g$. Then $p \mid by^2$, so $p \mid y$ since $\gcd(x, y, z) = 1$. Therefore, $p \mid (ax^2 + by^2)$, so $p^2 \mid c$ contradicting the hypothesis. Hence, $\gcd(x, c) = 1$ and by a similar argument $\gcd(y, c) = 1$. Thus, since

$$ax^2 + by^2 \equiv 0 \pmod{|c|} \tag{7.10}$$

it follows that

$$(axy^{-1})^2 \equiv -ab \pmod{|c|},$$

namely $-ab$ is a quadratic residue of $|c|$. Similarly, $-bc$ and $-ca$ are quadratic residues of $|a|$ and $|b|$, respectively. We have proved the necessity of the conditions.

Conversely, assume that $-ab$, $-bc$, and $-ca$ are quadratic residues of $|c|$, $|a|$, and $|b|$, respectively. Since a, b, c are not of the same sign, we may assume without loss of generality that two of them are negative and one of them is positive, say $a > 0$, $b < 0$, and $c < 0$. By hypothesis, there exists a $d \in \mathbb{Z}$ such that $-ab \equiv d^2 \pmod{|c|}$ and $ah \equiv 1 \pmod{|c|}$ for some $h \in \mathbb{Z}$. Now consider,

$$ax^2 + by^2 \equiv ah(ax^2 + by^2) \equiv h(a^2x^2 + aby^2) \equiv h(a^2x^2 - d^2y^2) \equiv$$

$$h(ax - dy)(ax + dy) \equiv (x - hdy)(ax + dy) \pmod{|c|}.$$

Therefore,

$$ax^2 + by^2 + cz^2 \equiv (x - hdy)(ax + dy) \pmod{|c|}.$$

Similarly, $ax^2 + by^2 + cz^2$ is a product of two linear factors modulo a and $|b|$. To continue, we need the following.

Claim 7.2 $ax^2 + by^2 + cz^2$ *factors into two linear factors modulo abc.*

It suffices to prove that $ax^2 + by^2 + cz^2$ factors into two linear factors modulo $|ab|$ since the same argument can then be used on ab and c. Suppose that

$$ax^2 + by^2 + cz^2 \equiv (a_1x + b_1y + c_1z)(a_2x + b_2y + c_2z) \pmod{a}$$

and

$$ax^2 + by^2 + cz^2 \equiv (a_3x + b_3y + c_3z)(a_4x + b_4y + c_4z) \pmod{|b|}.$$

Since a, b, c are pairwise relatively prime, then we may use the Chinese Remainder Theorem, to select m, n, r, s, t, u such that

$$m \equiv a_1, n \equiv b_1, r \equiv c_1, s \equiv a_2, t \equiv b_2, u \equiv c_2 \pmod{a},$$

$$m \equiv a_3, n \equiv b_3, r \equiv c_3, s \equiv a_4, t \equiv b_4, u \equiv c_4 \pmod{|b|}.$$

Therefore, the congruence,

$$ax^2 + by^2 + cz^2 \equiv (mx + ny + rz)(sx + ty + uz) \pmod{a}$$

holds modulo $|a|$ and $|b|$, so it holds modulo $|ab|$, which secures the claim.

Using Claim 7.2, we have that there are integers m, n, r, s, t, u such that

$$ax^2 + by^2 + cz^2 \equiv (mx + ny + rz)(sx + ty + uz) \pmod{abc}. \tag{7.11}$$

To continue further, we need the following.

Claim 7.3 *The congruence*

$$mx + ny + rz \equiv 0 \pmod{abc} \tag{7.12}$$

has a solution x, y, z *such that* $|x| \leq \sqrt{bc}$, $|y| \leq \sqrt{|ac|}$, *and* $|z| \leq \sqrt{|ab|}$.

Consider the set,

$$\mathcal{S} = \{(x, y, z) : 0 \leq x \leq \lfloor\sqrt{bc}\rfloor; 0 \leq y \leq \lfloor\sqrt{|ac|}\rfloor; 0 \leq z \leq \lfloor\sqrt{|ab|}\rfloor\}.$$

Then the cardinality of $\mathcal{S}$ is

$$(1 + \lfloor\sqrt{bc}\rfloor)(1 + \lfloor\sqrt{|ac|}\rfloor)(1 + \lfloor\sqrt{|ab|}\rfloor) > \sqrt{bc}\sqrt{|ac|}\sqrt{|ab|} = abc.$$

Thus, by the Pigeonhole Principle, there must be two triples (x_1, y_1, z_1) and (x_2, y_2, z_2) such that

$$mx_1 + ny_1 + rz_1 \equiv mx_2 + ny_2 + rz_2 \pmod{abc}.$$

In other words,

$$m(x_1 - x_2) + n(y_1 - y_2) + r(z_1 - z_2) \equiv 0 \pmod{abc},$$

where $|x_1 - x_2| \leq \sqrt{bc}$, $|y_1 - y_2| \leq \sqrt{|ac|}$, and $|z_1 - z_2| \leq \sqrt{|ab|}$, which secures the claim.

By Claim 7.3, the solution (x, y, z) to the congruence (7.12) satisfies $x^2 \leq bc$ where $x^2 = bc$ if and only if $b = c = -1$; $y^2 \leq -ac$ with $y^2 = -ac$ if and only if $a = 1$ and $c = -1$, and $z^2 \leq -ab$ with $z^2 = -ab$ if and only if $a = 1$ and $b = -1$. Given that $a > 0$, $b < 0$, and $c < 0$, then if

$$b \neq -1 \text{ or } c \neq -1, \tag{7.13}$$

we have $ax^2 + by^2 + cz^2 \leq ax^2 < abc$ and

$$ax^2 + by^2 + cz^2 \geq by^2 + cz^2 > b(-ac) + c(-ab) = -2abc.$$

Hence,

$$-2ab < ax^2 + by^2 + cz^2 < abc. \tag{7.14}$$

However, by (7.11), (x, y, z) also is a solution to

$$ax^2 + by^2 + cz^2 \equiv 0 \pmod{abc}.$$

Hence, by (7.14), either $ax^2 + by^2 + cz^2 = 0$ or $ax^2 + by^2 + cz^2 = -abc$. In the former case, we are done since this is the solution that we seek. In the latter case,

$$a(-by + xz)^2 + b(ax + yz)^2 + c(z^2 + ab)^2 = 0,$$

so we have a solution unless $-by + xz = ax + yz = z^2 + ab = 0$, in which case, $z^2 = -ab$ forcing $z = \pm 1$ since ab is squarefree by hypothesis. Thus, $a = 1 = b$ and $x = 1$, $y = -1$, and $z = 0$ is a solution.

We have proved the theorem for all but the case $b = c = -1$ since we arrived at the above proof via the assumption in (7.13). It follows from the hypothesis that -1 is a quadratic residue modulo a. Therefore, by Lemma 6.2 on page 246, $a = y^2 + z^2$ has a solution, so if we take $x = 1$, then $ax^2 + by^2 + cz^2 = 0$ since $b = c = -1$. $\qquad\square$

Implicit in the proof of Theorem 7.3 is the following.

Corollary 7.1 *Suppose that $a, b, c \in \mathbb{Z}$ are nonzero, not of the same sign, such that abc is squarefree. Then the following are equivalent.*

(a) $ax^2 + by^2 + cz^2 = 0$ *has a solution* $x, y, z \in \mathbb{Z}$ *not all zero.*

(b) $ax^2 + by^2 + cz^2$ *factors into linear factors modulo* $|abc|$.

(c) $-ab$, $-bc$, *and* $-ca$ *are quadratic residues of* $|c|$, $|a|$, *and* $|b|$, *respectively.*

Exercises

7.9. Determine if $-3x^2 + 5y^2 - 11z^2 = 0$ has a solution.

7.10. Determine if $2x^2 + 5y^2 - 11z^2 = 0$ has a solution.

7.11. Determine if $5x^2 + y^2 - z^2 = 0$ has a solution.

7.12. Verify how Corollary 7.1 follows from the proof of Theorem 7.3.

7.3 Bachet's Equation

> *One must divide one's time between politics and equations. But our equations are much more important to me.*
> **Albert Einstein (1879–1955), German-born theoretical physicist and founder of the theory of relativity**

The earliest published result on the equation

$$y^2 = x^3 + k \tag{7.15}$$

was given by Bachet in 1621, who found solutions when $k = -2$, such as $(x, y) = (3, 5)$. Thus, the equation is known as *Bachet's equation* — see Biography 7.2 on page 279. Fermat later claimed to have a method for solving Bachet's equation, but he never published it. In 1869, V.A. Lebesque proved the following for (7.15).

Theorem 7.4 | **Lebesque's Theorem**

If $k = 7$, then Equation (7.15) has no solutions $x, y \in \mathbb{Z}$.

Proof. If x is even, then $y^2 \equiv 3 \pmod 4$, which is impossible since -1 is not a quadratic residue modulo 4. If $x \equiv 3 \pmod 4$, then $y^2 \equiv 2 \pmod 4$, which is also impossible. Therefore, we must have $x \equiv 1 \pmod 4$, so

$$y^2 + 1 = x^3 + 8 = (x + 2)(x^2 - 2x + 4),$$

and $z = x^2 - 2x + 4 \equiv 3 \pmod 4$, so $y^2 \equiv -1 \pmod z$, again impossible. $\square$

There are more general results that can be proved using elementary techniques.

Theorem 7.5 | **Special Forms of Bachet's Equation**

The equation

$$y^2 = x^3 + k(b^2 - k^2 a^3) \tag{7.16}$$

has no integer solutions when each of the following holds.

(a) $a \equiv -1 \pmod 4$.

(b) b is even.

(c) k is squarefree.

(d) $k \equiv 3 \pmod 4$.

(e) $\gcd(k, b) = 1$.

(f) $b \not\equiv 0 \,(\text{mod } 3)$ if $k \equiv 2 \,(\text{mod } 3)$.

(g) If p is a prime such that the Legendre symbol $\left(\frac{k}{p}\right) = -1$, then $\gcd(a,b) \not\equiv 0$ (mod p).

Proof. Since from conditions (a), (b), and (d), we have that

$$y^2 \equiv x^3 + k(b^2 - k^2 a^3) \equiv x^3 + 3(0 - 3^2(-1)^3) \equiv x^3 - 1 \pmod{4},$$

then it is not possible for $x \equiv 0, 2, 3 \,(\text{mod } 4)$, so

$$x \equiv 1 \pmod{4}. \tag{7.17}$$

Now, we may write (7.16) in the form

$$y^2 - kb^2 = (x - ka)(x^2 + kax + k^3 a^2).$$

Set $\ell = x^2 + kax + k^3 a^2 \in \mathbb{N}$. If a prime $p \mid \gcd(x, k)$, then $p \mid y^2$, so $p^2 \mid kb^2$, which contradicts condition (e) since condition (c) tells us that k is squarefree. Therefore, $\gcd(x,k) = 1$, so $\gcd(k,\ell) = 1$. By conditions (a), (d), and (7.17), we have,

$$\ell \equiv x^3 + kax + k^2 a^2 \equiv 1^3 + 3(-1)(1) + 3^2(-1)^2 \equiv 3 \pmod{4}.$$

Since we also have, from condition (d), that $k \equiv 3 \,(\text{mod } 4)$, then the following Jacobi symbol equality holds,

$$\left(\frac{k}{\ell}\right) = -\left(\frac{\ell}{k}\right) = -\left(\frac{x^3 + kax + k^2 a^2}{k}\right) = -\left(\frac{x^3}{k}\right) =$$

$$-\left(\frac{y^2 - k}{k}\right) = -\left(\frac{y^2}{k}\right) = -\left(\frac{y}{k}\right)^2 = -1.$$

Therefore, ℓ is divisible by a prime p to an odd power such that $\left(\frac{k}{p}\right) = -1$. Therefore, if p does not divide y and p does not divide b, then

$$1 = \left(\frac{y^2}{p}\right) = \left(\frac{kb^2}{p}\right) = \left(\frac{k}{p}\right) = -1,$$

a contradiction that tells us

$$y \equiv b \equiv 0 \pmod{p}. \tag{7.18}$$

Thus, p divides $y^2 - kb^2$ to an even power, and so p divides $x - ka$ to an odd power. Hence,

$$x - ka \equiv 0 \pmod{p} \tag{7.19}$$

and

$$x^2 + kax + k^2 a^2 \equiv 0 \pmod{p}. \tag{7.20}$$

Multiplying (7.19) by $-x$ and subtracting the result from (7.20), we get

$$2kax + k^2a^2 \equiv 0 \pmod{p}.$$

However, multiplying (7.19) by ka tells us that

$$kax \equiv k^2a^2 \pmod{p},$$

whence, the former translates into

$$0 \equiv 2kax + k^2a^2 \equiv 2kax + kax \equiv 3kax \pmod{p}.$$

If $p = 3$, then since $\left(\frac{k}{p}\right) = -1$, we must have that $k \equiv 2 \pmod{3}$. However, this contradicts condition (f) since (7.18) holds. If $x \equiv 0 \pmod{p}$, then by (7.19), $ka \equiv 0 \pmod{p}$. Since $\gcd(k, p) = 1$, then $a \equiv 0 \pmod{p}$ and this contradicts condition (g) since (7.18) holds. This exhausts all possibilities so there are no solutions to Equation (7.16). $\qquad\square$

Example 7.7 *If $k = 3$, $b = 2$ and $a = 7$ in Theorem 7.5, then we have that*

$$y^2 = x^3 - 9249$$

has no solutions. Similarly, if $k = 3 = a$ and $b = 2$, then

$$y^2 = x^3 - 717$$

has no integer solutions.

There are other results on forms of Bachet's equation that require some techniques from algebraic number theory involving factorizations in quadratic fields. We leave those results for a second course in number theory.

Biography 7.2 Claude Gaspar Bachet De Méziriac (1581–1638) *was born on October 9, 1581, in Bourg-en-Bresse, France. He studied under the Jesuits at several locations including Rheims where his mentor was the Jesuit mathematician Jacques de Billy. He eventually resolved to live a life of leisure on his estate in Bourg-en-Bresse, where he was financially very well established. He spent most of his life on his estate except for a couple of years spent in Paris.*

In 1612, he published his work Problèmes Plaisants, *and in 1621, he published* Les Éléments Arithmétiques *which contained a translation from Greek to Latin of Diophantus' book* Arithmetica. *This is the famous book in which Fermat wrote his margin notes about what we now reference as Fermat's Last Theorem. Bachet also wrote books on mathematical puzzles, and arithmetic tricks, including the construction of magic squares, and these works provided the foundation for later books on recreational mathematics. He seems also to be the earliest writer to discuss the solution of indeterminate equations via continued fractions. In 1635, three years before his death, he was elected a member of the French Academy.*

Exercises

7.13. Prove that if $k = -16$ in Equation (7.15), then there are no solutions $x, y \in \mathbb{Z}$.

(*Hint: Prove that $x \equiv 1 \pmod 8$ and use Exercise 4.3 on page 187.*)

7.14. Prove that if $k = 45$ in Equation (7.15), then there are no solutions $x, y \in \mathbb{Z}$.

(*Hint: Prove that $x \equiv 3, 7 \pmod 8$ and look at the equation modulo 8.*)

7.15. Prove that if $k = 23$ in Equation (7.15), then there are no solutions $x, y \in \mathbb{Z}$.

(*Hint: Prove that $x \equiv 1 \pmod 4$ and use the fact that*

$$y^2 + 4 = (x + 3)(x^2 - 3x + 9).)$$

7.16. Prove that if $k = 339$ in Equation (7.15), then there are no solutions $x, y \in \mathbb{Z}$.

(*Hint: Prove that $x \equiv 1 \pmod 4$ and use the fact that*

$$y^2 + 4 = (x + 7)(x^2 - 7x + 49).)$$

7.17. Prove that $y^2 = x^3 - 9213$ has no solutions $x, y \in \mathbb{Z}$.

(*Hint: Use Theorem 7.5.*)

7.18. Prove that $y^2 = x^3 - 35925$ has no solutions $x, y \in \mathbb{Z}$.

(*Hint: Use Theorem 7.5.*)

7.19. Prove that $x^2 + y^2 = z^3$ has infinitely many solutions $x, y, z \in \mathbb{Z}$.

(*Hint: Look at $x = 3a^2 - 1$ and $y = a^3 - 3a$ for any $a \in \mathbb{N}$.*)

7.20. Prove that for any natural number $n > 1$, the equation

$$x^n + y^n = z^{n-1}$$

has infinitely many solutions $x, y, z \in \mathbb{Z}$.

(*Hint: Look at $x = (1 + m^n)^{n-2}$ and $y = m(1 + m^n)^{n-2}$ for any $m \in \mathbb{N}$.*)

7.21. Prove that for any natural number n, the equation

$$x^n + y^n = z^{n+1}$$

has infinitely many solutions $x, y, z \in \mathbb{Z}$.

(*Hint: Look at $x = 1 + m^n$ and $y = m(1 + m^n)$ for any $m \in \mathbb{N}$.*)

7.4   Fermat's Last Theorem

> *I consider that I understand an equation when I can predict the properties of its solutions, without actually solving it.*
>   **Paul Dirac (1902–1984), British theoretical physicist and founder of the field of quantum mechanics**

As noted in Biography 1.10 on page 38, the now-verified result known as *Fermat's Last Theorem* is the statement that the Diophantine equation

$$x^n + y^n = z^n \tag{7.21}$$

has no solutions $x, y, z \in \mathbb{Z}$ for any integer $n > 2$. Although the proof of this fact is well beyond the scope of this book, we can prove some instances of it by elementary means. The first result is used to illustrate Fermat's *method of infinite descent.* This method involves assuming the existence, in natural numbers, of a solution to a given problem and constructing new solutions using smaller natural numbers; and from the new ones other solutions using still smaller natural numbers, and so on. Since this cannot go on indefinitely for natural numbers, then the initial assumption must have been false. Fermat used this method to prove (7.21) for $n = 4$. We prove something slightly stronger since the stronger case is actually easier to verify. In order to prove it, we require a result on what are known as *Pythagorean triples*, namely a set of integers (x, y, z) such that

$$x^2 + y^2 = z^2, \tag{7.22}$$

which is said to be *primitive* if $\gcd(x, y, z) = 1$. (Also, see part (f) of Exercise 1.13 on page 13 for a result on Fibonacci numbers and Pythagorean triples.)

Theorem 7.6 | **Pythagorean Triples**

A primitive Pythagorean triple (x, y, z) is a solution to (7.22), where $x, y, z \in \mathbb{N}$ and x is even, if and only if

$$(x, y, z) = (2uv, v^2 - u^2, v^2 + u^2),$$

for relatively prime natural numbers u and v of opposite parity.

Proof. Given that x is even, then we may set $z - y = 2\ell$ and $z + y = 2m$. Hence, Equation (7.22) may be written as

$$x^2 = z^2 - y^2 = (z - y)(z + y),$$

so

$$\left(\frac{x}{2}\right)^2 = \left(\frac{z-y}{2}\right)\left(\frac{z+y}{2}\right) = \ell m.$$

If a prime $p \mid \gcd(\ell, m)$, then $p \mid (\ell - m)$ and $p \mid (\ell + m)$, so $p \mid y$ and $p \mid z$, contradicting that $\gcd(y, z) = 1$. Therefore, $\gcd(\ell, m) = 1$.

Claim 7.4 *There are natural numbers u and v such that $\ell = u^2$ and $m = v^2$.*

The result is vacuously true if $\ell = 1$ or $m = 1$, so we assume that $\ell > 1$ and $m > 1$. Let $\ell = \prod_{j=1}^{r} p_j^{a_j}$ and $m = \prod_{k=1}^{s} q_j^{b_j}$ be the canonical prime factorizations of ℓ and m, respectively. Since $\gcd(\ell, m) = 1$, then $p_j \neq q_k$ for any $j = 1, 2, \ldots, r$ and any $k = 1, 2, \ldots, s$. Let $x/2 = \prod_{i=1}^{t} r_j^{c_j}$ be the canonical prime factorization of $x/2$. Then

$$\prod_{j=1}^{r} p_j^{a_j} \prod_{k=1}^{s} q_j^{b_j} = \prod_{i=1}^{t} r_j^{2c_j}.$$

Thus, the primes $r_1, r_2, \ldots, r_t$ are just the primes $p_1, p_2, \ldots, p_r, q_1, q_2, \ldots, q_s$, and $2c_1, 2c_2, \ldots, 2c_t$ are the corresponding exponents $a_1, a_2, \ldots, a_r, b_1, b_2, \ldots, b_s$. Therefore, each a_j, b_k must be divisible by 2. Set

$$u = \prod_{j=1}^{r} p_j^{a_j/2} \text{ and } v = \prod_{k=1}^{s} q_j^{b_j/2}.$$

Hence,

$$\ell = u^2 \text{ and } m = v^2,$$

as required for the claim.

By Claim 7.4,

$$z = m + \ell = v^2 + u^2, \tag{7.23}$$

$$y = m - \ell = v^2 - u^2, \tag{7.24}$$

and

$$x^2 = 4\ell m = 4u^2 v^2,$$

which implies that

$$x = 2uv.$$

If a prime $p \mid \gcd(u, v)$, then by (7.23)–(7.24), $p \mid z$ and $p \mid y$, contradicting that $\gcd(y, z) = 1$. Thus, $\gcd(u, v) = 1$. Lastly, if u and v have the same parity, then z and y are both even by (7.23)–(7.24), an impossibility since x is even by hypothesis. This proves the necessity of the conditions.

Conversely, if the conditions hold, then

$$x^2 + y^2 = (2uv)^2 + (v^2 - u^2)^2 = (v^2 + u^2)^2 = z^2.$$

If p is a prime divisor of $\gcd(x, y, z)$, then $p \mid (z + y)$ and $p \mid (z - y)$. In other words, $p \mid 2u^2$ and $p \mid 2v^2$. Since $p \mid z$, which is odd, then $p > 2$. Hence, $p \mid u$ and $p \mid v$, contradicting the hypothesis that $\gcd(u, v) = 1$. Thus, $\gcd(x, y, z) = 1$, which completes the proof of the sufficiency. $\square$

Theorem 7.7 | **The Equation $x^4 + y^4 = z^2$** |

The Diophantine equation

$$x^4 + y^4 = z^2 \tag{7.25}$$

has no solutions $x, y, z \in \mathbb{Z}$.

Proof. Assume that (7.25) has a solution in natural numbers x_0, y_0, z_0, where we may assume without loss of generality that $\gcd(x_0, y_0) = 1$ since we may otherwise divide through (7.25) by the gcd and achieve the assumed state. Therefore, (x_0^2, y_0^2, z_0) is a primitive Pythagorean triple, so by Theorem 7.6, we may take x_0 to be even, so there exist relatively prime natural numbers u, v such that

$$x_0^2 = 2uv, \tag{7.26}$$

$$y_0^2 = v^2 - u^2, \tag{7.27}$$

and

$$z_0 = v^2 + u^2, \tag{7.28}$$

where u and v have different parity. If v is even, then

$$1 \equiv y_0^2 = v^2 - u^2 \equiv 0 - 1 \pmod{4},$$

which is impossible. Thus, $u = 2w$, and Equation (7.26) becomes $x_0^2 = 4wv$. In other words,

$$\left(\frac{x_0}{2}\right)^2 = wv.$$

However, by the same argument as in Claim 7.4, $w = q_1^2$ and $v = z_1^2$ for some $q_1, z_1 \in \mathbb{N}$ given that $\gcd(w, v) = 1 = \gcd(u, v)$ because the parity of u and v differ. Now consider the equation

$$u^2 + y_0^2 = v^2.$$

Since $\gcd(u, v) = 1$, then $\gcd(u, y_0, v) = 1$, so (u, y_0, v) is a primitive Pythagorean triple for which we may invoke Theorem 7.6 to get relatively prime natural numbers a, b with

$$u = 2ab,$$

$$y_0 = a^2 - b^2,$$

and

$$v = a^2 + b^2. \tag{7.29}$$

Since, $ab = u/2 = w = q_1^2$, then we use the argument as in Claim 7.4 to get that $a = x_1^2$ and $b = y_1^2$, so via (7.29),

$$z_1^2 = v = a^2 + b^2 = x_1^4 + y_1^4,$$

where

$$0 < z_1 \leq z_1^2 = v \leq v^2 < v^2 + u^2 = z_0.$$

What we have shown is that the assumption of one solution (x_0, y_0, z_0) leads to another solution (x_1, y_1, z_1) with $0 < z_1 < z_0$, so we may continue in this fashion to produce a solution (x_2, y_2, z_2) with $0 < z_2 < z_1$, and so on, leading to an infinitely descending sequence

$$z_0 > z_1 > z_2 > \cdots > z_n > \cdots$$

However, this is a contradiction since there are only finitely many natural numbers less than z_0. Hence, $x^4 + y^4 = z^2$ has no solutions $x, y, z \in \mathbb{N}$. $\square$

An immediate consequence of Theorem 7.7 is Fermat's result that we discussed above.

Corollary 7.2 *The equation*

$$x^4 + y^4 = z^4 \tag{7.30}$$

has no solutions in natural numbers x, y, z.

Proof. If (7.30) has solutions $x_0, y_0, z_0 \in \mathbb{N}$, then x_0, y_0, z_0^2 would satisfy (7.25) contradicting Theorem 7.7. $\square$

Exercises

☆ 7.22. Prove that $x^4 - y^4 = z^2$ has no solutions in natural numbers x, y, z.

(*Hint: Use an infinite descent type of argument by assuming there is a solution with minimal x value and deducing one smaller.*)

(*This result is also due to Fermat with a slight variation on the method of infinite descent suggested in the hint.*)

7.23. Prove that there does not exist a primitive Pythagorean triple $x, y, z \in \mathbb{Z}$ such that both $x^2 + y^2 = z^2$ and $xy = 2w^2$ for some $w \in \mathbb{N}$.

(*Hint: Use Exercise 7.22.*)

(*This result says that the area of a Pythagorean* (right) *triangle cannot be equal to the square of an integer.*)

7.24. Prove that the equation $x^4 - y^4 = 2z^2$ has no solutions in natural numbers x, y, z.

7.25. Show that the only natural number solutions to the equation $x^4 + y^4 = 2z^2$ are $x = y = z = 1$.

☆ 7.26. Prove that the equation

$$x^4 - y^4 = pz^2 \tag{7.31}$$

where p is a prime congruent to 3 modulo 8, has no solutions in natural numbers x, y, z.

(*Hint: Use an infinite descent type of argument by assuming there is a solution with minimal x value and deducing one smaller.*)

Appendix A: Fundamental Facts

In this appendix, we set down some fundamental facts, beginning with the fundamental notion of a set. Proofs may be found in standard introductory texts on the subject matter.

◆ Well-Definedness

A set of objects is *well-defined* provided that it is always possible to determine whether or not a particular element belongs to the set. The classical example of a collection that is *not* well-defined is described as follows. Suppose that there is a library with many books, and each of these books may be placed into one of two categories, those that list themselves in their own index and those that do not. The chief librarian decides to set up a Master Directory, which will keep track of those books that do not list themselves. Now, the question arises: Does the Master Directory list itself? If it does not, then it *should* since it only lists those that do not list themselves. If it does, then it *should not* for the same reason — a paradox! This is called the *Russell Paradox* or *Russell Antinomy*. The problem illustrated by the Russell Paradox is with *self-referential* collections of objects. We see that Russell's collection is not *well-defined*, so it is not a set. Russell's example may be symbolized as $S = \{x : x \notin S\}$. The term "unset" is often used to describe such a situation.

Definition A.1 | Sets |

A set *is a* well-defined *collection of* distinct *objects. The terms* set, *collection, and* aggregate *are synonymous. The* objects *in the set are called* elements *or* members. *We write* $a \in S$ *to denote membership of an element* a *in a set* S, *and if* a *is not in* S, *then we write* $a \notin S$.

This definition *avoids* the problems of the contradictions that arise in such discussions as the Russell Antinomy.

Set notation is given by putting elements between two braces. For instance, an important set is the set of *natural numbers*:

$$\mathbb{N} = \{1, 2, 3, 4, \ldots\}.$$

In general, we may specify a set by properties. For instance,

$$\{x \in \mathbb{N} : x > 3\}$$

specifies those natural numbers that satisfy the property of being bigger than 3, which is the same as $\{x \in \mathbb{N} : x \neq 1, 2, 3\}$.

(Note that the symbol $\mathbb{N}$ comes from the German *natuerlich* for natural.)

Definition A.2 | Subsets and Equality |

A set $\mathcal{T}$ is called a subset *of a set $\mathcal{S}$, denoted by $\mathcal{T} \subseteq \mathcal{S}$ if every element of $\mathcal{T}$ is in $\mathcal{S}$. On the other hand, if there is an element $t \in \mathcal{T}$ such that $t \notin \mathcal{S}$, then we write $\mathcal{T} \not\subseteq \mathcal{S}$ and say that $\mathcal{T}$ is* not *a subset of $\mathcal{S}$. We say that two sets $\mathcal{S}$ and $\mathcal{T}$ are* equal, *denoted by $\mathcal{T} = \mathcal{S}$ provided that $t \in \mathcal{T}$ if and only if $t \in \mathcal{S}$, namely both $\mathcal{T} \subseteq \mathcal{S}$, and $\mathcal{S} \subseteq \mathcal{T}$. If $\mathcal{T} \subseteq \mathcal{S}$, but $\mathcal{T} \neq \mathcal{S}$, then we write $\mathcal{T} \subset \mathcal{S}$ and call $\mathcal{T}$ a* proper subset *of $\mathcal{S}$. All sets contain the* empty set, *denoted by $\varnothing$, or $\{\}$, consisting of no elements. The set of all subsets of a given set $\mathcal{S}$ is called its* power set.

Definition A.3 | Complement, Intersection, and Union |

The intersection *of two sets $\mathcal{S}$ and $\mathcal{T}$ is the set of all elements common to both, denoted by $\mathcal{S} \cap \mathcal{T}$, namely*

$$\mathcal{S} \cap \mathcal{T} = \{a : a \in \mathcal{S} \text{ and } a \in \mathcal{T}\}.$$

The union *of the two sets consists of all elements that are in $\mathcal{S}$ or in $\mathcal{T}$ (possibly both), denoted by $\mathcal{S} \cup \mathcal{T}$, namely*

$$\mathcal{S} \cup \mathcal{T} = \{a : a \in \mathcal{S} \text{ or } a \in \mathcal{T}\}.$$

If $\mathcal{T} \subseteq \mathcal{S}$, then the complement *of $\mathcal{T}$ in $\mathcal{S}$, denoted by $\mathcal{S} \setminus \mathcal{T}$ is the set of all those elements of $\mathcal{S}$ that are not in $\mathcal{T}$, namely*

$$\mathcal{S} \setminus \mathcal{T} = \{s : s \in \mathcal{S} \text{ and } s \notin \mathcal{T}\}.$$

Two sets $\mathcal{S}$ and $\mathcal{T}$ are called disjoint *if $\mathcal{S} \cap \mathcal{T} = \varnothing$.*

For instance, if $\mathcal{S} = \mathbb{N}$, and $\mathcal{T} = \{1,2,3\}$, then $\mathcal{S} \cap \mathcal{T} = \mathcal{T} = \{1,2,3\}$, and $\mathcal{S} \cup \mathcal{T} = \mathbb{N}$. Also, $\mathcal{S} \setminus \mathcal{T} = \{x \in \mathbb{N} : x > 3\}$.

Definition A.4 | Set Partitions |

Let $\mathcal{S}$ be a set and let $\mathfrak{G} = \{\mathcal{S}_1, \mathcal{S}_2, \ldots\}$ be a set of nonempty subsets of $\mathcal{S}$. Then $\mathfrak{G}$ is called a partition *of $\mathcal{S}$ provided both of the following are satisfied.*

(a) $\mathcal{S}_j \cap \mathcal{S}_k = \varnothing$ for all $j \neq k$.

(b) $\mathcal{S} = \mathcal{S}_1 \cup \mathcal{S}_2 \cup \cdots \cup \mathcal{S}_j \cdots$, namely $s \in \mathcal{S}$ if and only if $s \in \mathcal{S}_j$ for some j.

For an example of partitioning, see the notion of *congruence* on page 73.

Definition A.5 | Binary Relations and Operations |

Let s_1, s_2 be elements of a set $\mathcal{S}$. Then we call (s_1, s_2) an ordered pair, *where s_1 is called the* first component *and s_2 is called the* second component. *If $\mathcal{T}$ is* *another set, then the* Cartesian product *of* S *with* $\mathcal{T}$*, denoted by* $S \times \mathcal{T}$*, is given by the set of ordered pairs:*

$$S \times \mathcal{T} = \{(s, t) : s \in S, t \in \mathcal{T}\}.$$

A relation R on $S \times \mathcal{T}$ is a subset of $S \times \mathcal{T}$ where $(s, t) \in R$ is denoted by sRt. A relation on $S \times S$ is called a binary relation*. A relation R on $(S \times S) \times S$ is called a* binary operation *on S if R associates with each $(s_1, s_2) \in S \times S$, a unique element $s_3 \in S$. In other words, if $(s_1, s_2)Rs_3$ and $(s_1, s_2)Rs_4$, then $s_3 = s_4$.*

For example, a relation on $S \times \mathcal{T} = \{1, 2, 3\} \times \{1, 2\}$ is $\{(1, 1), (1, 2)\}$. Notice that there does not exist a unique second element for 1 in this relation. We cannot discuss a binary operation here since $S \neq \mathcal{T}$. The next section provides us with an important notion of a binary operation.

◆ Functions

Definition A.6 *A* function *f (also called a* mapping *or* map*) from a set S to a set $\mathcal{T}$ is a relation on $S \times \mathcal{T}$, denoted by $f : S \to \mathcal{T}$, which assigns each $s \in S$ a unique $t \in \mathcal{T}$, called the* image *of s under f, denoted by $f(s) = t$. The set S is called the* domain *of f and $\mathcal{T}$ is called the* range *of f. If $S_1 \subseteq S$, then the* image *of S_1 under f, denoted by $f(S_1)$, is the set $\{t \in \mathcal{T} : t = f(s)$ for some $s \in S_1\}$. If $S = S_1$, then $f(S)$ is called the* image *of f, denoted by* img(S)*. If $\mathcal{T}_1 \subseteq \mathcal{T}$, the* inverse image *of $\mathcal{T}_1$ under f, denoted by $f^{-1}(\mathcal{T}_1)$, is the set $\{s \in S : f(s) \in \mathcal{T}_1\}$.*
A function $f : S \to \mathcal{T}$ is called injective *(also called* one-to-one*) if and only if for each $s_1, s_2 \in S$, $f(s_1) = f(s_2)$ implies that $s_1 = s_2$. A function f is* surjective *(also called* onto*) if $f(S) = \mathcal{T}$, namely if for each $t \in \mathcal{T}$, $t = f(s)$ for some $s \in S$. A function f is called* bijective *(or a* bijection*) if it is both injective and surjective. Two sets are said to be in a* one-to-one correspondence *if there exists a bijection between them.*

Each of the following may be verified for a given function $f : S \to \mathcal{T}$.

A.1. If $S_1 \subseteq S$, then $S_1 \subseteq f^{-1}(f(S_1))$.

A.2. If $\mathcal{T}_1 \subseteq \mathcal{T}$, then $f(f^{-1}(\mathcal{T}_1)) \subseteq \mathcal{T}_1$.

A.3. The identity map, $1_S : S \to S$, given by $1_S(s) = s$ for all $s \in S$, is a bijection.

A.4. f is injective if and only if there exists a function $g : \mathcal{T} \to S$ such that $gf = 1_S$, and g is called a *left inverse of f*.

A.5. f is surjective if and only if there exists a function $h : \mathcal{T} \to S$ such that $fh = 1_{\mathcal{T}}$, and h is called a *right inverse for f*.

A.6. If f has both a left inverse g and a right inverse h, then $g = h$ is a unique map called the *two-sided inverse of f*.

A.7. f is bijective if and only if f has a two-sided inverse.

Notice that in Definition A.5 a binary operation on $\mathcal{S}$ is just a function on $\mathcal{S} \times \mathcal{S}$. The number of elements in a set is of central importance.

Definition A.7 | Cardinality |

If $\mathcal{S}$ and $\mathcal{T}$ are sets, and there exists a one-to-one mapping from $\mathcal{S}$ to $\mathcal{T}$, then $\mathcal{S}$ and $\mathcal{T}$ are said to have the same cardinality. *A set $\mathcal{S}$ is finite if either it is empty or there is an $n \in \mathbb{N}$ and a bijection $f : \{1, 2, \ldots, n\} \mapsto \mathcal{S}$. The number of elements in a finite set $\mathcal{S}$ is sometimes called its* cardinality, *or* order, *denoted by $|\mathcal{S}|$. A set is said to be* countably infinite *if there is a bijection between the set and $\mathbb{N}$. If there is no such bijection and the set is infinite, then the set is said to be* uncountably infinite.

Example A.1 If $n \in \mathbb{N}$ is arbitrary and $n_0 \in \mathbb{N}$ is arbitrary but fixed, then the map $f : \mathbb{N} \mapsto n_0\mathbb{N}$ via $f(n) = n_0 n$ is bijective, so the multiples of $n_0 \in \mathbb{N}$ can be identified with $\mathbb{N}$. For instance, the case where $n_0 = 2$ shows that the even natural numbers may be identified with the natural numbers themselves.

Definition A.8 | Indexing Sets and Set Operations |

Let I be a set, which may be finite or infinite (possibly uncountably infinite), and let $\mathcal{U}$ be a universal set, *which means a set that has the property of containing all sets under consideration. We define*

$$\cup_{j \in I} \mathcal{S}_j = \{s \in \mathcal{U} : s \in \mathcal{S}_j \text{ for some } j \in I\},$$

and

$$\cap_{j \in I} \mathcal{S}_j = \{s \in \mathcal{U} : s \in \mathcal{S}_j \text{ for all } j \in I\}.$$

Here, I is called the indexing set, *$\cup_{j \in I} \mathcal{S}_j$ is called a* generalized set-theoretic union, *and $\cap_{j \in I} \mathcal{S}_j$ is called a* generalized set-theoretic intersection.

Example A.2 The reader may verify both of the following properties about generalized unions and intersections. In what follows, $\mathcal{T}, \mathcal{S}_j \subseteq \mathcal{U}$.

(a) $\mathcal{T} \cup (\cap_{j \in I} \mathcal{S}_j) = \cap_{j \in I}(\mathcal{T} \cup \mathcal{S}_j)$.

(b) $\mathcal{T} \cap (\cup_{j \in I} \mathcal{S}_j) = \cup_{j \in I}(\mathcal{T} \cap \mathcal{S}_j)$.

◆ Arithmetic

The natural numbers $\{1, 2, 3, 4, \ldots\}$ are denoted by $\mathbb{N}$ and the *integers* $\{\ldots, -3, -2, -1, 0, 1, 2, 3, \ldots\}$ are denoted by $\mathbb{Z}$. (Note that the symbol $\mathbb{Z}$ comes from the German *Zahl* for number.)

For this we need a larger set. The following are called the *rational numbers*.

$$\mathbb{Q} = \{a/b : a, b \in \mathbb{Z}, \text{ and } b \neq 0\}.$$

(Note that the symbol $\mathbb{Q}$ comes from *quotient*, and was introduced, as were $\mathbb{N}$ and $\mathbb{Z}$, by Bourbaki in the 1930s — see Biography A.1 on the next page.)

Rational numbers have *periodic* decimal expansions. In other words, they have patterns that repeat *ad infinitum*. For instance, $1/2 = 0.5000\ldots$ and $1/3 = 0.333\ldots$. However, there are numbers whose decimal expansions have *no* repeated pattern, such as

$$\sqrt{2} = 1.41421356237\ldots,$$

so it is not a quotient of integers. These numbers, having decimal expansions that are not periodic, are called irrational numbers, denoted by $\mathfrak{J}$. It is possible that a sequence of rational numbers may *converge* to an irrational one. For instance, define

$$q_0 = 2, \text{ and } q_{j+1} = 1 + \frac{1}{q_j} \text{ for } j \geq 0.$$

Then

$$\lim_{j \to \infty} q_j = \frac{1 + \sqrt{5}}{2},$$

called the *Golden Ratio*, denoted by $\mathfrak{g}$ which we introduced on page 4. The *real numbers* consist of the set-theoretic union:

$$\mathbb{R} = \mathbb{Q} \cup \mathfrak{J}.$$

To complete the hierarchy of numbers (at least for our purposes), the *complex numbers* employ $\sqrt{-1}$, as follows:

$$\mathbb{C} = \{a + b\sqrt{-1} : a, b \in \mathbb{R}\}.$$

We now provide the *Fundamental Laws of Arithmetic* as a fingertip reference for the convenience of the reader.

The Laws of Arithmetic:

◆ **The Laws of Closure** If $a, b \in \mathbb{R}$, then $a + b \in \mathbb{R}$ and $ab \in \mathbb{R}$.

◆ **The Commutative Laws** If $a, b \in \mathbb{R}$, then $a + b = b + a$, and $ab = ba$.

◆ **The Associative Laws** If $a, b, c \in \mathbb{R}$, then $(a + b) + c = a + (b + c)$, and $(ab)c = a(bc)$.

◆ **The Distributive Law** If $a, b, c \in \mathbb{R}$, then $a(b + c) = ab + ac$.

◆ **The Cancellation Law** Let $a, b, c \in \mathbb{R}$. If $a + c = b + c$, then $a = b$ for any $c \in \mathbb{R}$. Also, if $ac = bc$, then $a = b$ for any $c \in \mathbb{R}$, with $c \neq 0$.

(Compare the above with the congruence cancellation law on page 75.)

Note that as a result of the distributive law, we may view $-a$ for any $a \in \mathbb{R}$ as $(-1) \cdot a$, or -1 *times* a.

Biography A.1 Nicolas Bourbaki *"Bourbaki" is the collective pen name of a group of some of the most respected mathematicians. The precise membership of Bourbaki, which has changed over the years, is a closely guarded secret but it is known that most of the members are French. Since 1939, Bourbaki has been publishing a monumental work, the* Eléments de mathématique *or* Elements of mathematics, *of which over thirty volumes have so far appeared. In this Bourbaki attempts to classify all of mathematics starting from certain carefully chosen logical and set-theoretic concepts. The emphasis throughout the Eléments is on the interrelationships to be found between the various structures present in mathematics. Thus it may be said that, for Bourbaki, pure mathematics is to be considered strictly as the study of pure structure.*

We now look at inverses under multiplication.

◆ **The Multiplicative Inverse** If $z \in \mathbb{R}$ with $z \neq 0$, then the *multiplicative inverse* of z is that number $1/z = z^{-1}$ (since $z \cdot \frac{1}{z} = 1$, the multiplicative identity). In fact, division may be considered the inverse of multiplication.

(Compare the above with the modular multiplicative inverse on page 80.)

Now we look at square roots and the relationship with exponentiation.

If $a < 0$, then $\sqrt{a} \notin \mathbb{R}$. For instance, $\sqrt{-1} \notin \mathbb{R}$ and $\sqrt{-5} \notin \mathbb{R}$. Consider, $\sqrt{25} = 5 \in \mathbb{R}$. A common error is to say that $\sqrt{25} = \pm 5$, but this is **false**. The error usually arises from the confusion of the solutions to $x^2 = 25$ with the solutions to $\sqrt{5^2} = x$. Solutions to $x^2 = 25$ are certainly $x = \pm 5$, but the **only** solution to $\sqrt{5^2} = x$ is $x = 5$, the unique *positive* integer such that $x^2 = 25$. A valid way of avoiding confusion with $\sqrt{x^2}$ is the following development.

Definition A.9 | Absolute Value |

If $x \in \mathbb{R}$, *then*
$$|x| = \begin{cases} x & \text{if } x \geq 0, \\ -x & \text{if } x < 0, \end{cases}$$
called the absolute value *of* x.

With Definition A.9 in mind, we see that if $x > 0$, then

$$\sqrt{x^2} = (x^2)^{1/2} = (x)^{2 \cdot 1/2} = x^1 = x = |x|,$$

and if $x < 0$, then

$$\sqrt{x^2} = \sqrt{(-x)^2} = (-x)^{2 \cdot 1/2} = (-x)^1 = -x = |x|.$$

Hence,

$$\sqrt{x^2} = |x|.$$

We may define *exponentiation* by observing that for any $x \in \mathbb{R}$, $n \in \mathbb{N}$,

$$x^n = x \cdot x \cdots x,$$

multiplied n times. Note that by convention $x^0 = 1$ for any nonzero real number x (and 0^0 is undefined). In what follows, the notation $\mathbb{R}^+$ means all of the *positive* real numbers. For rational exponents, we have the following.

Definition A.10 │Rational Exponents│

Let $n \in \mathbb{N}$. If n is even and $a \in \mathbb{R}^+$, then $\sqrt[n]{a} = b$ means that unique value of $b \in \mathbb{R}^+$ such that $b^n = a$. If n is even and $a \in \mathbb{R}$ with a negative, then $\sqrt[n]{a}$ is undefined. If n is odd, then $\sqrt[n]{a} = b$ is that unique value of $b \in \mathbb{R}$ such that $b^n = a$. In each case, a is called the base for the exponent.

Based on Definition A.10, the symbol $a^{\frac{m}{n}}$ for $a \in \mathbb{R}^+$ and $m, n \in \mathbb{N}$ is given by

$$a^{\frac{m}{n}} = \left(a^{\frac{1}{n}}\right)^m.$$

Also,

$$a^{-\frac{m}{n}} = \frac{1}{a^{\frac{m}{n}}}.$$

In general, we have the following laws.

Theorem A.1 │Laws for Exponents│

Let $a, b \in \mathbb{R}^+$, and $n, m \in \mathbb{N}$.

(a) $a^n b^n = (ab)^n$.

(b) $a^m a^n = a^{m+n}$.

(c) $(a^m)^n = a^{mn}$.

(d) $(a^m)^{\frac{1}{n}} = \sqrt[n]{a^m} = a^{\frac{m}{n}} = (a^{\frac{1}{n}})^m$.

Corollary A.1 *Let $a, n \in \mathbb{N}$. Then $\sqrt[n]{a} \in \mathbb{Q}$ if and only if $\sqrt[n]{a} \in \mathbb{Z}$.*

Note that we cannot have a *negative* base in Theorem A.1. The reason for this assertion is given in the following discussion. If we were to allow $-5 = \sqrt{25}$, then by Theorem A.1,

$$-5 = \sqrt{25} = 25^{1/2} = (5^2)^{1/2} = 5^{2 \cdot 1/2} = 5^1 = 5,$$

which is a contradiction. From another perspective, suppose that we allowed for negative bases in Theorem A.1. Then

$$5 = \sqrt{25} = \sqrt{(-5)^2} = ((-5)^2)^{1/2} = (-5)^{2 \cdot 1/2} = (-5)^1 = -5,$$

again a contradiction. Hence, only positive bases are allowed for the laws in Theorem A.1 to hold. It is worthy of note that even the great Euler (see Biography 1.17 on page 56) made the error of assuming that $\sqrt{a}\sqrt{b} = \sqrt{ab}$ regardless of whether a and b are both positive or not! Indeed there is a recent article [27] that looks at his error and the implications it has. The reader is highly recommended to read it since it is quite informative, accessible to the novice reader, and highlights what has been emphasized in this discussion.

Since we have the operations of addition and multiplication, it would be useful to have a notation that would simplify calculations.

◆ The Sigma Notation

We can write $n = 1 + 1 + \cdots + 1$ for the sum of n copies of 1. We use the Greek letter upper case *sigma* to denote *summation*. For instance, $\sum_{i=1}^{n} 1 = n$ would be a simpler way of stating the above. Also, instead of writing the sum of the first one hundred natural numbers as $1 + 2 + \cdots + 100$, we may write it as $\sum_{i=1}^{100} i$. In general, if we have numbers $a_m, a_{m+1}, \cdots, a_n$ $(m \leq n)$, we may write their sum as

$$\sum_{i=m}^{n} a_i = a_m + a_{m+1} + \cdots a_n,$$

and by convention

$$\sum_{i=m}^{n} a_i = 0 \text{ if } m > n.$$

The letter i is the *index of summation* (and any letter may be used here), n is *the upper limit of summation*, m is *the lower limit of summation*, and a_i is a *summand*. In the previous example, $\sum_{i=1}^{n} 1$, there is no i in the summand since we are adding the *same* number n times. The upper limit of summation tells us how many times that is (when $i = 1$). Similarly, we can write, $\sum_{j=1}^{4} 3 = 3 + 3 + 3 + 3 = 12$. This is the simplest application of the sigma notation. Another example is $\sum_{i=1}^{10} i = 55$.

Theorem A.2 | Properties of the Summation (Sigma) Notation

Let $h, k, m, n \in \mathbb{Z}$ *with* $m \leq n$ *and* $h \leq k$. *If* R *is a ring, then:*

(a) If $a_i, c \in R$, then $\sum_{i=m}^{n} ca_i = c \sum_{i=m}^{n} a_i$.

(b) If $a_i, b_i \in R$, then $\sum_{i=m}^{n} (a_i + b_i) = \sum_{i=m}^{n} a_i + \sum_{i=m}^{n} b_i$.

(c) If $a_i, b_j \in R$, then

$$\sum_{i=m}^{n} \sum_{j=h}^{k} a_i b_j = \left(\sum_{i=m}^{n} a_i \right) \left(\sum_{j=h}^{k} b_j \right) = \sum_{j=h}^{k} \sum_{i=m}^{n} a_i b_j = \left(\sum_{j=h}^{k} b_j \right) \left(\sum_{i=m}^{n} a_i \right).$$

A close cousin of the summation symbol is the product symbol defined as follows.

◆ The Product Symbol

The multiplicative analogue of the summation notation is the *product symbol* denoted by Π, upper case Greek *pi*. Given $a_m, a_{m+1}, \ldots, a_n \in R$, where R is a given ring and $m \leq n$, their product is denoted by:

$$\prod_{i=m}^{n} a_i = a_m a_{m+1} \cdots a_n,$$

and by convention $\prod_{i=m}^{n} a_i = 1$ if $m > n$.

The letter i is the *product index*, m is the *lower product limit* n is the *upper product limit*, and a_i is a *multiplicand* or *factor*.

For example, if $x \in \mathbb{R}^+$, then

$$\prod_{j=0}^{n} x^j = x^{\sum_{j=0}^{n} j} = x^{n(n+1)/2},$$

(see Theorem 1.1 on page 2).

Above, we defined the product notation. For instance, $\prod_{i=1}^{7} i = 1 \cdot 2 \cdot 3 \cdot 4 \cdot 5 \cdot 6 \cdot 7 = 5040$. This is an illustration of the following concept.

In the above, we have used the symbols $>$ (greater than) and $<$ (less than). We now formalize this notion of ordering as follows.

Definition A.11 | Ordering

If $a, b \in \mathbb{R}$, then we write $a < b$ if $a - b$ is negative and say that a is strictly less than b. Equivalently, $b > a$ means that b is strictly bigger than a. (Thus, to say that $b - a$ is positive is equivalent to saying that $b - a > 0$.) We also write $a \leq b$ to mean that $a - b$ is not positive, namely $a - b = 0$ or $a - b < 0$. Equivalently, $b \geq a$ means that $b - a$ is nonnegative, namely $b - a = 0$ or $b - a > 0$.

Now we state the principle governing order.

◆ **The Law of Order** If $a, b \in \mathbb{R}$, then *exactly one* of the following must hold: $a < b$, $a = b$, or $a > b$.

A basic rule, which follows from the Law of Order, is the following.

◆ **The Transitive Law**

Let $a, b, c \in \mathbb{R}$. If $a < b$ and $b < c$, then $a < c$.

What now follows easily from this is the connection between order and the operations of addition and multiplication, namely if $a < b$, then $a + c < b + c$ for any $c \in \mathbb{R}$, and $ac < bc$ for any $c \in \mathbb{R}^+$. However, if $c < 0$, then $ac > bc$.

To conclude our discussion on the basics of arithmetic, we need to understand how the order of operations holds in a given calculation. This is given in what follows.

◆ **Laws for Order of Operations.** Each of the following must be carried out from left to right and in the order listed when doing any numerical calculations.

P: Start by working inside parentheses, innermost first.

E: Simplify any exponent expression next.

MD: Then work all multiplications from left to right, as they appear.

AS: Finally work all additions and subtractions from left to right.

To do the operations in the right order, remember PEMDAS, which stands for: *Parentheses, Exponents, Multiplication-Division, Addition-Subtraction.*

A mnemonic, or memory aid, for *PEMDAS* is **P**owerful **E**arthquakes **M**ay **D**eliver **A**fter-**S**hocks.

For example, a common error is to confuse $1/ab$ with $1/(ab)$. According to the Laws for Order of Operations, the expression on the left dictates that we first divide 1 by a, which is $\frac{1}{a}$, then multiply the result by b to get $\frac{1}{a} \cdot b = \frac{b}{a} = b/a$. The expression $1/(ab)$, however, requires that we first do the multiplication in the brackets to get ab, then execute 1 divided by ab to get $\frac{1}{ab} = 1/(ab)$ that is *not* the same as $\frac{b}{a}$ in general. For instance, $1/3 \cdot 2 = 2/3 = \frac{2}{3}$, whereas $1/(3 \cdot 2) = 1/6 = \frac{1}{6}$.

Let's do a complete calculation. For instance,

$$1/2 \cdot 3^2 - 5 \cdot 7/(2 \cdot 9) = 1/2 \cdot 3^2 - 5 \cdot 7/18 = \frac{1}{2} \cdot 9 - \frac{5 \cdot 7}{18} = \frac{9}{2} - \frac{35}{18} = \frac{81 - 35}{18} = \frac{46}{18} = \frac{23}{9}.$$

In the text, we will be in need of some elementary facts concerning matrix theory. We now list these facts, without proof, for the convenience of the reader. The proofs, background, and details may be found in any text on elementary linear algebra.

◆ **Basic Matrix Theory**

If $m, n \in \mathbb{N}$, then an $m \times n$ matrix (read "m by n matrix") is a rectangular array of entries with m rows and n columns. We will assume, for the sake of simplicity, that the entries come from $\mathbb{R}$. If A is such a matrix, and $a_{i,j}$ denotes the entry in the ith row and j^{th} column, then

$$A = (a_{i,j}) = \begin{pmatrix} a_{1,1} & a_{1,2} & \cdots a_{1,n} \\ a_{2,1} & a_{2,2} & \cdots a_{2,n} \\ \vdots & \vdots & \vdots \\ a_{m,1} & a_{m,2} & \cdots a_{m,n} \end{pmatrix}.$$

Two $m \times n$ matrices $A = (a_{i,j})$, and $B = (b_{i,j})$ are equal if and only if $a_{i,j} = b_{i,j}$ for all i and j. The matrix $(a_{j,i})$ is called the *transpose* of A, denoted by

$$A^t = (a_{j,i}).$$

Addition of two $m \times n$ matrices A and B is done in the natural way.

$$A + B = (a_{i,j}) + (b_{i,j}) = (a_{i,j} + b_{i,j}),$$

and if $r \in \mathbb{R}$, then $rA = r(a_{i,j}) = (ra_{i,j})$, called *scalar multiplication*.

Under the above definition of addition and scalar multiplication, the set of all $m \times n$ matrices with entries from $\mathbb{R}$, form a set, denoted by $\mathcal{M}_{m \times n}(\mathbb{R})$. When $m = n$, this set is in fact a ring given by the following — see Remark 2.5 on page 79.

If $A = (a_{i,j})$ is an $m \times n$ matrix and $B = (b_{j,k})$ is an $n \times r$ matrix, then the *product* of A and B is defined as the $m \times r$ matrix:

$$AB = (a_{i,j})(b_{j,k}) = (c_{i,k}),$$

where

$$c_{i,k} = \sum_{\ell=1}^{n} a_{i,\ell} b_{\ell,k}.$$

Multiplication, if defined, is associative, and distributive over addition. If $m = n$, then $\mathcal{M}_{n \times n}(\mathbb{R})$ is a ring, with identity given by the $n \times n$ matrix:

$$I_n = \begin{pmatrix} 1 & 0 & \cdots & 0 \\ 0 & 1 & \cdots & 0 \\ \vdots & \vdots & \vdots & \vdots \\ 0 & 0 & \cdots & 1 \end{pmatrix},$$

called *the $n \times n$ identity matrix*, where 1 is, of course, the identity of $\mathbb{R}$.

Another important aspect of matrices that we will need throughout the text is motivated by the following. We maintain the assumption that the entries are from $\mathbb{R}$. Let $(a, b), (c, d) \in \mathcal{M}_{1 \times 2}(\mathbb{R})$. If we set up these row vectors into a single 2×2 matrix

$$A = \begin{pmatrix} a & b \\ c & d \end{pmatrix},$$

then $ad - bc$ is called the *determinant* of A, denoted by $\det(A)$. More generally, we may define the determinant of any $n \times n$ matrix in $\mathcal{M}_{n \times n}(\mathbb{R})$ for any $n \in \mathbb{N}$. The determinant of any $r \in \mathcal{M}_{1 \times 1}(\mathbb{R})$ is just $\det(r) = r$. Thus, we have the definitions for $n = 1, 2$, and we may now give the general definition inductively. The definition of the determinant of a 3×3 matrix

$$A = \begin{pmatrix} a_{1,1} & a_{1,2} & a_{1,3} \\ a_{2,1} & a_{2,2} & a_{2,3} \\ a_{3,1} & a_{3,2} & a_{3,3} \end{pmatrix}$$

is defined in terms of the above definition of the determinant of a 2×2 matrix, namely $\det(A)$ is given by

$$a_{1,1} \det \begin{pmatrix} a_{2,2} & a_{2,3} \\ a_{3,2} & a_{3,3} \end{pmatrix} - a_{1,2} \det \begin{pmatrix} a_{2,1} & a_{2,3} \\ a_{3,1} & a_{3,3} \end{pmatrix} + a_{1,3} \det \begin{pmatrix} a_{2,1} & a_{2,2} \\ a_{3,1} & a_{3,2} \end{pmatrix}.$$

Therefore, we may inductively define the determinant of any $n \times n$ matrix in this fashion. Assume that we have defined the determinant of an $n \times n$ matrix. Then we define the determinant of an $(n+1) \times (n+1)$ matrix $A = (a_{i,j})$ as follows. First, we let $A_{i,j}$ denote the $n \times n$ matrix obtained from A by deleting the ith row and j^{th} column. Then we define the *minor* of $A_{i,j}$ at position (i,j) to be $\det(A_{i,j})$. The *cofactor* of $A_{i,j}$ is defined to be

$$\mathrm{cof}(A_{i,j}) = (-1)^{i+j} \det(A_{i,j}).$$

We may now define the determinant of A by

$$\det(A) = a_{i,1}\mathrm{cof}(A_{i,1}) + a_{i,2}\mathrm{cof}(A_{i,2}) + \cdots + a_{i,n+1}\mathrm{cof}(A_{i,n+1}). \qquad \text{(A.1)}$$

This is called the *expansion of a determinant by cofactors* along the ith row of A. Similarly, we may expand along a column of A.

$$\det(A) = a_{1,j}\mathrm{cof}(A_{1,j}) + a_{2,j}\mathrm{cof}(A_{2,j}) + \cdots + a_{n+1,j}\mathrm{cof}(A_{n+1,j}),$$

called the *cofactor expansion along the jth column of A*. Both expansions can be shown to be identical. Hence, a determinant may be viewed as a function that assigns a real number to an $n \times n$ matrix, and the above gives a method for finding that number. Other useful properties of determinants that we will have occasion to use in the text are given in the following.

Theorem A.3 $\boxed{\textbf{Properties of Determinants}}$

Let $A = (a_{i,j})$, $B = (b_{i,j}) \in \mathcal{M}_{n \times n}(\mathbb{R})$. *Then each of the following hold.*

(a) $\det(A) = \det(a_{i,j}) = \det(a_{j,i}) = \det(A^t)$.

(b) $\det(AB) = \det(A)\det(B)$.

(c) *If matrix A is achieved from matrix B by interchanging two rows (or two columns), then $\det(A) = -\det(B)$.*

(d) *If $\mathcal{S}_n$ is the symmetric group on n symbols, then*

$$det(A) = \sum_{\sigma \in \mathcal{S}_n} (\mathrm{sgn}(\sigma)) a_{1,\sigma(1)} a_{2,\sigma(2)} \cdots a_{n,\sigma(n)},$$

where $\mathrm{sgn}(\sigma)$, *is 1 or -1 according as σ is even or odd.*

If $A \in \mathcal{M}_{n \times n}(\mathbb{R})$, then A is said to be *invertible*, or *nonsingular* if there is a unique matrix denoted by

$$A^{-1} \in M_{n \times n}(\mathbb{R})$$

such that

$$AA^{-1} = I_n = A^{-1}A.$$

Here are some properties of invertible matrices.

Theorem A.4 | **Properties of Invertible Matrices**

Let $n \in \mathbb{N}$, and A invertible in $\mathcal{M}_{n \times n}(\mathbb{R})$. Then each of the following holds.

(a) $(A^{-1})^{-1} = A$.

(b) $(A^t)^{-1} = (A^{-1})^t$, *where "t" denotes the transpose.*

(c) $(AB)^{-1} = B^{-1}A^{-1}$.

In order to provide a formula for the inverse of a given matrix, we need the following concept.

Definition A.12 | **Adjoint**

If $A = (a_{i,j}) \in \mathcal{M}_{n \times n}(\mathbb{R})$, then the matrix

$$A^a = (b_{i,j})$$

given by

$$b_{i,j} = (-1)^{i+j}\det(A_{j,i}) = \text{cof}(A_{j,i}) = \left[(-1)^{i+j}\det(A_{i,j})\right]^t$$

is called the adjoint *of A.*

Some properties of adjoints related to inverses, including a formula for the inverse, are as follows. Recall that a *unit* in $\mathbb{R}$ means an element for which there exists a multiplicative inverse. In other words, an element $u \in \mathbb{R}$ is a unit if there exists an element $u^{-1} \in \mathbb{R}$ such that $uu^{-1} = 1$, in other words, all nonzero elements of $\mathbb{R}$ are units. Thus, the following says, in particular, that the only *non*-invertible matrices are those with zero determinant.

Theorem A.5 | **Properties of Adjoints**

If $A \in \mathcal{M}_{n \times n}(\mathbb{R})$, then each of the following holds.

(a) $AA^a = \det(A)I_n = A^a A$.

(b) *A is invertible in $\mathcal{M}_{n \times n}(R)$ if and only if $\det(A)$ is a unit in $\mathbb{R}$, in which case $A^{-1} = A^a/\det(A)$.*

Example A.3 If $n = 2$, then the inverse of a nonsingular matrix

$$A = \begin{pmatrix} a & b \\ c & d \end{pmatrix}$$

is given by

$$A^{-1} = \begin{pmatrix} \frac{d}{\det(A)} & \frac{-b}{\det(A)} \\ \frac{-c}{\det(A)} & \frac{a}{\det(A)} \end{pmatrix}.$$

✦ Polynomials and Polynomial Rings

If R is a ring (see Remark 2.5 on page 79), then a *polynomial* $f(x)$ in an *indeterminant* x with *coefficients* in R is an infinite formal sum

$$f(x) = \sum_{j=0}^{\infty} a_j x^j = a_0 + a_1 x + \cdots + a_n x^n + \cdots,$$

where the *coefficients* a_j are in R for $j \geq 0$ and $a_j = 0$ for all but a finite number of those values of j. The set of all such polynomials is denoted by $R[x]$. If $a_n \neq 0$, and $a_j = 0$ for $j > n$, then a_n is called the *leading coefficient* of $f(x)$. If the leading coefficient $a_n = 1_R$, in the case where R is a commutative ring with identity 1_R, then $f(x)$ is said to be *monic*.

We may add two polynomials from $\mathbb{R}[x]$, $f(x) = \sum_{j=0}^{\infty} a_j x^j$ and $g(x) = \sum_{j=0}^{\infty} b_j x^j$, by

$$f(x) + g(x) = \sum_{j=0}^{\infty} (a_j + b_j) x^j \in R[x],$$

and multiply them by

$$f(x)g(x) = \sum_{j=0}^{\infty} c_j x^j,$$

where

$$c_j = \sum_{i=0}^{j} a_i b_{j-i}.$$

Also, $f(x) = g(x)$ if and only if $a_j = b_j$ for all $j = 0, 1, \ldots$. Under the above operations $R[x]$ is a ring, called the *polynomial ring over R in the indeterminant x*. Furthermore, if R is commutative, then so is $R[x]$, and if R has identity 1_R, then 1_R is the identity for $R[x]$. Notice that with these conventions, we may write $f(x) = \sum_{j=0}^{n} a_j x^j$, for some $n \in \mathbb{N}$, where a_n is the leading coefficient since we have tacitly agreed to "ignore" zero terms.

If $\alpha \in R$, we write $f(\alpha)$ to represent the element $\sum_{j=0}^{n} a_j \alpha^j \in R$, called the *substitution* of α for x. When $f(\alpha) = 0$, then α is called a *root* of $f(x)$. The substitution gives rise to a mapping $\overline{f} : R \mapsto R$ given by $\overline{f} : \alpha \mapsto f(\alpha)$, which is determined by $f(x)$. Thus, $\overline{f}$ is called a *polynomial function* over R.

✦ Characteristic of a Ring

The characteristic of a ring R is the smallest $n \in \mathbb{N}$ (if there is one) such that $n \cdot r = 0$ for all $r \in R$. If there is no such n, then R is said to have characteristic 0. Any field containing $\mathbb{Q}$ has characteristic zero, while any field containing the finite field $\mathbb{F}_p$ for a prime p has characteristic p (see the discussion following Definition A.15 below).

Definition A.13 Degrees of Polynomials

If $f(x) \in R[x]$, with $f(x) = \sum_{j=0}^{d} a_j x^j$, and $a_d \neq 0$, then $d \geq 0$ is called the degree of $f(x)$ over R, denoted by $\deg_R(f)$. If no such d exists, we write $\deg_R(f) = -\infty$, in which case $f(x)$ is the zero polynomial in $R[x]$ (for instance, see Example A.5 below). If F is a field of characteristic zero, then

$$\deg_{\mathbb{Q}}(f) = \deg_F(f)$$

for any $f(x) \in \mathbb{Q}[x]$. If F has characteristic p, and $f(x) \in \mathbb{F}_p[x]$, then

$$\deg_{\mathbb{F}_p}(f) = \deg_F(f).$$

In either case, we write $\deg(f)$ for $\deg_F(f)$, without loss of generality, and call this the degree of $f(x)$.

With respect to roots of polynomials, the following is important.

Definition A.14 Discriminant of Polynomials

Let $f(x) = a \prod_{j=1}^{n} (x - \alpha_j) \in F[x]$, $\deg(f) = n > 1$, $a \in F$ a field in $\mathbb{C}$, where $\alpha_j \in \mathbb{C}$ are all the roots of $f(x) = 0$ for $j = 1, 2, \ldots, n$. Then the discriminant of f is given by

$$\mathrm{disc}(f) = a^{2n-2} \prod_{1 \leq i < j \leq n} (\alpha_j - \alpha_i)^2.$$

From Definition A.14, we see that f has a multiple root in $\mathbb{C}$ (namely for some $i \neq j$ we have $\alpha_i = \alpha_j$, also called a *repeated root*) if and only if $\mathrm{disc}(f) = 0$.

Example A.4 If $f(x) = ax^2 + bx + c$ where $a, b, c \in \mathbb{Z}$, then $\mathrm{disc}(f) = b^2 - 4ac$ and if $f(x) = x^3 - c$, then $\mathrm{disc}(f) = -27c^2$.

Definition A.15 Division of Polynomials

We say that a polynomial $g(x) \in R[x]$ divides $f(x) \in R[x]$, if there exists an $h(x) \in R[x]$ such that $f(x) = g(x)h(x)$. We also say that $g(x)$ is a factor of $f(x)$.

Definition A.16 | Irreducible Polynomials over Rings |

A polynomial $f(x) \in R[x]$ is called irreducible (over R) if $f(x)$ is not a unit in R and any factorization $f(x) = g(x)h(x)$, with $g(x), h(x) \in R[x]$ satisfies the property that one of $g(x)$ or $h(x)$ is in R, called a constant polynomial. In other words, $f(x)$ cannot be the product of two nonconstant polynomials. If $f(x)$ is not irreducible, then it is said to be reducible.

Note that it is possible that a reducible polynomial $f(x)$ could be a product of two polynomials of the *same degree* as that of f. For instance, $f(x) = (1 - x) = (2x + 1)(3x + 1)$ in $R = \mathbb{Z}/6\mathbb{Z}$.

In general, it is important to make the distinction between degrees of a polynomial over various rings, since the base ring under consideration may alter the makeup of the polynomial.

For the following example, recall that a *finite field* is a field with a finite number of elements $n \in \mathbb{N}$, denoted by $\mathbb{F}_n$. In general, if K is a finite field, then $K = \mathbb{F}_{p^m}$ for some prime p and $m \in \mathbb{N}$, also called *Galois fields*. The field $\mathbb{F}_p$ is called the *prime subfield* of K. In general, a prime subfield is a field having no proper subfields, so $\mathbb{Q}$ is the prime subfield of any field of characteristic 0 and $\mathbb{Z}/p\mathbb{Z} = \mathbb{F}_p$ is the prime field of any field $K = \mathbb{F}_{p^m}$. In the following result, the term *cyclic* in reference to a multiplicative abelian group G means that a group generated by some $g \in G$ *coincides* with G. Note that any group of prime order is cyclic and the product of two cyclic groups of relatively prime order is also a cyclic group. Also, if $\mathcal{S}$ is a nonempty subset of a group G, then the intersection of all subgroups of G containing $\mathcal{S}$ is called the subgroup *generated* by $\mathcal{S}$.

Theorem A.6 | Multiplicative Subgroups of Fields |

If F is any field and F^* is a finite subgroup of the multiplicative subgroup of nonzero elements of F, then F^* is cyclic. In particular, if $F = \mathbb{F}_{p^n}$ is a finite field, then F^* is a finite cyclic group.

(See the discussion surrounding the above and congruences on page 81.)

Example A.5 The polynomial $f(x) = 2x^2 + 2x + 2$ is of degree two over $\mathbb{Q}$. However, over $\mathbb{F}_2$, $\deg_{\mathbb{F}_2}(f) = -\infty$, since f is the zero polynomial in $\mathbb{F}_2[x]$.

Some facts concerning irreducible polynomials will be needed in the text as follows.

Theorem A.7 | Irreducible Polynomials Over Finite Fields |

The product of all monic irreducible polynomials over a finite field $\mathbb{F}_q$ whose degrees divide a given $n \in \mathbb{N}$ is equal to $x^{q^n} - x$.

Based upon Theorem A.7, the following may be used as an algorithm for testing polynomials for irreducibility over prime fields and thereby generate irreducible polynomials.

Corollary A.2 *The following are equivalent.*

(a) f is irreducible over $\mathbb{F}_p$, where p is prime, and $\deg_{\mathbb{F}_p}(f) = n$.

(b) $\gcd(f(x), x^{p^i} - x) = 1$ for all natural numbers $i \leq \lfloor n/2 \rfloor$.

The following is also a general result concerning irreducible polynomials over any field.

Theorem A.8 | **Irreducible Polynomials Over Arbitrary Fields**

Let F be a field and $f(x) \in F[x]$. Denote by $(f(x))$ the principal ideal in $F[x]$ generated by $f(x)$ (see Definition A.20 on page 303). Then the following are equivalent.

(a) f is irreducible over F.

(b) $F[x]/(f(x))$ is a field.

Another useful result is the following.

Theorem A.9 | **Polynomials, Traces, and Norms**

Suppose that $f(x) \in R[x]$ is a monic, irreducible polynomial (over R where R is an integral domain), $\deg(f) = d \in \mathbb{N}$, and α_j for $j = 1, 2, \ldots, d$ are all of the roots of $f(x)$ in $\mathbb{C}$. Then

$$f(x) = x^d - Tx^{d-1} + \cdots \pm N,$$

where

$$T = \sum_{j=1}^{d} \alpha_j \text{ and } N = \prod_{j=1}^{d} \alpha_j,$$

where T is called the trace *and N is called the* norm *(of any of the roots of $f(x)$).*

(Compare the above with the discussion of norm and trace in Remark 5.1 on page 222.)

Now that we have the notion of irreducibility for polynomials, we may state a unique factorization result for polynomials over fields.

Theorem A.10 | **Unique Factorization for Polynomials**

If F is a field, then every nonconstant polynomial $f(x) \in F[x]$ can be factored in $F[x]$ into a product of irreducible polynomials $p(x)$, each of which is unique up to order and units (nonzero constant polynomials) in F.

The Euclidean Algorithm applies to polynomials in a way that allows us to talk about common divisors of polynomials in a fashion similar to that for integers.

Definition A.17 $\boxed{\textbf{The GCD of Polynomials}}$

If $f_i(x) \in F[x]$ *for* $i = 1, 2$, *where* F *is a field, then the* greatest common divisor *of* $f_1(x)$ *and* $f_2(x)$ *is the unique* monic polynomial $g(x) \in F[x]$ *satisfying both:*

(a) *For* $i = 1, 2$, $g(x)|f_i(x)$.

(b) *If there is a* $g_1(x) \in F[x]$ *such that* $g_1(x)|f_i(x)$ *for* $i = 1, 2$, *then* $g_1(x)|g(x)$.

If $g(x) = 1$, *we say that* $f_1(x)$ *and* $f_2(x)$ *are* relatively prime, *or* coprime *denoted by*
$$\gcd(f_1(x), f_2(x)) = 1.$$

There is also a Euclidean result for polynomials over a field.

Theorem A.11 $\boxed{\textbf{Euclidean Algorithm for Polynomials}}$

If $f(x), g(x) \in F[x]$, *where* F *is a field, and* $g(x) \neq 0$, *there exist unique* $q(x), r(x) \in F[x]$ *such that*

$$f(x) = q(x)g(x) + r(x),$$

where $\deg(r) < \deg(g)$. *(Note that if* $r(x) = 0$, *the zero polynomial, then* $\deg(r) = -\infty$.)

Finally, if $f(x)$ *and* $g(x)$ *are relatively prime, there exist* $s(x), t(x) \in F[x]$ *such that*

$$1 = s(x)f(x) + t(x)g(x).$$

We will need the following important polynomial in the main text.

Definition A.18 $\boxed{\textbf{Cyclotomic Polynomials}}$

If $n \in \mathbb{N}$, *then the* n^{th} cyclotomic polynomial *is given by*

$$\Phi_n(x) = \prod_{\substack{\gcd(n,j)=1 \\ 1 \leq j < n}} (x - \zeta_n^j).$$

Also, the degree of $\Phi_n(x)$ *is* $\phi(n)$, *the Euler Totient (see Definition 2.7).*

Note that despite the form of the cyclotomic polynomial given in Definition A.18, it can be shown that $\Phi_n(x) \in \mathbb{Z}[x]$. The reader may think of the term *cyclotomic* as "circle dividing," since the nth roots of unity divide the unit circle into n equal arcs. Also, the ζ_n^j are sometimes called *De Moivre Numbers*.

Biography A.2 Abraham De Moivre (1667–1754) *was a French-born Huguenot who left for England when Louis XIV revoked the Edict of Nantes in 1685. He was one of the pioneers of the theory of probability in the early eighteenth century. He became acquainted with Newton and Halley when he went to England. However, as a Frenchman, he was unable to secure a university position there and remained mostly self-supporting through fees for tutorial services. Yet he produced a considerable amount of research, perhaps the most famous of which is his* Doctrine of Chances *first published in 1718. This and subsequent editions had more than fifty problems on probability. Perhaps the most famous theorem with De Moivre's name attached to it is the one that says: For a, b coordinates in the complex plane, r the radius and ϕ the angle that the radius vector makes with the real axis, $(a + bi)^n = r^n(\cos(n\phi) + i\sin(n\phi))$.*

The following section is of importance for us in the main text as a tool for the description of numerous cryptographic devices (see page 79).

✦ Action on Rings

Definition A.19 | Morphisms of Rings

If R and S are two rings and $f : R \to S$ is a function such that $f(ab) = f(a)f(b)$, and $f(a + b) = f(a) + f(b)$ for all $a, b \in R$, then f is called a ring homomorphism. *If, in addition, $f : R \to S$ is an injection as a map of sets, then f is called a* ring monomorphism. *If a ring homomorphism f is a surjection as a map of sets, then f is called a* ring epimorphism. *If a ring homomorphism f is a bijection as a map of sets, then f is called a* ring isomorphism, *and R is said to be* isomorphic *to S, denoted by $R \cong S$. Lastly,* $\ker(f) = \{s \in S : f(s) = 0\}$ *is called the* kernel *of f. Also, f is injective if and only if $\ker(f) = \{0\}$.*

There is a fundamental result that we will need in the text. In order to describe it, we need the following notion.

Definition A.20 | Ideal, Cosets, and Quotient Rings

An ideal *I in a commutative ring R with identity is a subring of R satisfying the additional property that $rI \subseteq I$ for all $r \in R$. If I is an ideal in R then a* coset *of I in R is a set of the form $r + I = \{r + \alpha : \alpha \in I\}$ where $r \in R$. The set*

$$R/I = \{r + I : r \in R\}$$

becomes a ring under multiplication and addition of cosets given by

$$(r + I)(s + I) = rs + I, \text{ and } (r + I) + (s + I) = (r + s) + I,$$

for any $r, s \in R$ (and this can be shown to be independent of the representatives r and s). R/I is called the quotient ring *of R by I, or the* factor ring *of R by I, or the* residue class ring modulo I. *The cosets are called the* residue classes modulo I. *A mapping*

$$f : R \mapsto R/I,$$

which takes elements of R to their coset representatives in R/I, is called the natural map *of R to R/I, and it is easily seen to be an epimorphism. The cardinality of R/I is denoted by $|R : I|$.*

Example A.6 Consider the ring of integers modulo $n \in \mathbb{N}$, $\mathbb{Z}/n\mathbb{Z}$. Then $n\mathbb{Z}$ is an ideal in $\mathbb{Z}$, and the quotient ring is the residue class ring modulo n. In particular, we will need to use elementary results of this nature in §1.8.

Remark A.1 *Since rings are also groups, then the above concept of cosets and quotients specializes to groups. In particular, we have the following. Note that an index of a subgroup H in a group G can be defined similarly to the above situation for rings as follows. The* index of H in G, *denoted by $|G : H|$, is the cardinality of the set of distinct right (respectively left) cosets of H in G. Our principal interest is when this cardinality is finite (so this allows us to access the definition of cardinality given earlier). Then* Lagrange's Theorem for groups *says that*

$$|G| = |G : H| \cdot |H|,$$

so if G is a finite group, then $|H| \mid |G|$. In particular, a finite abelian group G has subgroups of all orders dividing $|G|$.

Now we are in a position to state the important result for rings.

Theorem A.12 | **Fundamental Isomorphism Theorem for Rings**

If R and S are commutative rings with identity, and

$$\phi : R \to S$$

is a homomorphism of rings, then

$$\frac{R}{\ker(\phi)} \cong \text{img}(\phi).$$

Example A.7 If $\mathbb{F}_q$ is a finite field where $q = p^n$ (p prime) and $f(x) \in \mathbb{F}_p[x]$ is an irreducible polynomial of degree n (see page 298), then

$$\mathbb{F}_q \cong \frac{\mathbb{F}_p[x]}{(f(x))}.$$

The situation in Example A.7 is related to the following definition and theorem.

Definition A.21 | Maximal and Proper Ideals

Let R be a commutative ring with identity. An ideal $I \neq R$ is called maximal *if whenever $I \subseteq J$, where J is an ideal in R, then $I = J$ or $I = R$. (An ideal $I \neq R$ is called a* proper *ideal.)*

Theorem A.13 | Rings Modulo Maximal Ideals

If R is a commutative ring with identity, then M is a maximal ideal in R if and only if R/M is a field.

Example A.8 If F is a field and $r \in F$ is a fixed nonzero element, then

$$I = \{f(x) \in F[x] : f(r) = 0\}$$

is a maximal ideal and

$$F \cong F[x]/I.$$

Another aspect of rings that we will need in the text is the following. If $\mathcal{S} = \{R_j : j = 1, 2, \ldots, n\}$ is a set of rings, then let R be the set of n-tuples $(r_1, r_2, \ldots, r_n)$ with $r_j \in R_j$ for $j = 1, 2, \ldots, n$, with the *zero element* of R being the n-tuple, $(0, 0, \ldots, 0)$. Define addition in R by

$$(r_1, r_2, \ldots, r_n) + (r_1', r_2', \ldots, r_n') = (r_1 + r_1', r_2 + r_2', \ldots, r_n + r_n'),$$

for all $r_j, r_j' \in R_j$ with $j = 1, 2, \ldots, n$, and multiplication by

$$(r_1, r_2, \ldots, r_n)(r_1', r_2', \ldots, r_n') = (r_1 r_1', r_2 r_2', \ldots, r_n r_n').$$

This defines a structure on R called the *direct sum* of the rings R_j, $j = 1, 2, \ldots, n$, denoted by

$$\oplus_{j=1}^{n} R_j = R_1 \oplus \cdots \oplus R_n, \tag{A.2}$$

which is easily seen to be a ring. Similarly, when the R_j are groups, then this is a direct sum of groups, which is again a group.

In the text, we will have occasion to refer to such items as vector spaces, so we remind the reader of the definition. The reader is referred to pages 77–81, where we discussed the axioms for algebraic objects such as groups, rings, and fields. In particular, for the sake of completeness, note that any set satisfying all of the axioms of Theorem 2.1 on page 77, except (g), is called a *division ring*.

✦ Vector Spaces

A *vector space* consists of an additive abelian group V and a field F together with an operation called *scalar multiplication* of each element of V by each element of F on the left, such that for each $r, s \in F$ and each $\alpha, \beta \in V$ the following conditions are satisfied:

A.1. $r\alpha \in V$.

A.2. $r(s\alpha) = (rs)\alpha$.

A.3. $(r + s)\alpha = (r\alpha) + (s\alpha)$.

A.4. $r(\alpha + \beta) = (r\alpha) + (r\beta)$.

A.5. $1_F\alpha = \alpha$.

The set of elements of V are called *vectors* and the elements of F are called *scalars*. The generally accepted abuse of language is to say that V is a *vector space over F*. If V_1 is a subset of a vector space V that is a vector space in its own right, then V_1 is called a *subspace of V*.

Example A.9 For a given prime p, $m, n \in \mathbb{N}$, the finite field $\mathbb{F}_{p^n}$ is an n-dimensional vector space over $\mathbb{F}_{p^m}$ with p^{mn} elements.

Definition A.22 | Bases, Dependence, and Finite Generation

If $\mathcal{S}$ is a subset of a vector space V, then the intersection of all subspaces of V containing $\mathcal{S}$ is called the subspace generated by $\mathcal{S}$, *or* spanned by $\mathcal{S}$. *If there is a finite set $\mathcal{S}$, and $\mathcal{S}$ generates V, then V is said to be* finitely generated. *If $\mathcal{S} = \varnothing$, then $\mathcal{S}$ generates the zero vector space. If $\mathcal{S} = \{m\}$, a singleton set, then the subspace generated by $\mathcal{S}$ is said to be the* cyclic subspace generated by m.

A subset $\mathcal{S}$ of a vector space V is said to be linearly independent *provided that for distinct $s_1, s_2, \ldots, s_n \in \mathcal{S}$, and $r_j \in V$ for $j = 1, 2, \ldots, n$,*

$$\sum_{j=1}^{n} r_j s_j = 0 \text{ implies that } r_j = 0 \text{ for } j = 1, 2, \ldots, n.$$

If $\mathcal{S}$ is not linearly independent, then it is called linearly dependent. *A linearly independent subset of a vector space that spans V is called a* basis *for V.*

In the text, we will have need of the following notion, especially as it pertains to the infinite binary case.

✦ Sequences and Series

Definition A.23 *A* sequence *is a function whose domain is* $\mathbb{N}$, *with images denoted by* a_n, *called the* n^{th} *term of the sequence. The entire sequence is denoted by* $\{a_n\}_{n=1}^{\infty}$, *or simply* $\{a_n\}$, *called an* infinite sequence *or simply a* sequence. *If* $\{a_n\}$ *is a sequence, and* $L \in \mathbb{R}$ *such that*

$$\lim_{n \to \infty} a_n = L,$$

then the sequence is said to converge (*namely when the limit exists*) *whereas sequences that have no such limit are said to* diverge. *If the terms of the sequence are nondecreasing,* $a_n \le a_{n+1}$ *for all* $n \in \mathbb{N}$, *or nonincreasing,* $a_n \ge a_{n+1}$ *for all* $n \in \mathbb{N}$, *then* $\{a_n\}$ *is said to be* monotonic. *A sequence* $\{a_n\}$ *is called* bounded above *if there exists an* $M \in \mathbb{R}$ *such that* $a_n \le M$ *for all* $n \in \mathbb{N}$. *The value* M *is called an* upper bound *for the sequence. A sequence* $\{a_n\}$ *is called* bounded below *if there is a* $B \in \mathbb{R}$ *such that* $B \le a_n$ *for all* $n \in \mathbb{N}$, *and* B *is called a* lower bound *for the sequence. A sequence* $\{a_n\}$ *is called* bounded *if it bounded above and bounded below.*

Some fundamental facts concerning sequences are contained in the following.

Theorem A.14 | **Properties of Sequences** | *Let* $\{a_n\}$ *and* $\{b_n\}$ *be sequences. Then*

(a) *If* $\{a_n\}$ *is bounded and monotonic, then it converges.*

(b) *If* $\lim_{n \to \infty} a_n = \lim_{n \to \infty} b_n = L \in \mathbb{R}$, *and* $\{c_n\}$ *is a sequence such that there exists a natural number* N *with* $a_n \le c_n \le b_n$ *for all* $n > N$, *then* $\lim_{n \to \infty} c_n = L$.

(c) *If* $\lim_{n \to \infty} |a_n| = 0$, *then* $\lim_{n \to \infty} a_n = 0$.

Note that part (c) of Theorem A.14 is a corollary to part (b).

Now we look at series. If a_n is a function on $n \in \mathbb{N}$, then $\sum_{j=1}^{n} a_j = A_n$ is called an n *partial sum* of the infinite series $\sum_{j=1}^{\infty} a_n$, which is said to converge is the sequence $\{A_j\}_{j=1}^{\infty}$ converges, and to diverge if that sequence diverges. When $\sum_{j=1}^{\infty} |a_n|$ is convergent, we say that $\sum_{j=1}^{\infty} a_n$ is *absolutely convergent*.

Theorem A.15 | **Properties of Series**

(a) *If* $\sum_{n=1}^{\infty} a_n$ *converges, then* $\lim_{n \mapsto \infty} a_n = 0$.

(b) *If $a_n \geq 0$, then $\prod_{n=1}^{\infty}(1 + a_n)$ and $\sum_{n=1}^{\infty} a_n$ are both convergent or both divergent.*

(c) *If $\lim_{n \mapsto \infty} \left| \frac{a_{n+1}}{a-n} \right| < 1$, then $\sum_{j=1}^{\infty} a_n$ is absolutely convergent. (This is called the* ratio test.)

(d) *If $a_n \geq 0$ and $\lim_{n \mapsto \infty} a_n^{1/n} < 1$, then $\sum_{n=1}^{\infty} a_n$ is convergent. (This is called Cauchy's test.)*

Some important examples are given as follows.

Example A.10 *We have the formula for an infinite geometric series,*

$$\sum_{n=0}^{\infty} x^n = \frac{1}{1-x},$$

which converges for $|x| < 1$.

Example A.11 *We have the series for the natural exponential function,*

$$\sum_{n=0}^{\infty} x^n/n! = e^x.$$

The following will be a valuable tool in §1.7 when we look at some partition theory.

Definition A.24 | Taylor and Maclaurin Series |

If a function f has derivatives $f^{(n)}(c)$ of all orders at $x = c$, then the series

$$\sum_{n=0}^{\infty} \frac{f^{(n)}(c)}{n!}(x - c)^n = f(c) + f'(c)(x - c) + \cdots + \frac{f^{(n)}(c)}{n!}(x - c)^n + \cdots$$

is called the Taylor *series for $f(x)$ at c. Also, if $c = 0$, then the series is called the* Maclaurin *series for f.*

Biography A.3 Colin Maclaurin (1698–1746) *was born in Kilmodan, Cowla, Argyllshire, Scotland, where his father was minister of the small parish. However, his father died when he was only six weeks old, and his mother died when he was nine years old. His was raised by his uncle Daniel Maclaurin who was minister at Kilfinnan on Loch Fyne. Colin entered the University of Glasgow in 1709 when he was eleven years old. By the age of 14 he was awarded his Master's degree, after which he took a year at the university to study divinity. He left the university in 1714, to return to live with his uncle at Kilfinnan. There he worked hard, and it paid high rewards since he was appointed professor of mathematics at Marischal College in the University of Aberdeen in August of 1717. By 1725, with the assistance of a supportive letter from Newton himself, Maclaurin was appointed to the University of Edinburgh on November 3, 1725, and it is at Edinburgh where he spent the rest of his life. Maclaurin's work included geometry, astronomy, and algebra. In 1740 he stood tall beside two of the greatest mathematicians of the day when he was jointly awarded a second prize from the* Académie des Sciences *in Paris for a study of tides, along with Euler and Daniel Bernoulli. In 1742 he wrote his two-volume work,* Treatise of Fluxions, *which was the first mathematically rigorous display of Newton's methods, essentially written to counter Berkeley's attack on the lack of a rigorous structure for the calculus. In 1748, his* Treatise on Algebra *was published posthumously. He died in Edinburgh, Scotland on June 14, 1746.*

Example A.12 *The Maclaurin series for* $f(x) = (1+x)^k$ *is*

$$1 + kx + \frac{k(k-1)x^2}{2} + \cdots + \frac{k(k-1)\cdots(k-n+1)x^n}{n!} + \cdots$$

called the Binomial Series. *The binomial series converges for* $|x| < 1$.

Biography A.4 Brook Taylor (1685–1731) *was born in Edmonton, Middlesex, England. He entered St. John's College, Cambridge on April 3, 1703, and graduated with an LL.B. in 1709. On April 3, 1712, he was elected to the Royal Society, and in 1714 he was elected Secretary to the Royal Society, the latter of which he held until October 21, 1718. In 1715, he published two renowned book,* Methodus Incrementorum Directa et Inversa *and* Linear Perspective. *In the former, he introduced a new branch of mathematics called* calculus of finite differences, *invented the method in calculus known as* integration by parts, *and discovered the series above that bears his name. Also included in that important volume was the method of singular solutions to differential equations, a change of variables formula, and methodology for associating the derivative of a function to the derivative of the inverse function. There was even some applications to Taylor's love of music, a treatment of vibrating strings. The aforementioned second book from 1715, contained the first overall discussion of vanishing points. This work may be considered to be the beginnings of the theory of projective geometry. It is generally agreed that his health problems and family tragedies prevented him from developing his ideas to greater fruition, and that his contribution to mathematics is exponentially greater than the mere attachment of his name to the above series. He died on December 29, 1731 in Somerset House, London, England.*

Example A.13 *The power series for* $1/(1+x)$ *is given by*

$$\frac{1}{1+x} = 1 - x + x^2 - x^3 + x^4 - x^5 + \cdots + (-1)^n x^n + \cdots,$$

which converes for $|x| < 1$, *and that for* $(1+x)^k$ *is given by*

$$(1+x)^k = 1 + kx + \frac{k(k-1)x^2}{2!} + \frac{k(k-1)(k-2)x^3}{3!} + \cdots,$$

which converges for $|x| < 1$.

Appendix B: Complexity

In this appendix, we look at the notion of complexity. In particular, the amount of time required on a computer to perform an algorithm is measured in terms of what are *bit operations*, by which we mean addition, subtraction, or multiplication of two binary digits, the division of a two-bit integer by a one-bit integer, or the shifting of a binary digit by one place. The number of bit operations required to perform an algorithm is its *computational complexity*. To describe the order of magnitude of this complexity, we need the following.

Definition B.1 | Big O Notation |

Suppose that f and g are positive real-valued functions. If there exists a real number c such that $f(x) < cg(x)$ for all sufficiently large x, then we write $f(x) = O(g(x))$, or simply $f = O(g)$. Typically, mathematicians also write $f << g$ to denote $f = O(g)$.

Remark B.1 *The Big O notation was introduced by Edmund Landau (1877–1938), whose most famous work is Vorlesungen über Zahlentheorie, published in 1927.*

Also note that in the above, "sufficiently large" means that there exists some bound $B \in \mathbb{R}^+$ such that $g(x) < cf(x)$ for all $x > B$. We just do not know explicitly the value of B. Often f is defined on $\mathbb{N}$ rather than $\mathbb{R}$, and occasionally over any subset of $\mathbb{R}$.

Furthermore, the notation $<<$ was introduced by I.M. Vinogradov, a Russian mathematician who proved, in 1937, that every sufficiently large positive integer is the sum of at most four primes. This is related to Goldbach's Conjecture, which says that every even $n \in \mathbb{N}$ with $n > 2$ is a sum of two primes.

A simple illustration of the use of Big O is to determine the number of bits in a base b integer. If n is a t-bit base b integer, then $b^{t-1} \leq n < b^t$. Therefore, $t = \lfloor \log_b n \rfloor + 1$, so $t = O(\log n)$. Here, $\log n$ means $\log_e n$, the logarithm to the base e, the *natural* or *canonical* base.

A simple illustration of the use of Big O in determining computational complexity is that of the addition or subtraction of two n-bit integers, using conventional methods as illustrated above. The number of bit operations required can be shown to be $O(n)$. On the other hand, the multiplication of two n-bit integers will vary depending upon the algorithm used. It can take as much as $O(n^2)$ bit operations, or as little as $O(n^{1.5})$ bit operations. Also, it can be shown that computation of $n!$ is bounded by $cn^2 \log^2 n$ bit operations (although more sophisticated algorithms can reduce this number). Furthermore, the actual amount of time required to carry out bit operations on a computer varies, depending upon the computer being used and current computer technology. Nevertheless, the advantage of the Big O notation for measuring complexity is independent of the particular computer being used. In other words, despite the relative differences of the various machines, the order of magnitude complexity of an algorithm remains the same.

A fundamental *time estimate* in performing an algorithm A is *polynomial time* (or simply *polynomial*) namely an algorithm is polynomial if its computational complexity is $O(n^c)$ for some constant $c \in \mathbb{R}^+$, where n is the bitlength of the input to the algorithm. For example, if $c = 0$, the algorithm is constant; if $c = 1$, it is linear; if $c = 2$, it is quadratic, and so on. Examples of polynomial time algorithms are the ordinary arithmetic operations of addition, subtraction, multiplication, and division. However, the computational complexity for computing $n!$ is *not* polynomial. On the other hand, algorithms with computational complexity $O(c^{f(n)})$ where c is constant and f is a polynomial on $n \in \mathbb{N}$ are *exponential time algorithms* or simply *exponential*. Now we examine some properties of this important Big O notation.

Theorem B.1 $\boxed{\text{Properties of the Big O Notation}}$

Suppose that f, g are positive real-valued functions.

B.1. If $c \in \mathbb{R}^+$, then $cO(g) = O(g)$.

B.2. $O(\max\{f, g\}) = O(f) \pm O(g)$.

B.3. $O(fg) = O(f)O(g)$.

There are more sophisticated algorithms for reducing the computational complexity of the operations described above. For instance, there is an algorithm (using fast Fourier transforms — beyond the scope of this book) which shows that two n-bit integers can be multiplied using $O(n \log n)$ bit operations. However, there are complexity calculations which are within our scope. For instance, see Corollary 1.4 on page 21, known as Lamé's Theorem and the discussion of the computational complexity of the gcd in Remark 1.1 on page 22.

What is implicit in this appendix is the newest branch of number theory, *computational number theory*. During the past two decades, computational number theory has become a discipline in its own right. In particular, the study of cryptology (or secrecy systems) that we study throughout the text as applications of fundamental number theory, may be credited with the favour which has fallen upon computational number theory. The reason is that cryptography (the design and implementation of secrecy codes or systems) involves cryptosystems (methods for encoding and decoding messages) whose security is based upon the (presumed) difficulty of number-theoretic problems such as factoring.

Appendix C: Primes ≤ 9547 and Least Primitive Roots

This is a table of all primes $p \leq 9547$, and their least primitive roots a.

p	a	p	a	p	a	p	a	p	a	p	a
2	1	173	2	401	3	647	5	919	7	1193	3
3	2	179	2	409	21	653	2	929	3	1201	11
5	2	181	2	419	2	659	2	937	5	1213	2
7	3	191	19	421	2	661	2	941	2	1217	3
11	2	193	5	431	7	673	5	947	2	1223	5
13	2	197	2	433	5	677	2	953	3	1229	2
17	3	199	3	439	15	683	5	967	5	1231	3
19	2	211	2	443	2	691	3	971	6	1237	2
23	5	223	3	449	3	701	2	977	3	1249	7
29	2	227	2	457	13	709	2	983	5	1259	2
31	3	229	6	461	2	719	11	991	6	1277	2
37	2	233	3	463	3	727	5	997	7	1279	3
41	6	239	7	467	2	733	6	1009	11	1283	2
43	3	241	7	479	13	739	3	1013	3	1289	6
47	5	251	6	487	3	743	5	1019	2	1291	2
53	2	257	3	491	2	751	3	1021	10	1297	10
59	2	263	5	499	7	757	2	1031	14	1301	2
61	2	269	2	503	5	761	6	1033	5	1303	6
67	2	271	6	509	2	769	11	1039	3	1307	2
71	7	277	5	521	3	773	2	1049	3	1319	13
73	5	281	3	523	2	787	2	1051	7	1321	13
79	3	283	3	541	2	797	2	1061	2	1327	3
83	2	293	2	547	2	809	3	1063	3	1361	3
89	3	307	5	557	2	811	3	1069	6	1367	5
97	5	311	17	563	2	821	2	1087	3	1373	3
101	2	313	10	569	3	823	3	1091	2	1381	2
103	5	317	2	571	3	827	2	1093	5	1399	13
107	2	331	3	577	5	829	2	1097	3	1409	3
109	6	337	10	587	2	839	11	1103	5	1423	3
113	3	347	2	593	3	853	2	1109	2	1427	2
127	3	349	2	599	7	857	3	1117	2	1429	6
131	2	353	3	601	7	859	2	1123	2	1433	3
137	3	359	7	607	3	863	5	1129	11	1439	7
139	2	367	6	613	2	877	2	1151	17	1447	3
149	2	373	2	617	3	881	3	1153	5	1451	2
151	6	379	2	619	2	883	2	1163	5	1453	2
157	5	383	5	631	3	887	5	1171	2	1459	3
163	2	389	2	641	3	907	2	1181	7	1471	6
167	5	397	5	643	11	911	17	1187	2	1481	3

p	a	p	a	p	a	p	a	p	a	p	a
1483	2	1777	5	2087	5	2389	2	2711	7	3037	2
1487	5	1783	10	2089	7	2393	3	2713	5	3041	3
1489	14	1787	2	2099	2	2399	11	2719	3	3049	11
1493	2	1789	6	2111	7	2411	6	2729	3	3061	6
1499	2	1801	11	2113	5	2417	3	2731	3	3067	2
1511	11	1811	6	2129	3	2423	5	2741	2	3079	6
1523	2	1823	5	2131	2	2437	2	2749	6	3083	2
1531	2	1831	3	2137	10	2441	6	2753	3	3089	3
1543	5	1847	5	2141	2	2447	5	2767	3	3109	6
1549	2	1861	2	2143	3	2459	2	2777	3	3119	7
1553	3	1867	2	2153	3	2467	2	2789	2	3121	7
1559	19	1871	14	2161	23	2473	5	2791	6	3137	3
1567	3	1873	10	2179	7	2477	2	2797	2	3163	3
1571	2	1877	2	2203	5	2503	3	2801	3	3167	5
1579	3	1879	6	2207	5	2521	17	2803	2	3169	7
1583	5	1889	3	2213	2	2531	2	2819	2	3181	7
1597	11	1901	2	2221	2	2539	2	2833	5	3187	2
1601	3	1907	2	2237	2	2543	5	2837	2	3191	11
1607	5	1913	3	2239	3	2549	2	2843	2	3203	2
1609	7	1931	2	2243	2	2551	6	2851	2	3209	3
1613	3	1933	5	2251	7	2557	2	2857	11	3217	5
1619	2	1949	2	2267	2	2579	2	2861	2	3221	10
1621	2	1951	3	2269	2	2591	7	2879	7	3229	6
1627	3	1973	2	2273	3	2593	7	2887	5	3251	6
1637	2	1979	2	2281	7	2609	3	2897	3	3253	2
1657	11	1987	2	2287	19	2617	5	2903	5	3257	3
1663	3	1993	5	2293	2	2621	2	2909	2	3259	3
1667	2	1997	2	2297	5	2633	3	2917	5	3271	3
1669	2	1999	3	2309	2	2647	3	2927	5	3299	2
1693	2	2003	5	2311	3	2657	3	2939	2	3301	6
1697	3	2011	3	2333	2	2659	2	2953	13	3307	2
1699	3	2017	5	2339	2	2663	5	2957	2	3313	10
1709	3	2027	2	2341	7	2671	7	2963	2	3319	6
1721	3	2029	2	2347	3	2677	2	2969	3	3323	2
1723	3	2039	7	2351	13	2683	2	2971	10	3329	3
1733	2	2053	2	2357	2	2687	5	2999	17	3331	3
1741	2	2063	5	2371	2	2689	19	3001	14	3343	5
1747	2	2069	2	2377	5	2693	2	3011	2	3347	2
1753	7	2081	3	2381	3	2699	2	3019	2	3359	11
1759	6	2083	2	2383	5	2707	2	3023	5	3361	22

p	a	p	a	p	a	p	a	p	a	p	a
3371	2	3691	2	4019	2	4357	2	4703	5	5039	11
3373	5	3697	5	4021	2	4363	2	4721	6	5051	2
3389	3	3701	2	4027	3	4373	2	4723	2	5059	2
3391	3	3709	2	4049	3	4391	14	4729	17	5077	2
3407	5	3719	7	4051	10	4397	2	4733	5	5081	3
3413	2	3727	3	4057	5	4409	3	4751	19	5087	5
3433	5	3733	2	4073	3	4421	3	4759	3	5099	2
3449	3	3739	7	4079	11	4423	3	4783	6	5101	6
3457	7	3761	3	4091	2	4441	21	4787	2	5107	2
3461	2	3767	5	4093	2	4447	3	4789	2	5113	19
3463	3	3769	7	4099	2	4451	2	4793	3	5119	3
3467	2	3779	2	4111	12	4457	3	4799	7	5147	2
3469	2	3793	5	4127	5	4463	5	4801	7	5153	5
3491	2	3797	2	4129	13	4481	3	4813	2	5167	6
3499	2	3803	2	4133	2	4483	2	4817	3	5171	2
3511	7	3821	3	4139	2	4493	2	4831	3	5179	2
3517	2	3823	3	4153	5	4507	2	4861	11	5189	2
3527	5	3833	3	4157	2	4513	7	4871	11	5197	7
3529	17	3847	5	4159	3	4517	2	4877	2	5209	17
3533	2	3851	2	4177	5	4519	3	4889	3	5227	2
3539	2	3853	2	4201	11	4523	5	4903	3	5231	7
3541	7	3863	5	4211	6	4547	2	4909	6	5233	10
3547	2	3877	2	4217	3	4549	6	4919	13	5237	3
3557	2	3881	13	4219	2	4561	11	4931	6	5261	2
3559	3	3889	11	4229	2	4567	3	4933	2	5273	3
3571	2	3907	2	4231	3	4583	5	4937	3	5279	7
3581	2	3911	13	4241	3	4591	11	4943	7	5281	7
3583	3	3917	2	4243	2	4597	5	4951	6	5297	3
3593	3	3919	3	4253	2	4603	2	4957	2	5303	5
3607	5	3923	2	4259	2	4621	2	4967	5	5309	2
3613	2	3929	3	4261	2	4637	2	4969	11	5323	5
3617	3	3931	2	4271	7	4639	3	4973	2	5333	2
3623	5	3943	3	4273	5	4643	5	4987	2	5347	3
3631	15	3947	2	4283	2	4649	3	4993	5	5351	11
3637	2	3967	6	4289	3	4651	3	4999	3	5381	3
3643	2	3989	2	4297	5	4657	15	5003	2	5387	2
3659	2	4001	3	4327	3	4663	3	5009	3	5393	3
3671	13	4003	2	4337	3	4673	3	5011	2	5399	7
3673	5	4007	5	4339	10	4679	11	5021	3	5407	3
3677	2	4013	2	4349	2	4691	2	5023	3	5413	5

p	a	p	a	p	a	p	a	p	a	p	a
5417	3	5737	5	6079	17	6379	2	6763	2	7103	5
5419	3	5741	2	6089	3	6389	2	6779	2	7109	2
5431	3	5743	10	6091	7	6397	2	6781	2	7121	3
5437	5	5749	2	6101	2	6421	6	6791	7	7127	5
5441	3	5779	2	6113	3	6427	3	6793	10	7129	7
5443	2	5783	7	6121	7	6449	3	6803	2	7151	7
5449	7	5791	6	6131	2	6451	3	6823	3	7159	3
5471	7	5801	3	6133	5	6469	2	6827	2	7177	10
5477	2	5807	5	6143	5	6473	3	6829	2	7187	2
5479	3	5813	2	6151	3	6481	7	6833	3	7193	3
5483	2	5821	6	6163	3	6491	2	6841	22	7207	3
5501	2	5827	2	6173	2	6521	6	6857	3	7211	2
5503	3	5839	6	6197	2	6529	7	6863	5	7213	5
5507	2	5843	2	6199	3	6547	2	6869	2	7219	2
5519	13	5849	3	6203	2	6551	17	6871	3	7229	2
5521	11	5851	2	6211	2	6553	10	6883	2	7237	2
5527	5	5857	7	6217	5	6563	5	6899	2	7243	2
5531	10	5861	3	6221	3	6569	3	6907	2	7247	5
5557	2	5867	5	6229	2	6571	3	6911	7	7253	2
5563	2	5869	2	6247	5	6577	5	6917	2	7283	2
5569	13	5879	11	6257	3	6581	14	6947	2	7297	5
5573	2	5881	31	6263	5	6599	13	6949	2	7307	2
5581	6	5897	3	6269	2	6607	3	6959	7	7309	6
5591	11	5903	5	6271	11	6619	2	6961	13	7321	7
5623	5	5923	2	6277	2	6637	2	6967	5	7331	2
5639	7	5927	5	6287	7	6653	2	6971	2	7333	6
5641	14	5939	2	6299	2	6659	2	6977	3	7349	2
5647	3	5953	7	6301	10	6661	6	6983	5	7351	6
5651	2	5981	3	6311	7	6673	5	6991	6	7369	7
5653	5	5987	2	6317	2	6679	7	6997	5	7393	5
5657	3	6007	3	6323	2	6689	3	7001	3	7411	2
5659	2	6011	2	6329	3	6691	2	7013	2	7417	5
5669	3	6029	2	6337	10	6701	2	7019	2	7433	5
5683	2	6037	5	6343	3	6703	5	7027	2	7451	2
5689	11	6043	5	6353	3	6709	2	7039	3	7457	3
5693	2	6047	5	6359	13	6719	11	7043	2	7459	2
5701	2	6053	2	6361	19	6733	2	7057	5	7477	2
5711	19	6067	2	6367	3	6737	3	7069	2	7481	6
5717	2	6073	10	6373	2	6761	3	7079	7	7487	5

p	a	p	a	p	a	p	a	p	a	p	a
7489	7	7817	3	8179	2	8543	5	8863	3	9227	2
7499	2	7823	5	8191	17	8563	2	8867	2	9239	19
7507	2	7829	2	8209	7	8573	2	8887	3	9241	13
7517	2	7841	12	8219	2	8581	6	8893	5	9257	3
7523	2	7853	2	8221	2	8597	2	8923	2	9277	5
7529	3	7867	3	8231	11	8599	3	8929	11	9281	3
7537	7	7873	5	8233	10	8609	3	8933	2	9283	2
7541	2	7877	2	8237	2	8623	3	8941	6	9293	2
7547	2	7879	3	8243	2	8627	2	8951	13	9311	7
7549	2	7883	2	8263	3	8629	6	8963	2	9319	3
7559	13	7901	2	8269	2	8641	17	8969	3	9323	2
7561	13	7907	2	8273	3	8647	3	8971	2	9337	5
7573	2	7919	7	8287	3	8663	5	8999	7	9341	2
7577	3	7927	3	8291	2	8669	2	9001	7	9343	5
7583	5	7933	2	8293	2	8677	2	9007	3	9349	2
7589	2	7937	3	8297	3	8681	15	9011	2	9371	2
7591	6	7949	2	8311	3	8689	13	9013	5	9377	3
7603	2	7951	6	8317	6	8693	2	9029	2	9391	3
7607	5	7963	5	8329	7	8699	2	9041	3	9397	2
7621	2	7993	5	8353	5	8707	5	9043	3	9403	3
7639	7	8009	3	8363	2	8713	5	9049	7	9413	3
7643	2	8011	14	8369	3	8719	3	9059	2	9419	2
7649	3	8017	5	8377	5	8731	2	9067	3	9421	2
7669	2	8039	11	8387	2	8737	5	9091	3	9431	7
7673	3	8053	2	8389	6	8741	2	9103	6	9433	5
7681	17	8059	3	8419	3	8747	2	9109	10	9437	2
7687	6	8069	2	8423	5	8753	3	9127	3	9439	22
7691	2	8081	3	8429	2	8761	23	9133	6	9461	3
7699	3	8087	5	8431	3	8779	11	9137	3	9463	3
7703	5	8089	17	8443	2	8783	5	9151	3	9467	2
7717	2	8093	2	8447	5	8803	2	9157	6	9473	3
7723	3	8101	6	8461	6	8807	5	9161	3	9479	7
7727	5	8111	11	8467	2	8819	2	9173	2	9491	2
7741	7	8117	2	8501	7	8821	2	9181	2	9497	3
7753	10	8123	2	8513	5	8831	7	9187	3	9511	3
7757	2	8147	2	8521	13	8837	2	9199	3	9521	3
7759	3	8161	7	8527	5	8839	3	9203	2	9533	2
7789	2	8167	3	8537	3	8849	3	9209	3	9539	2
7793	3	8171	2	8539	2	8861	2	9221	2	9547	2

Appendix D: Indices

In Definition 3.3 on page 153, we introduced the notion of the index of an integer. In the following table, we give sample indices for certain integers modulo certain primes less than 100.

Table 7.1

b	a	p	$\mathrm{ind}_a(b)$	b	a	p	$\mathrm{ind}_a(b)$	b	a	p	$\mathrm{ind}_a(b)$
2	2	3	1	6	2	19	14	7	2	29	12
2	2	5	1	7	2	19	6	8	2	29	3
3	2	5	3	8	2	19	3	9	2	29	10
4	2	5	2	9	2	19	8	10	2	29	23
2	3	7	2	10	2	19	17	11	2	29	25
3	3	7	1	11	2	19	12	12	2	29	7
4	3	7	4	12	2	19	15	13	2	29	18
5	3	7	5	13	2	19	5	14	2	29	13
6	3	7	3	14	2	19	7	15	2	29	27
2	2	11	1	15	2	19	11	16	2	29	4
3	2	11	8	16	2	19	4	17	2	29	21
4	2	11	2	17	2	19	10	18	2	29	11
5	2	11	4	18	2	19	9	19	2	29	9
6	2	11	9	2	5	23	2	20	2	29	24
7	2	11	7	3	5	23	16	21	2	29	17
8	2	11	3	4	5	23	4	22	2	29	26
9	2	11	6	5	5	23	1	23	2	29	20
10	2	11	5	6	5	23	18	24	2	29	8
2	2	13	1	7	5	23	19	25	2	29	16
3	2	13	4	8	5	23	6	26	2	29	19
2	3	17	14	9	5	23	10	27	2	29	15
3	3	17	1	10	5	23	3	28	2	29	14
4	3	17	12	11	5	23	9	2	3	31	2
5	3	17	5	12	5	23	20	24	2	37	29
6	3	17	15	13	5	23	14	17	6	41	33
7	3	17	11	14	5	23	21	24	3	43	40
8	3	17	10	15	5	23	17	34	5	47	34
9	3	17	2	16	5	23	8	13	2	53	24
10	3	17	3	17	5	23	7	4	2	59	2
11	3	17	7	18	5	23	12	14	2	61	50
12	3	17	13	19	5	23	15	11	2	67	59
13	3	17	4	20	5	23	5	29	7	71	68
14	3	17	9	21	5	23	13	19	5	73	62
15	3	17	6	22	5	23	11	53	3	79	77
16	3	17	8	2	2	29	1	21	2	83	80
2	2	19	1	3	2	29	5	10	3	89	86
3	2	19	13	4	2	29	2	88	5	97	92
4	2	19	2	5	2	29	22	17	5	97	89

Appendix E: The ABC Conjecture

The conjecture highlighted in this appendix has been used effectively in the literature to prove other results. We give a brief overview of the work done, and related conjectures proved under the assumption of the validity of the following deep, and seemingly intractable conjecture due to Oesterlé, and later refined by Masser (see [29, pp. 35–39]).

◆ **The ABC Conjecture**

Suppose that $a, b, c \in \mathbb{N}$ such that $a + b = c$ with $\gcd(a, b, c) = 1$. Let $G = G(a, b, c)$ be the product of the primes dividing a, b, c, each to the first power. Then for all $\varepsilon > 0$, there exists a constant $k = k(\varepsilon)$ such that $c < kG^{1+\varepsilon}$.

In [17], Granville was able to prove the following conjectures under the assumption of the validity of the ABC Conjecture. Recall the definition of powerful numbers studied in Exercises 1.44–1.46 on page 39.

Conjecture 7.1 *There are only finitely many $n \in \mathbb{N}$ such that $n - 1$, n, $n + 1$ are all powerful.*

Conjecture 7.2 *The largest prime factor of $1 + x^2 y^3$ tends to infinity as $x + |y|$ tends to infinity.*

Granville attributed Conjecture 7.2 to K. Mahler who gave an analogous one in 1953. To see how easily one can achieve results using the ABC Conjecture, we make the following simple observation. By taking $a = 1$, $b = n^2 - 1$, and $c = n^2$ in the ABC Conjecture, we get $G \leq \sqrt{bn} < n^{3/2}$, so $n^2 < kn^{3/2+\varepsilon}$, which bounds n. Hence, conjecture 7.1 holds. Given that conjecture 7.1 remains open, and we can prove it in a sentence using the ABC Conjecture points to the depth of the latter.

The following conjecture was made by P.G. Walsh, a former student of this author, in [50].

Conjecture 7.3 *The equation $x^n - m^3 y^2 = \pm 1$ is solvable in integers $x, y > 1$, $m \geq 1$, and $n > 2$ if and only if $(x, m, y, n) \in \{(2, 1, 3, 3), (23, 2, 39, 3)\}$.*

Conjecture 7.3 is a generalization of Catalan's Conjecture (see Exercise 3.6 on page 144). It is an easy exercise to prove that Conjecture 7.3 follows from the validity of the ABC Conjecture (see [29, Exercise 1.6.11, p. 39]). Given that all powerful numbers are of the form $x^2 y^3$ (see Exercise 1.44), then conjecture 7.3 upgrades Catalan's Conjecture from pure powers to powerful numbers.

We know that Fermat's Last Theorem is true, and it took centuries to achieve a proof. Yet, the ABC Conjecture seems beyond our reach at this time. For relationships between Fermat's Last Theorem and the ABC Conjecture see [17].

Appendix F: Primes is in P

The following is an unconditional deterministic polynomial-time algorithm for primality testing presented in [1] by M. Agrawal, N. Kayal, and N. Saxena. The following is adapted from [32]. For notation in what follows, see Appendix A, especially Definition 2.7 on page 90 and Definition 3.1 on page 139, as well as results on polynomial rings especially as they pertain to finite fields starting on on page 298.

In what follows, $\mathbb{Z}_n$ for a given integer $n > 1$ denotes $\mathbb{Z}/n\mathbb{Z}$, and if $h(X) \in \mathbb{Z}_n[X]$, then the notation, $f(X) \equiv g(X) \,(\mathrm{mod}\, h(X), n))$, is used to represent the equation $f(X) = g(X)$ in the quotient ring $\mathbb{Z}_n[X]/(h(X))$. In particular, for suitably chosen r and a values, we will be looking at an equation of the following type:

$$(X + a)^n \equiv X^n + a \quad (\mathrm{mod}\, X^r - 1, n). \tag{F.1}$$

Algorithm F.1 —

Unconditional Deterministic Polynomial-Time Primality Test

Input an integer $n > 1$, and execute the following steps.

1. If $n = a^b$ for some $a \in \mathbb{N}$ and $b > 1$, then terminate with output

   "n is composite."

2. Find the smallest $r \in \mathbb{N}$ such that $\mathrm{ord}_r(n) > 4 \log_2^2 n$.

3. If $1 < \gcd(a, n) < n$ for some $a \leq r$, then output

   "n is composite."

4. If $n \leq r$, then output
   "n is prime."

5. Set $a = 1$ and execute the following:

   (i) Compute $Y(a) \equiv (X + a)^n - X^n - a \,(\mathrm{mod}\, X^r - 1, n)$.

   (ii) If $Y(a) \not\equiv 0 \,(\mathrm{mod}\, X^r - 1, n)$, output

   "n is composite."

   Otherwise, go to step (iii).

   (iii) If $Y(a) \equiv 0 \,(\mathrm{mod}\, X^r - 1, n)$, set $a = a + 1$. If $a < \lfloor 2\sqrt{\phi(r)} \cdot \log_2(n) \rfloor$, go to step (i). Otherwise, go to step 6.

6. Output
   "n is prime."

◆ **Analysis**

The reason the authors of [1] considered equations of type (F.1) was that they were able to prove the following.

Polynomial Primality Criterion

If $a \in \mathbb{Z}$, $n \in \mathbb{N}$ with $n > 1$, and $\gcd(a, n) = 1$, then n is prime if and only if

$$(X + a)^n \equiv X^n + a \pmod{n}. \tag{F.2}$$

The satisfaction of polynomial congruence (F.2) is a simple test but the time taken to test the congruence is too expensive. To save time, they looked at the congruence modulo a polynomial, whence congruence (F.1). However, by looking at such congruences, they introduced the possibility that composite numbers might satisfy (F.1), which indeed they do. Yet, the authors were able to (nearly) restore the characterization given in the above polynomial primality criterion by showing that for a suitably chosen r, if (F.1) is satisfied for several values of a, then n must be a prime power. Since the number of a values and the suitably chosen r value are bounded by a polynomial in $\log_2(n)$, they achieved a deterministic polynomial time algorithm for primality testing.

The authors of [1] were able to establish the following facts about their algorithm. The reader will need the concepts of ceiling and floor functions — see §2.5.

Facts Concerning Algorithm F.1

1. The algorithm outputs "n is prime" if and only if n is prime. (Hence, it outputs "n is composite" if and only if n is composite.)

2. There exists and $r \leq \lceil 16 \log_2^5(n) \rceil$ such that $\text{ord}_r(n) > 4 \log_2^2(n)$.

3. The asymptotic time complexity of the algorithm is $O(\log_2^{10.5+\varepsilon}(n))$ for any $\varepsilon > 0$.

4. It is conjectured that the time complexity of the algorithm can be improved to the best-case scenario where $r = O(\log_2^2(n))$, which would mean that the complexity of the algorithm would be

$$O(\log_2^{6+\varepsilon}(n)) \text{ for any } \varepsilon > 0.$$

Two conjectures support the authors' conjecture in part 4 above. They are given as follows.

Artin's Conjecture

If $n \in \mathbb{N}$ is not a perfect square, then the number of primes $q \leq m$ for which $\text{ord}_q(n) = q - 1$ is asymptotically $A(n) \cdot m/\ln(m)$, where $A(n)$ is *Artin's constant* given by

$$A(n) = \prod_{j=1}^{\infty} \left(1 - \frac{1}{p_k(p_k - 1)} \right) = 0.3739558136\ldots,$$

with p_k being the kth prime.

If Artin's conjecture becomes effective for $m = O(\log_2^2(n))$, then it follows that there is an $r = O(\log_2^2(n))$ with the desired properties.

The other conjecture that supports their contention is given as follows.

Sophie Germane's Prime Density Conjecture

The number of primes $q \leq m$ such that $2q+1^a$ is also a prime is asymptotically $2C_2 m / \ln^2(m)$, where C_2 is the *twin prime constant* given by

$$C_2 = \prod_{p \geq 3} \frac{p(p-2)}{(p-1)^2} \approx 0.6601611816\ldots.$$

---

^{a}Such primes are called *Sophie Germane primes*.

If the Sophie Germane conjecture holds, then $r = O(\log_2^{2+\varepsilon}(n))$ for any $\varepsilon > 0$ such that $\text{ord}_r(n) \geq 4\log_2^2(n)$. Hence, the algorithm, with this r value, yields a time complexity of $O(\log_2^{6+\varepsilon}(n))$ for any $\varepsilon > 0$.

The authors of [1] leave one more conjecture, the affirmative solution of which would improve the complexity of algorithm F.1 to $O(\log_2^{3+\varepsilon}(n))$ for any $\varepsilon > 0$.

Conjecture F.4 *If r is a prime not dividing $n > 1$ and if*

$$(X - 1)^n \equiv X^n - 1 \pmod{X^r - 1, n},$$

then either n is prime or $n^2 \equiv 1 \pmod{r}$.

The result given in Algorithm F.1 is a major breakthrough and the simplicity of the approach is noteworthy. The algorithm uses essentially only elementary properties of polynomial rings over finite fields and a generalization of Fermat's little theorem in that context, quite impressive indeed.

Solutions to Odd-Numbered Exercises

Section 1.1

1.1 $\mathfrak{g}^2 = ((1 + \sqrt{5})/2)^2 = (6 + 2\sqrt{5})/4 = (3 + \sqrt{5})/2 = (1 + \sqrt{5})/2 + 1 = \mathfrak{g} + 1$.

1.3 Assume that the Principle of Mathematical Induction holds. Let $\mathcal{S} \neq \varnothing$, and $\mathcal{S} \subseteq \mathbb{N}$. Suppose that $\mathcal{S}$ has no least element. Then $1 \notin \mathcal{S}$, so $2 \notin \mathcal{S}$, and similarly $3 \notin \mathcal{S}$, and so on, which implies that $\mathcal{S} = \varnothing$ by induction, a contradiction.

Conversely, assume the Well-Ordering Principle holds. Also, assume that $1 \in \mathcal{S}$, and that $k \in \mathcal{S}$, whenever $k - 1 \in \mathcal{S}$. If $\mathcal{S} \neq \mathbb{N}$, then the Well-Ordering Principle says that there is a least $n \in \mathbb{N} \setminus \mathcal{S}$. Thus, $n - 1 \in S$. However, by assumption $n \in \mathcal{S}$, a contradiction. Therefore, $\mathcal{S} = \mathbb{N}$, so the Principle of Mathematical Induction holds.

1.5 If we set $x = \sqrt{1 + \sqrt{1 + \sqrt{1 + \cdots}}}$, then $x^2 = 1 + x$, whose roots are exactly $\mathfrak{g}$ and $\mathfrak{g}'$, but $\mathfrak{g}' < 0 < \mathfrak{g}$, so $x = g$.

1.7 We use induction on n. If $n = 1$, then $\sum_{j=1}^{1} F_j^2 = 1 = F_1 F_2$. Assume that the result holds for all integers k such that $1 \leq k \leq n$. Therefore,

$$\sum_{j=1}^{n+1} F_j^2 = F_n F_{n+1} + F_{n+1}^2 = F_{n+1}(F_n + F_{n+1}) = F_{n+1} F_{n+2}.$$

Therefore, by induction the above statement holds for all $n \in \mathbb{N}$.

1.9 We use induction on n. If $n = 2$, then

$$F_{n-1} F_{n+1} - F_n^2 = F_1 F_3 - F_2^2 = 1 \cdot 2 - 1^2 = 1 = (-1)^n,$$

which is the induction step.

Now assume that

$$F_{n-1} F_{n+1} - F_n^2 = (-1)^n,$$

which is the induction hypothesis and prove the result for $n + 1$. We have,

$$F_n F_{n+2} - F_{n+1}^2 = F_n(F_{n+1} + F_n) - F_{n+1}^2 = F_n F_{n+1} + F_n^2 - F_{n+1}^2 =$$

$$F_{n+1}(F_n - F_{n+1}) + F_n^2 = F_{n+1}(-F_{n-1}) + F_n^2 = -(F_{n-1} F_{n+1} - F_n^2),$$

and this equals $(-1)^{n+1}$ by the induction hypothesis, so we have secured the result by induction for all $n \in \mathbb{N}$.

1.11 We use induction on n. If $n = 1$, then

$$\sum_{j=1}^{n} b^{n-j} R_j^2 = R_1^2 = \frac{R_1 R_2}{a} = \frac{R_n R_{n+1}}{a},$$

since $a R_1 = R_2$, which is the induction step. Now assume that

$$\sum_{j=1}^{n-1} b^{n-1-j} R_j^2 = \frac{R_{n-1} R_n}{a}.$$

Then multiplying through by b and adding R_n^2 to both sides we get,

$$\sum_{j=1}^{n} b^{n-j} R_j^2 = \frac{bR_{n-1}R_n}{a} + R_n^2 = \frac{(aR_n + bR_{n-1})R_n}{a} = \frac{R_{n+1}R_n}{a}.$$

1.13 (a) We prove this by induction on n. If $n = 1$, then

$$\begin{pmatrix} 0 & 1 \\ b & a \end{pmatrix}^n \begin{pmatrix} R_1 \\ R_2 \end{pmatrix} = \begin{pmatrix} R_2 \\ bR_1 + aR_2 \end{pmatrix} = \begin{pmatrix} R_2 \\ R_3 \end{pmatrix},$$

which establishes the induction step. The induction hypothesis is

$$\begin{pmatrix} 0 & 1 \\ b & a \end{pmatrix}^n \begin{pmatrix} R_1 \\ R_2 \end{pmatrix} = \begin{pmatrix} R_{n+1} \\ R_{n+2} \end{pmatrix}.$$

Thus,

$$\begin{pmatrix} 0 & 1 \\ b & a \end{pmatrix}^{n+1} \begin{pmatrix} R_1 \\ R_2 \end{pmatrix} = \begin{pmatrix} 0 & 1 \\ b & a \end{pmatrix} \begin{pmatrix} R_{n+1} \\ R_{n+2} \end{pmatrix} = \begin{pmatrix} R_{n+2} \\ R_{n+3} \end{pmatrix},$$

and we have the result.

(b) This is immediate from part (a) since

$$\begin{pmatrix} 1 & 0 \end{pmatrix} \begin{pmatrix} 0 & 1 \\ b & a \end{pmatrix}^{n+1} \begin{pmatrix} R_1 \\ R_2 \end{pmatrix} = \begin{pmatrix} 1 & 0 \end{pmatrix} \begin{pmatrix} R_{n+2} \\ R_{n+3} \end{pmatrix} = R_{n+2}.$$

(c) We use induction on n. If $n = 1$,

$$\begin{pmatrix} 0 & 1 \\ 1 & 1 \end{pmatrix}^{n+1} = \begin{pmatrix} 0 & 1 \\ 1 & 1 \end{pmatrix}^2 = \begin{pmatrix} F_1 & F_2 \\ F_2 & F_3 \end{pmatrix},$$

which is the induction step. The induction hypothesis is for any $n > 1$,

$$\begin{pmatrix} 0 & 1 \\ 1 & 1 \end{pmatrix}^n = \begin{pmatrix} F_{n-1} & F_n \\ F_n & F_{n+1} \end{pmatrix}.$$

Therefore,

$$\begin{pmatrix} 0 & 1 \\ 1 & 1 \end{pmatrix}^{n+1} = \begin{pmatrix} 0 & 1 \\ 1 & 1 \end{pmatrix} \begin{pmatrix} F_{n-1} & F_n \\ F_n & F_{n+1} \end{pmatrix} =$$

$$\begin{pmatrix} F_n & F_{n+1} \\ F_{n-1} + F_n & F_n + F_{n+1} \end{pmatrix} = \begin{pmatrix} F_n & F_{n+1} \\ F_{n+1} & F_{n+2} \end{pmatrix}.$$

(d) Taking determinants in part (c) yields the result.

(e) Since

$$\begin{pmatrix} 0 & 1 \\ 1 & 1 \end{pmatrix}^{2n+1} = \begin{pmatrix} 0 & 1 \\ 1 & 1 \end{pmatrix}^{n+1} \begin{pmatrix} 0 & 1 \\ 1 & 1 \end{pmatrix}^n,$$

then

$$\begin{pmatrix} F_{2n} & F_{2n+1} \\ F_{2n+1} & F_{2n+2} \end{pmatrix} = \begin{pmatrix} F_n & F_{n+1} \\ F_{n+1} & F_{n+2} \end{pmatrix} \begin{pmatrix} F_{n-1} & F_n \\ F_n & F_{n+1} \end{pmatrix} =$$

$$\begin{pmatrix} F_n F_{n-1} + F_{n+1}F_n & F_n^2 + F_{n+1}^2 \\ F_{n+1}F_{n-1} + F_{n+2}F_n & F_n F_{n+1} + F_{n+2}F_{n+1} \end{pmatrix}.$$

Reading the entries in both sides in column 2 and row 1 we get,

$$F_{2n+1} = F_n^2 + F_{n+1}^2.$$

(f) We have,

$$(F_{n+1}^2 - F_n^2)^2 + (2F_n F_{n+1})^2 = F_{n+1}^4 - 2F_n^2 F_{n+1}^2 + F_n^4 + 4F_n^2 F_{n+1}^2 =$$

$$F_{n+1}^4 + 2F_n^2 F_{n+1}^2 + F_n^4 = (F_{n+1}^2 + F_n^2)^2 = F_{2n+1}^2,$$

where the last equality comes from part (e).

1.15 This follows from the fact that $(n - (n - r))!(n - r)! = r!(n - r)!$.

1.17 Use $x = y = 1$ in the Binomial Theorem and the result follows.

1.19 (a) Since $R_{(b,n+1)} = bR_{(b,n)} + 1$ is a recursive definition for a repunit, then an induction yields the result. For $n = 1$, $R_{(b,1)} = 1 = (b^1 - 1)/(b - 1) = 1$. If $R_{(b,n)} = (b^n - 1)/(b - 1)$, then $R_{(b,n+1)} = bR_{(b,n)} + 1 = b(b^n - 1)/(b - 1) + 1 = (b^{n+1} - 1)/(b - 1)$.

(b) Since

$$R_{(10,p)} = \frac{10^p - 1}{10 - 1},$$

then by the Binomial Theorem this equals

$$\frac{1 + \sum_{i=1}^p 9^i \binom{p}{i} - 1}{9} = \sum_{i=1}^p 9^{i-1} \binom{p}{i}.$$

Thus, $p | R_{(10,p)}$.

Section 1.2

1.21 If $a \mid b$, then $b = xa$ for some integer x, and if $b \mid c$, then $c = by$ for some integer y. Thus, $c = by = a(xy)$, so $a \mid c$.

1.23 If $a \mid b$, then there exists an integer d such that $b = ad$. Therefore, $bc = (ac)d$, so $ac \mid bc$. Conversely, if $ac \mid bc$, then there is an integer f such that $bc = acf$. Since $c \neq 0$, we may divide through by it to get $b = af$, so $a \mid b$.

1.25 If $g = \gcd(a, b) = |a|$, then $g \mid b$. Conversely, if $|a| \mid b$, then $b = |a|c$ for some $c \in \mathbb{Z}$ and $\gcd(a, b) = \gcd(a, |a|c) = |a| \gcd(\pm 1, c) = |a|$, where the penultimate equality follows from Exercise 1.24.

1.27 Prove this by induction. If $n = 0$, then $10^7 + 23 = 57 \cdot 175439$. Assume that $57 \mid 10^{36n+7} + 23$. Since

$$10^{36(n+1)+7} + 23 = 10^{36}(10^{36n+7} + 23) + 23(1 - 10^{36}),$$

then we need only show that $57 \mid (10^{36} - 1)$, and a computer calculation (with say *Maple*) shows that it does.

1.29 Clearly if $g = \gcd(a_1, a_2, \ldots, a_n) \nmid c$, then no solutions are possible. To show that this necessary condition is also sufficient, we use induction on n. If $n = 2$, this is Example 1.15, which is the induction step. Assume that the condition is sufficient for all $k < n$, where $a_1 x_1 + a_2 x_2 + \cdots + x_k a_k = c$. Then consider

$$a_1 x_1 + \cdots + x_n a_n = a_1 x_1 + a_2 x_2 + \cdots + a_{n-2} x_{n-2} + \gcd(a_{n-1}, a_n)z, \quad \text{(S1)}$$

where the last term holds since Example 1.15 tells us that the linear combinations $a_{n-1}x_{n-1} + a_n x_n$ are the set $z \gcd(a_{n-1}, a_n)$ as z ranges over all integers and from a solution of $a_{n-1}x_{n-1} + a_n x_n = \gcd(a_{n-1}, a_n)$, we can find a general solution to $a_{n-1}x_{n-1} + a_n x_n = \gcd(a_{n-1}, a_n)z$. Hence, since there are $n - 1$ terms in Equation (S1), we are done by induction.

1.31 Suppose that $n \mid m$, so $kn = m$ for some $k \in \mathbb{Z}$. Use induction on k. The result is clear if $k = 1$, so assume $F_n \mid F_{nt}$ for all $t < k$. By the convolution formula given in Exercise 1.10 on page 12,

$$F_m = F_{n+(k-1)n} = F_{n-1}F_{(k-1)n} + F_n F_{(k-1)n+1},$$

and by the induction hypothesis, $F_n \mid F_m$, so we have the solution of the first part.

Let $g = \gcd(m, n)$, and let $d = \gcd(F_m, F_n)$. Then by the first part, $F_g \mid F_m$ and $F_g \mid F_n$, so $F_g \mid d$. It remains to show that $d \mid F_g$. By Theorem 1.10 on page 22, there exist $x, y \in \mathbb{Z}$ such that $g = mx + ny$. Also, by the aforementioned convolution formula,

$$F_g = F_{mx+ny} = F_{mx-1}F_{ny} + F_{mx}F_{ny+1}.$$

However, $d \mid F_n \mid F_{ny}$, and $d \mid F_m \mid F_{mx}$ both by the first part. Hence $d \mid F_g$.

1.33 We use induction on k. If $k = 1$, then

$$A_k B_{k-1} - A_{k-1}B_k = 1 = (-1)^{k-1},$$

which is the induction step. The induction hypothesis is

$$A_{k-1}B_{k-2} - A_{k-2}B_{k-1} = (-1)^{k-2}.$$

Therefore,

$$A_k B_{k-1} - A_{k-1}B_k = (q_k A_{k-1} + A_{k-2})B_{k-1} - A_{k-1}(q_k B_{k-1} + B_{k-2}) =$$

$$A_{k-2}B_{k-1} - A_{k-1}B_{k-2} = (-1)^{k-1}.$$

1.35 (a) Let $\ell' = \operatorname{lcm}(an, bn)$. Then $\ell \mid \ell'$. If $n\ell \mid (na)$ and $n\ell \mid (nb)$, then $\ell' \mid (n\ell)$, so $n\ell = \ell'$.

(b) We have, $\ell = \operatorname{lcm}(c(a/c), (b/c)) = c \cdot \operatorname{lcm}(a/c, b/c)$, where the last equality follows from part (a). Therefore, $\ell/c = \operatorname{lcm}(a/c, b/c)$, as required.

Section 1.3

1.37 If $p \mid \operatorname{lcm}(p_1, p_2, \ldots, p_k)$, then $p \mid \prod_{j=1}^k p_j$, so by Lemma 1.2 on page 32, $p \mid p_j$ for some natural number $j \le k$. Hence, $p = p_j$.

1.39 Since $p^m \| a$ and $p^n \| b$, then there exist $c, d \in \mathbb{Z}$ such that $a = p^m c$ and $b = p^n d$ with $\gcd(p, c) = \gcd(p, d) = 1$. Thus, $ab = p^{m+n}cd$ with $\gcd(cd, p) = 1$, so $p^{m+n} \| ab$.

1.41 There exists $c \in \mathbb{Z}$ such that $a = p^n b$ with $\gcd(p, b) = 1$. Therefore, $a^m = p^{mn}b^m$ with $\gcd(b^m, p) = 1$, from which we have $p^{mn} \| a^m$.

1.43 If $p \mid a^n$, then by Lemma 1.2 on page 32, $p \mid a$, so by exponentiation by n, we get that $p^n \mid a^n$.

1.45 If we have three consecutive powerful numbers $a, a+1, a+2$, none of them can have remainder 2 when divided by 4, that is, none of the $a, a+1, a+2$ can be of the form $4n+2$ for some $n \in \mathbb{N}$. Hence, the only choice of a is $a = 4n-1$, for some $n \in \mathbb{N}$.

1.47 If $1 + 2^a = x^2$ $(a > 3)$, then $2^a = x^2 - 1 = (x-1)(x+1)$, so $x - 1 = 2^{a'}$ and $x + 1 = 2^{a-a'}$ for some $a' \in \mathbb{Z}$ with $0 \le a' \le a - a'$. By subtracting the last two equations, we get $2 = 2^{a-a'} - 2^{a'}$. If $a' > 0$, then $1 = 2^{a-a'-1} - 2^{a'-1}$. If $a' > 1$, then $1 = 2(2^{a-a'-2} - 2^{a'-2})$, forcing $2|1$, which is absurd. Therefore, $a' = 1$, so $1 = 2^{a-a'-1} - 2^{a'-1} = 2^{a-2} - 1$, forcing $a = 3$, a contradiction. Hence, $a' = 0$, but $2 = 2^{a-a'} - 2^{a'} = 2^a - 1$, forcing $2^a = 3$, which is impossible. Hence, such an x cannot exist.

1.49 Since $a \mid c$, then there exists an integer d such that $c = ad$, and since $b \mid c$, then $b \mid ad$. Given that $\gcd(a,b) = 1$, then by Euclid's Lemma 1.1 on page 18, $b \mid d$, so $d = bf$ for some $f \in \mathbb{Z}$. Thus, $c = ad = abf$, so $ab \mid c$.

1.51 Since for any $n > 2$, $m = n! - 1 > 1$, then it has a prime divisor, p, say. If $p \le n$, then $p \mid n!$, so $p \mid 1$, a contradiction, so $p > n$. Also, if $p \ge n!$, then $p > m$, contradicting that $p \mid m$. Hence, $n < p < n!$, as required.

Section 1.4

1.53 $x = 1$ for $n_1 = -1 = n_2$.

1.55 $x = 497$ for $n_1 = 55$, $n_2 = 45$, and $n_3 = 38$.

1.57 $x = 576$ for $n_1 = 116$, $n_2 = 25$, and $n_3 = 22$.

1.59 $x = 42239$ for $n_1 = 21119$, $n_2 = 2223$, $n_3 = 1280$, and $n_4 = 797$.

Section 1.5

1.61 Since $p \mid (n^2 + 3)$ and $\gcd(n, p) = 1$, then by Thue's Theorem there exist $x \in \mathbb{N}$ and $y \in \mathbb{Z}$ with $x, |y| < \sqrt{p}$ such that $p \mid (nx - y)$. Therefore, there exist $u, v \in \mathbb{Z}$ such that $n^2 + 3 = pu$ and $nx - y = pv$. Hence,

$$pu = n^2 + 3 = \left(\frac{pv + y}{x}\right)^2 + 3,$$

and by rewriting, we get $p(x^2 u - pv^2 - 2vy) = y^2 + 3$, so $p \mid (y^2 + 3x^2)$. However, since $x, |y| < \sqrt{p}$, then $y^2 + 3x^2 < 4p$. There are three possibilities. First, we could have $p = y^2 + 3x^2$, in which case we have our representation with $a = |y|$, and $b = x$. Secondly, we could have $2p = y^2 + 3x^2$, in which case x and y have the same parity. This means that $y^2 + 3x^2$ is divisible by 4, so $2 \mid p$, a contradiction. Therefore, $2p \ne y^2 + 3x^2$. The remaining case is $3p = y^2 + 3x^2$, so $3 \mid y$ and we have our representation $p = 3(y/3)^2 + x^2$ with $a = x$ and $b = |y/3|$. This is existence. Now we show uniqueness.

Suppose that there exist $a_0, b_0 \in \mathbb{N}$ such that $p = a_0^2 + 3b_0^2$, with $p \mid (nb_0 - a_0)$ and $p = a^2 + 3b^2$ with $p \mid (nb - a)$. Then by Equation (1.22),

$$p^2 = (a_0^2 + 3b_0^2)(a^2 + 3b^2) = (aa_0 + 3bb_0)^2 + 3(a_0 b - ab_0)^2.$$

Therefore, $0 \leq aa_0 + 2bb_0 \leq p$. Also, since $p \mid (nb_0 - a_0)$ and $p \mid (nb - a)$, then there exist $w, z \in \mathbb{Z}$ such that $nb - a = pw$ and $nb_0 - a_0 = pz$. Hence,

$$aa_0 + 3bb_0 = (nb - pw)(nb_o - pz) + 3bb_0 = n^2 bb_o - nbpz - pwnb_0 + p^2 wz + 3bb_0 =$$

$$bb_0(n^2 + 3) + p(pwz - nbz - wnb_0) = bb_0 pu + p(pwz - nbz - wnb_0),$$

so $p \mid (aa_0 + 3bb_0)$, which implies that $aa_0 + 3bb_0 = p$. It follows that $a_0 b - ab_0 = 0$. Since we clearly have that $\gcd(a, b) = 1 = \gcd(a_0, b_0)$, then as above $a = a_0$ and $b = b_0$.

1.63 We prove one case only since the other is similar.

$$(xu - Dyv)^2 + D(xv + yu)^2 =$$

$$x^2 u^2 + D^2 y^2 v^2 - 2xuDyv + Dx^2 v^2 + 2xuDyv + Dy^2 u^2 =$$
$$(x^2 + Dy^2)u^2 + Dv^2(x^2 + Dy^2) = (x^2 + Dy^2)(u^2 + Dy^2).$$

Section 1.6

1.65 They are $5, 13$, and 563. It is not known if there are any others and this has been checked up to some high bounds.

1.67 We use induction on n. If $n = 1$, the result is clear. Assume that

$$p \mid \left[\left(\sum_{j=1}^{n-1} a_j \right)^p - \sum_{j=1}^{n-1} a_j^p \right],$$

the induction hypothesis. Therefore, by the Binomial Theorem

$$\left(\sum_{j=1}^{n} a_j \right)^p = \sum_{j=0}^{p} \binom{p}{j} \left(\sum_{j=1}^{n-1} a_j \right)^j a_n^{p-j}.$$

Thus, by the fact that $p \mid \binom{p}{j}$ for all natural numbers $j < p$, we have,

$$\sum_{j=0}^{p} \binom{p}{j} \left(\sum_{j=1}^{n-1} a_j \right)^j a_n^{p-j} = a_n^p + \left(\sum_{j=1}^{n-1} a_j \right)^p + px,$$

for some integer x. Therefore, by the induction hypothesis, the latter equals

$$a_n^p + \sum_{j=1}^{n-1} a_j^p + py = \sum_{j=1}^{n} a_j^p + py$$

for some integer y. This proves that

$$p \mid \left[\left(\sum_{j=1}^{n} a_j \right)^p - \sum_{j=1}^{n} a_j^p \right],$$

as required.

1.69 Assume that $p \mid (2^{2^n} - 1)$. By Fermat's Little Theorem, $p \mid (2^{p-1} - 1)$. Therefore, by Exercise 1.68, $2^{n+1} \leq p - 1$, so by the division algorithm, there exist integers q, r such that $p - 1 = 2^{n+1}q + r$ with $0 \leq r < 2^{n+1}$. Thus,

$$2^{p-1} - 1 = 2^{2^{n+1}q+r} - 1 = (2^{2^{n+1}})^q 2^r - 1 = (1 + px)^q 2^r - 1.$$

By the Binomial Theorem, $(1 + px)^q = \sum_{j=0}^{q} \binom{q}{j}(px)^j = 1 + pz$ for some integer z. Hence, p divides $2^r + 2^r pz - 1$, forcing $p \mid (2^r - 1)$, which cannot occur unless $r = 0$ by the minimality of 2^{n+1}. Therefore, $2^{n+1} \mid (p - 1)$, namely $p = 2^{n+1}m + 1$ for some $m \in \mathbb{N}$.

1.71 If $m > 2$, then since $(m - 1) \mid (m^n - 1)$, and $n > 1$, then $m^n - 1$ cannot be prime. Therefore $m = 2$, so $M_n = 2^m - 1$.

Section 1.7

1.73 As noted in the hint, we first establish that

$$p_m(n) = p_{m-1}(n) + p_m(n - m). \tag{S2}$$

If a given partition of n contains m, then it is of the form, $n = m + z_1 + z_2 + \cdots + z_\ell$ where $1 \leq z_j \leq m$ for $j = 1, 2, \ldots, \ell$. Clearly, $z_1 + z_2 + \cdots + z_\ell$ is a partiton of $n - m$. Also, each of the $p_m(n - m)$ paritions of $n - m$ into summands less than m happens to be exactly one partition of n, so that partition must be $z_1 + z_2 + \cdots z_\ell + m$. If a given partition of n does *not* contain m, then that partition is counted exactly one time by $p_{m-1}(n)$. Hence, we have established Equation (S2).

Now if $n - m < m$, then $\ell = 1$ so we are done. If $n - m > m$, then by Equation (S2), $p_m(n-m) = p_{m-1}(n-m) + p_{m-1}(n-2m)$, so $p_m(n) = p_{m-1}(n) + p_{m-1}(n-m) + p_{m-1}(n-2m)$. If $n - 2m < m$, then $\ell = 2$ and we are done. If $n - 2m > m$, then we continue as above, so by this inductive process, we have the full result.

1.75 This is just a combinatorial interpretation of Exercise 1.74 since it says that the coefficient of x^n in $\prod_{j=1}^{\infty}(1 - x^j)$ is

$$\sum (-1)^{d(n)} \tag{S3}$$

where the sum ranges over all partitions of n into distinct parts. However, Equation (S3) is $E(n) - U(n)$, so the result follows.

Section 1.8

1.77 By Example 1.14, for distinct primes p and q,

$$\gcd(2^p - 1, 2^q - 1) = 2^{\gcd(p,q)} - 1 = 2 - 1 = 1.$$

1.79 Let $s_j = 2r_j$ in the Lucas-Lehmer Theorem and the result follows.

1.81

$$V_m V_n + 12 U_n U_m =$$
$$(\alpha^m + \beta^m)(\alpha^n + \beta^n) + (\alpha - \beta)^2 \frac{(\alpha^n - \beta^n)(\alpha^m - \beta^m)}{(\alpha - \beta)^2} =$$
$$\alpha^{n+m} + \alpha^m \beta^n + \alpha^n \beta^m + \beta^{m+n} + \alpha^{n+m} - \alpha^n \beta^m - \alpha^m \beta^n + \beta^{m+n} =$$
$$2(\alpha^{n+m} + \beta^{n+m}) = 2V_{n+m}.$$

1.83 We have

$$4U_{n+1} - U_n = (\alpha + \beta)\frac{\alpha^{n+1} - \beta^{n+1}}{\alpha - \beta} - \alpha\beta\frac{\alpha^n - \beta^n}{\alpha - \beta} =$$

$$\frac{\alpha^{n+2} - \alpha\beta^{n+1} + \beta\alpha^{n+1} - \beta^{n+2} - \alpha^{n+1}\beta + \alpha\beta^{n+1}}{\alpha - \beta} =$$

$$\frac{\alpha^{n+2} - \beta^{n+2}}{\alpha - \beta} = U_{n+2}.$$

Also,

$$4V_{n+1} - V_n = (\alpha + \beta)(\alpha^{n+1} + \beta^{n+1}) - \alpha\beta(\alpha^n + \beta^n) =$$

$$\alpha^{n+2} + \alpha\beta^{n+1} + \beta\alpha^{n+1} + \beta^{n+2} - \alpha^{n+1}\beta - \alpha\beta^{n+1} =$$

$$\alpha^{n+2} + \beta^{n+2} = V_{n+2}.$$

Section 1.9

1.85 If n is composite, then $\pi(n) = \pi(n-1)$, so the inequality implies $n - 1 > n$, a contradiction. Conversely if n is prime, then

$$\pi(n) = \pi(n-1) + 1,$$

so

$$n\pi(n-1) = n(\pi(n) - 1) < (n-1)\pi(n),$$

since $\pi(n) < n$ for any $n \in \mathbb{N}$.

1.87 From the result, it follows that if t is sufficiently large, say $t \geq m_n$, then

$$\pi(2t) - \pi(t) \geq n$$

for any given $n \in \mathbb{N}$. Hence, there are at least n primes between t and $2t$.

1.89 By Bertrand's postulate, there exists a prime between 2^j and 2^{j+1} for all $j \in \mathbb{N}$. In other words, there are j primes less than 2^{j+1} for any natural number j. By letting $j = n - 1$ and counting in $p_1 = 2$, we get that

$$p_n \leq 2^n.$$

Section 2.1

2.1 Let $\mathcal{S} = \{s_0, s_1, \ldots, s_{n-1}\}$, where $s_j = j$ for $j = 0, 1, \ldots, n - 1$. Since $\mathcal{R}$ is a complete residue system modulo n, then for any nonnegative $j < n$, there is a unique element of $\mathcal{R}$ congruent to s_j modulo n. Hence, $n \leq m$. However, $\mathcal{S}$ is, itself, a complete residue system by part (b) of Proposition 2.5 on page 76, so each r_j is congruent to a unique element of $\mathcal{S}$. Thus, $n \geq m$, and we have shown that $m = n$.

2.3 Since a is even, then $a = 2n$ for some n so $a^2 = (2n)^2 = 4n^2 \equiv 0 \pmod{4}$.

2.5 This is proved in exactly the same fashion as in Example 2.4 on page 80 since any prime dividing n must divide $a^2 - 1 = (a-1)(a+1)$.

2.7 Let $z \in \mathbb{Z}$ be arbitrarily chosen. Then by part (a) of Proposition 2.5 on page 76, z is in one of the n disjoint congruences classes modulo n. Since $\mathcal{R}$ has elements from n disjoint congruences classes (given that $r_i \equiv r_j \pmod{n}$ if and only if $i = j$), then z is congruent to exactly one of them. Hence, $\{r_1, r_2, \ldots, r_n\}$ is a complete residue system modulo n.

2.9 By Exercise 2.7 it suffices to prove that no two elements of the set are congruent modulo mn. If for some natural numbers i, j, k, ℓ we have

$$mr_i + ns_j \equiv mr_k + ns_\ell \pmod{mn},$$

then since $\gcd(m, n) = 1$, and $m(r_i - r_k) \equiv n(s_\ell - s_j) \pmod{mn}$, then $m \mid (s_\ell - s_j)$, so $\ell = j$ by Exercise 2.7. Therefore, $r_i \equiv r_k \pmod{n}$, so by the same reasoning, $i = k$.

Section 2.2

2.11 Since
$$(n-1)! \equiv (n-2)(n-1) \equiv -(n-2)! \pmod{n},$$
then by Wilson's Theorem, n is prime if and only if $(n-1)! \equiv -1 \pmod{n}$ if and only if $(n-2)! \equiv 1 \pmod{n}$.

2.13 $x = 20$.

2.15 $x = 49$.

2.17 $x = 62$.

2.19 $x = 1213$.

2.21 $(p-1)! \equiv -1 \pmod{p}$ and $n^p \equiv n \pmod{p}$ if and only if
$$(p-1)!n^p \equiv -n \pmod{p}$$
by the multiplicative property given in Proposition 2.1 on page 74.

Section 2.3

2.23 By Fermat's Little Theorem,
$$\sum_{j=1}^{p-1} j^{p-1} \equiv \sum_{j=1}^{p-1} 1 \equiv p - 1 \equiv -1 \pmod{p}.$$

2.25 If $n = p^a m$ with $p \nmid m$, then by Theorem 2.9,
$$\phi(p^a m) = (p^a - p^{a-1}) \mid \phi(m) \mid (p^a m - 1).$$
Thus, $p \mid 1$ or $a = 1$, so n is squarefree.

2.27 $\sum_{j=1}^{\infty} \mu(j!) = 1$ since $\mu(j!) = 0$ for any $j \geq 4$ given that $4 \mid j!$ for any such j, and $\mu(1!) + \mu(2!) + \mu(3!) = 1 - 1 + 1$.

2.29 This is essentially Theorem 2.15 on page 97 restated since $(-1)^2 = |\mu(d)|$ when $\mu(d) = -1$.

2.31 $(\mu * u)(n) = \sum_{d\mid n} \mu(d) u(n/d) = \sum_{d\mid n} \mu(d) = I(n)$, by Theorem 2.14 on page 96.

Section 2.4

2.33 (a) $\tau(23) = 2$. (b) $\tau(133) = 4$. (c) $\tau(276) = 12$. (d) $\tau(1011) = 4$. (e) $\tau(5^{10}) = 11$. (f) $\tau(3001) = 2$.

2.35 For $n = 2$ in Thabit's rule, $2^n pq = 2^2 \cdot 5 \cdot 11 = 220$ and $2^2 \cdot 71 = 284$, so $(220, 284)$ is the smallest amicable pair.

2.37 Using Thabit's rule, $(2^7 \cdot 191 \cdot 383, 2^7 \cdot 73727) = (9363584, 9437056)$, which is our amicable pair.

2.39 Two are:
$$(1980, 2016, 2556), (9180, 9504, 11556).$$

2.41 By Exercise 2.40, n is of the prescribed form. We need only show that $n \equiv 1 \pmod 4$. Since $p \equiv 1 \pmod 4$, then $p^a \equiv 1 \pmod 4$. Since m is odd, then $m \equiv \pm 1 \pmod 4$. Hence, $m^2 \equiv 1 \pmod 4$. Also see Exercise 2.2 on page 83 that we could have invoked directly.

2.43 Since divisors of kn include those of the form kd where $d \mid n$, then for $k > 1$,
$$\sigma(kn) \geq 1 + \sum_{d|n} kd = 1 + k\sum_{d|n} d = 1 + k\sigma(n) \geq 1 + 2kn > 2kn.$$

2.45 By Theorem 2.19 on page 104, every even perfect number is of the form,
$$n = 2^{k-1}(2^k - 1).$$
Also, by Theorem 1.1 on page 2,
$$\sum_{j=1}^{2^k-1} j = 2^{k-1}(2^k - 1),$$
which is the desired result.

2.47 By Theorem 2.19 on page 104, every even perfect number is of the form,
$$n = 2^{k-1}(2^k - 1),$$
so
$$8n + 1 = 2^{2k+2} - 2^{k+2} + 1 = (2^{k+1} - 1)^2.$$

2.49 By the Möbius inversion formula,
$$n = \sum_{d|n} \sigma(d)\mu(n/d) = \sum_{(n/d)|n} \sigma(n/d)\mu(d) = \sum_{d|n} \sigma(n/d)\mu(d).$$

2.51 Since 2^k is superperfect, then $\sigma(\sigma(2^k)) = 2 \cdot 2^k = 2^{k+1}$. Also, by Theorem 2.18 on page 102, $\sigma(2^k) = 2^{k+1} - 1$. Therefore, $\sigma(2^{k+1} - 1) = 2^{k+1}$. Hence, $2^{k+1} - 1$ is prime, whence a Mersenne prime.

2.53 If n is perfect, then $s_1(n) = \sigma(n) - n = 2n - n = n$. Consider this to be the induction step. Assume that $s_k(n) = n$. Therefore, $s_{k+1}(n) = \sigma(s_k(n)) - s_k(n) = \sigma(n) - n = 2n - n = n$, so we are done by induction.

2.55 By Euler's Theorem 2.10 on page 93, $a^{\phi(n)} \equiv 1 \pmod n$, so $r \leq \phi(n)$. By the division algorithm, there exist integers q, s with $\phi(n) = qr + s$ where $0 \leq s < r$. Therefore,
$$1 \equiv a^{\phi(n)} \equiv a^{qr+s} \equiv (a^r)^q a^s \equiv a^s \pmod n,$$
so by the minimality of r, $s = 0$ is forced. Hence, $r|\phi(n)$.

Section 2.5

2.57 By Exercise 2.56,

$$-1 < x - \lfloor x \rfloor - 1 = x + n - 1 - \lfloor x \rfloor - n \leq \lfloor x + n \rfloor - \lfloor x \rfloor - n \leq$$
$$x + n - \lfloor x \rfloor - n = x - \lfloor x \rfloor < 1,$$

so

$$\lfloor x + n \rfloor - \lfloor x \rfloor - n = 0.$$

2.59 We let $n = \lfloor x \rfloor$ and $m = \lfloor y \rfloor$, so we get

$$\lfloor x \rfloor + \lfloor -x \rfloor = n + \lfloor -n - z \rfloor = n + \lfloor -n - 1 + 1 - z \rfloor,$$

and by Exercise 2.58,

$$= n - n - 1 + \lfloor 1 - z \rfloor = \begin{cases} 0 & \text{if } z = 0, \\ -1 & \text{if } z > 0. \end{cases}$$

2.61 (a) 1 (b) 11 (c) -1 (d) 0 (e) 3 (f) 1.

2.63 (a) -1 (b) 4 (c) 13 (d) 0.

2.65 (a) 0 (b) 6/7 (c) 26/35 (d) 19/42.

Section 2.6

2.67 If $p > 2$, then $x^2 \equiv 1 \pmod{p}$ has exactly two incongruent solutions, namely 1 and $p - 1$. It follows from Theorem 2.23 that $x^2 \equiv 1 \pmod{p_1 p_2 \cdots p_k}$ has exactly 2^k incongruent solutions. Thus, for any $k \geq n$,

$$2^k > n.$$

2.69 $x \equiv 240 \pmod{7^3}$.

2.71 $x \equiv 3, 7 \pmod{8}$.

2.73 $x \equiv 1367 \pmod{17^3}$.

Section 2.7

2.75 $n = 9221$ is prime. Since $x_0 \equiv 2^{(n-1)/4} \equiv 5921 \pmod{n}$ and

$$x_1 \equiv 2^{(n-1)/2} \equiv -1 \pmod{n},$$

then the MSR test declares it to be probably prime.

2.77 Since $2^{2^n} \equiv -1 \pmod{\mathcal{F}_n}$, where $\mathcal{F}_n = 2^{2^n} + 1$, then

$$2^{\mathcal{F}_n - 1} \equiv 2^{2^{2^n}} \equiv \left(2^{2^n}\right)^{2^{2^n - n}} \equiv (-1)^{2^{2^n - n}} \equiv 1 \pmod{\mathcal{F}_n},$$

which makes $\mathcal{F}_n$ a strong pseudoprime to base 2.

Section 2.8

2.79 ILOG WKH ZHDSRQV.

2.81 assume the worst.

2.83 IBMMBP ASD RBWDZ.

2.85 security agency.

2.87 CF RV DH EZ CM TE.

2.89 EHHUAVUIGJMG.

2.91 decipher.

Section 3.1

3.1 Let $c = \mathrm{ord}_n(m^b)$. Then $m^{bc} \equiv 1 \,(\mathrm{mod}\ n)$, implies $bc \geq ba$, so $c \geq a$. Since

$$1 \equiv m^{ab} \equiv (m^b)^a \pmod{n},$$

then $a \geq c$. Hence, $a = c = \mathrm{ord}_n(m^b)$.

3.3 If $p \equiv 1 \,(\mathrm{mod}\ 4)$, then

$$(p-g)^{(p-1)/2} \equiv (-1)^{(p-1)/2} g^{(p-1)/} \equiv g^{(p-1)/2} \equiv -1 \pmod{p},$$

where the last congruence follows from Example 2.4 on page 80 since $g^{(p-1)/2}$ is a multiplicative self-inverse modulo p. Hence, $p - g$ is a primitive root modulo p. Conversely, if $p - g$ is a primitive root modulo p, and $p \equiv -1 \,(\mathrm{mod}\ 4)$, then

$$(p-g)^{(p-1)/2} \equiv (-1)^{(p-1)/2} g^{(p-1)/2} \equiv -g^{(p-1)/2} \equiv (-1)(-1) \equiv 1 \pmod{p},$$

contradicting that $p - g$ is a primitive root modulo p, so $p \equiv 1 \,(\mathrm{mod}\ 4)$.

3.5 Since g is a primitive root modulo p, then

$$-1 \equiv g^{(p-1)/2} \equiv g_1^{-(p-1)/2} \pmod{p},$$

so $g_1^{(p-1)/2} \equiv (-1)^{-1} \equiv -1 \,(\mathrm{mod}\ p)$, from which we deduce that g_1 is a primitive root modulo p.

3.7 Since p has a primitive root g, then by Theorem 3.1 on page 142

$$(p-1)! \equiv g \cdot g^2 \cdots g^{p-1} \equiv g^{p(p-1)/2} \equiv g^{(p-1)/2} \pmod{p},$$

where the penultimate congruence comes from Theorem 1.1 on page 2. Hence, since g is a primitive root, then $g^{(p-1)/2} \equiv -1 \,(\mathrm{mod}\ p)$.

Section 3.2

3.9 (a) 2; (b) none since it is a product of two distinct odd primes; (c) 15; (d) 3; (e) 5; (f) 61.

3.11 Use induction on j. For $j = 1$, this is the definition of a Fibonacci primitive root. Assume that $f^j \equiv F_j f + F_{j-1} \,(\mathrm{mod}\ p)$. Then multiplying through by f, we get

$$f^{j+1} \equiv F_j f^2 + F_{j-1} f \equiv F_j(f+1) + F_{j-1} f \equiv$$
$$f(F_j + F_{j-1}) + F_j \equiv f F_{j+1} + F_j \pmod{p},$$

which is the desired result.

3.13 Since $\phi(\phi(26)) = 4$, there are four of them which are 7, 11, 15, and 19.

3.15 This follows from Theorem 3.2 on page 143 since $\phi(\phi(2p^n)) = \phi(\phi(p^n))$.

Section 3.3

3.17 (a) $\mathrm{ind}_2(11) = 12$; (b) $\mathrm{ind}_5(13) = 14$; (c) $\mathrm{ind}_7(25) = 56$; (d) $\mathrm{ind}_5(26) = 67$; (e) $\mathrm{ind}_5(29) = 13$; $\mathrm{ind}_5(22) = 3$.

3.19 (a) $x = 2$; (b) There is no solution since, for a primitive root 3 modulo 7 we would have

$$\mathrm{ind}_3(4) + 4\,\mathrm{ind}_3(x) \equiv \mathrm{ind}_3(5) \pmod 6.$$

However, $\mathrm{ind}_3(4) = 4$ and $\mathrm{ind}_3(5) = 5$, so

$$4 + 4\,\mathrm{ind}_3(x) \equiv 5 \pmod 6,$$

a contradiction.

(c) $x \in \{5, 6\}$; (d) $x \in \{2, 6, 7, 8, 10\}$; (e) $x = 11$; (f) $x = 13$.

3.21 Since m is a primitive root modulo $n > 2$, $m^{\phi(n)/2} \equiv -1 \pmod n$. Therefore, $\phi(n)/2 \equiv -1 \pmod{\phi(n)}$ by Proposition 3.1 on page 140. In other words, $\mathrm{ind}_m(-1) \equiv \phi(n)/2 \pmod{\phi(n)}$.

3.23 We prove this by induction on c. We are given $x^e \equiv b \pmod p$, and we assume that $x^e \equiv b \pmod{p^c}$ has a solution. We need only establish that

$$x^e \equiv b \pmod{p^{c+1}}$$

has a solution. If $g = \gcd(e, \phi(p^c))$, then

$$b^{\phi(p^c)/g} \equiv 1 \pmod{p^c},$$

by Theorem 3.10. Thus, there is an integer f such that

$$b^{\phi(p^c)/g} = 1 + fp^c.$$

Therefore, by Theorem 2.9 on page 91,

$$(1 + p^c f)^p = b^{p\phi(p^c)/g} = b^{\phi(p^{c+1})/g}.$$

However, by the Binomial Theorem,

$$(1 + p^c f)^p = \sum_{j=0}^{p} p^{cj} f^j \binom{p}{j} = 1 + p^{c+1} h,$$

for some $h \in \mathbb{Z}$ since $cj \geq c+1$ for all $j \geq 2$. Hence,

$$b^{\phi(p^{c+1})/g} \equiv 1 \pmod{p^{c+1}}. \tag{S4}$$

Since $p \nmid e$, then $g = \gcd(e, \phi(p^{c+1}))$ since

$$\gcd(e, (p-1)p^{c-1}) = \gcd(e, p-1) = \gcd(e, (p-1)p^c),$$

so by Theorem 3.10, (S4) says that b is an eth power residue modulo p^{c+1}.

3.25 If $c = 2$, then this is covered by Theorem 3.10 on page 155 since 4 possesses a primitive root by the Primitive Root Theorem, so we may let $c \geq 3$. By Exercise 3.24, there is a unique nonnegative integer $j < 2^c$ such that

$$b \equiv \pm 5^j \pmod{2^c}.$$

We seek as solution $b \equiv x^e \pmod{2^c}$. Let $k < 2^c$ be a nonnegative integer and note that the following criterion holds,

$b \equiv (\pm 5^k)^e \pmod{2^c}$ if and only if $\gcd(e,2) \mid j$ and $ek \equiv j \pmod{2^{c-2}}$.

The latter congruence is solvable if and only if $g = \gcd(2^{c-2}, e) \mid j$ by Theorem 2.3 on page 84. Since e is odd then this holds. But, and since $\text{ord}_{2^c}(5) = 2^{c-2}$, $g \mid j$ if and only if

$$(5^j)^{2^{c-2}/g} \equiv 1 \pmod{2^c}.$$

Thus we have the following criterion for b to be an eth power residue modulo 2^c.

$b \equiv (\pm 5^k)^e \pmod{2^c}$ if and only if $\gcd(e,2) \mid j$ and $(5^j)^{2^{c-2}/g} \equiv 1 \pmod{2^c}$. (S5)

Since e is odd, both conditions hold, so

$$b \equiv (\pm 5^k)^e \pmod{2^c}.$$

3.27 Using the notation in the solution of Exercise 3.25 above,

$$ek \equiv j \pmod{2^{c-2}}$$

has either no solutions or it has $g = \gcd(e, 2^{c-2})$ solutions. Thus, the $\gcd(2,e)g$ pairs (d,k) provide distinct values of $x \equiv (-1)^d 5^k \pmod{2^c}$ for which

$$x^e \equiv b \pmod{2^c}$$

where b is an eth power residue modulo 2^c. Since there exist 2^{c-1} incongruent odd positive integers less than 2^c, then the number of eth power residues is given by

$$\frac{2^{c-1}}{\gcd(2,e)\gcd(e,2^{c-2})}.$$

Section 3.4

3.29 4355, 9660, 3156, 9603, 2176, 7349, 78, 60.

3.31 $f(15) = (7, 17, 11, 25, 1, 5, 13, 3, 9, 21, 19, 15)$.

3.33 $f(6) = (62, 104, 111, 118, 55)$.

3.35 This is proved by induction with the case where $j = 0$ being clear since

$$s_0 \equiv a^0 s_0 + b(a^0 - 1)/(a - 1) \equiv s_0 + 0 \equiv s_0 \pmod{n}.$$

Assume the induction hypothesis,

$$s_j \equiv a^j s_0 + b\frac{(a^j - 1)}{(a - 1)} \pmod{n}.$$

Thus,

$$s_{j+1} \equiv as_j + b \equiv a^{j+1}s_0 + \frac{b(a(a^j-1))}{(a-1)} + b \equiv a^{j+1}s_0 + \frac{b(a(a^j-1)+a-1)}{(a-1)} \equiv$$

$$a^{j+1}s_0 + \frac{b(a^{j+1}-1)}{(a-1)} \pmod{n},$$

as required.

Section 3.5

3.37 Let m, m' be generators of $\mathbb{F}_p^*$ and let $\beta \in \mathbb{F}_p^*$ be arbitrary. Set

$$x = \log_m(\beta), \quad y = \log_{m'}(\beta), \text{ and } z = \log_m(m').$$

Then

$$m^x = \beta = (m')^y = (m^z)^y.$$

Thus, $x \equiv zy \pmod{p-1}$, so

$$\log_{m'}(\beta) = y \equiv xz^{-1} \equiv (\log_m(\beta))(\log_m(m'))^{-1} \pmod{p-1}.$$

Hence, any algorithm which computes logs to base m can be used to compute logs to any other base m' that is a generator of $\mathbb{F}_p^*$.

3.39 If $e = \phi(n)/2 + 1$, then by Euler's Theorem 2.10 on page 93,

$$m^{\phi(n)/2+1} = (m^{\phi(p)})^{\phi(q)/2}m \equiv m \pmod{p},$$

and

$$m^{\phi(n)/2+1} = (m^{\phi(q)})^{\phi(p)/2}m \equiv m \pmod{q},$$

so

$$m^{\phi(n)/2+1} \equiv m \pmod{n},$$

namely $m^e \equiv m \pmod{n}$ for all $m \in \mathfrak{M}$, clearly not a desirable outcome.

3.41 **fight to the death**, where $d = 115$.

3.43 **evacuate now**, where $d = 281$.

3.45 $d = 31271$, and $c^d \equiv m = 86840 \pmod{n}$.

3.47 $d = 254693$, and $c^d \equiv m = 73343 \pmod{n}$.

3.49 $p+q = n-\phi(n)+1 = 632119 - 630496 + 1 = 1624$ and $p-q = \sqrt{(p+q)^2 - 4n} = 330$, so $2q = p+q - (p-q) = 1294 = 2 \cdot 647$. Thus, $p = 977 = n/647$.

3.51 By the same reasoning as in the solution of Exercise 3.49 above we get $p = 1777$ and $q = 3037$.

3.53 $p = 6079$ and $q = 7103$.

Section 4.1

4.1 (a) -1 (b) -1.

4.3 By part (2) of Theorem 4.4 on page 182,

$$\left(\frac{-2}{p}\right) = \left(\frac{2}{p}\right)\left(\frac{-1}{p}\right),$$

so if $p \equiv 1 \pmod 8$, then by Theorem 4.6 on page 184,

$$\left(\frac{-2}{p}\right) = (1)(1) = 1.$$

Similarly, if $p \equiv 3 \pmod 8$,

$$\left(\frac{-2}{p}\right) = (-1)(-1) = 1;$$

if $p \equiv 7 \pmod 8$,

$$\left(\frac{-2}{p}\right) = (-1)(1) = -1;$$

and if $p \equiv 5 \pmod 8$, then

$$\left(\frac{-2}{p}\right) = (1)(-1) = -1.$$

4.5 This is an immediate consequence of Theorem 4.1 on page 178.

4.7 By Euler's criterion.

$$x \equiv a^{(p-1)/2} \equiv \left(\frac{a}{p}\right) \equiv 1 \pmod p,$$

and

$$x \equiv a^{(q-1)/2} \equiv \left(\frac{a}{q}\right) \equiv 1 \pmod q.$$

Therefore,

$$\left((\pm a)^{(p+1)/4}\right)^2 \equiv a^{(p+1)/2} \equiv a^{(p-1)/2} \cdot a \equiv a \pmod p,$$

and

$$\left((\pm a)^{(q+1)/4}\right)^2 \equiv a^{(q+1)/2} \equiv a^{(q-1)/2} \cdot a \equiv a \pmod q,$$

which establishes the result.

4.9 Since $\left(\frac{4}{p}\right) = 1$, then

$$\left(\frac{a}{p}\right)\left(\frac{f(x)}{p}\right) = \left(\frac{4}{p}\right)\left(\frac{a}{p}\right)\left(\frac{f(x)}{p}\right) = \left(\frac{4a^2x^2 + 4abx + 4ac}{p}\right) =$$
$$\left(\frac{(2ax+b)^2 - \Delta}{p}\right).$$

As x ranges over $0, 1, \ldots, p-1$, so does $2ax + b$ modulo p. Thus,

$$\sum_{x=0}^{p-1} \left(\frac{a}{p}\right)\left(\frac{f(x)}{p}\right) = \sum_{j=0}^{p-1} \left(\frac{j^2 - \Delta}{p}\right).$$

However, $\Delta \equiv 0 \pmod{p}$, so

$$\sum_{x=0}^{p-1} \left(\frac{a}{p}\right)\left(\frac{f(x)}{p}\right) = \sum_{j=1}^{p-1}\left(\frac{j^2}{p}\right) = \sum_{j=1}^{p-1} 1 = p-1.$$

Therefore, multiplying the above through by $\left(\frac{a}{p}\right)$, we get

$$\sum_{x=0}^{p-1} \left(\frac{f(x)}{p}\right) = (p-1)\left(\frac{a}{p}\right).$$

4.11 Assuming only finitely many p_j for $j = 1, 2, \ldots, s$, then for $n = (\prod_{j=1}^{s} p_j)^2 + 4$, we have that

$$\left(\prod_{j=1}^{s} p_j\right)^2 \equiv -4 \pmod{q},$$

for any prime q dividing n, which is necessarily odd. Thus, $(-4/q) = 1$, so by part (3) of Theorem 4.4 on page 182, $q \equiv 1 \pmod 4$. Since q cannot be of the form $8m + 5$, then q is of the form $8\ell + 1$ for all primes dividing n. However, products of primes of that form are also of the form so

$$n = 8k + 1 \text{ for some integer } k.$$

But by Exercise 2.2 on page 83, n is of the form

$$n = \left(\prod_{j=1}^{s} p_j\right)^2 + 4 = 8r + 1 + 4 = 8r + 5.$$

Hence, $8r + 5 = 8k + 1$, which leads to the contradiction, $8(k - r) = 4$.

4.13 We use the Legendre symbol on the Diophantine equation to get,

$$\left(\frac{-2}{p}\right) = \left(\frac{-2y^2}{p}\right) = \left(\frac{p-x^2}{p}\right) = \left(\frac{-x^2}{p}\right) = \left(\frac{-1}{p}\right).$$

Therefore,

$$\left(\frac{-2}{p}\right) = \left(\frac{-1}{p}\right),$$

so if $p \equiv 3 \pmod 8$, then by part (3) of Theorem 4.4, $(-1/p) = -1$ and by Exercise 4.3, $(-2/p) = 1$, a contradiction. If $p \equiv -3 \pmod 8$, then as above $(-1/p) = 1$ and $(-2/p) = -1$, again a contradiction.

4.15 If $r_1, r_2, \ldots, r_\ell$ are the quadratic residues less than $p/2$, then as noted in the proof of Gauss' Lemma, those bigger than $p/2$ are $p - r_j$ for $j = 1, 2, \ldots, \ell$. Therefore, by Theorem 4.1 on page 178, $2\ell = (p-1)/2$, so

$$\ell = (p-1)/4.$$

Section 4.2

4.17 (a) 1 (b) -1.

4.19 Since $p \equiv 1 \pmod 4$ and $p \equiv 3, 5 \pmod 7$, then by the quadratic reciprocity law,

$$\left(\frac{7}{p}\right) = \left(\frac{p}{7}\right) = -1.$$

Thus, by Euler's criterion Theorem 4.2 on page 179,

$$7^{(p-1)/2} \equiv -1 \pmod p.$$

Hence, $\operatorname{ord}_p(7) = p - 1$, so 7 is a primiive root modulo p.

4.21 (a) -1 (b) 1 (c) 1 (d) 1.

4.23 Let $m = 2^s m_1$ and $n = 2^t n_1$, where m_1, n_1 are odd and s, t are nonnegative integers. Then,

$$\left(\frac{a}{mn}\right) = \left(\frac{a}{2}\right)^{s+t}\left(\frac{a}{m_1 n_1}\right) = \left(\frac{a}{2}\right)^{s+t}\left(\frac{a}{m_1}\right)\left(\frac{a}{n_1}\right) =$$

$$\left[\left(\frac{a}{2}\right)^s\left(\frac{a}{m_1}\right)\right]\cdot\left[\left(\frac{a}{2}\right)^t\left(\frac{a}{n_1}\right)\right] = \left(\frac{a}{m}\right)\left(\frac{a}{n}\right).$$

4.25 First assume that $a < 0$, and $n = 2^b d_1$ where $b \geq 0$, and d_1 is odd. Then,

$$\left(\frac{a}{n}\right) = \left(\frac{a}{2}\right)^b\left(\frac{a}{d_1}\right) = \left(\frac{a}{2}\right)^b\left(\frac{-1}{d_1}\right)\left(\frac{|a|}{d_1}\right),$$

where the last equality follows from properties of the Jacobi symbol. Also by properties of the Jacobi symbol, this equals

$$\left(\frac{a}{2}\right)^b\left(\frac{-1}{d_1}\right)\left(\frac{d_1}{|a|}\right)(-1)^{\frac{-a-1}{2}\cdot\frac{d_1-1}{2}} = \left(\frac{a}{2}\right)^b\left(\frac{d_1}{|a|}\right)(-1)^{\frac{-a-1}{2}\cdot\frac{d_1-1}{2}+\frac{d_1-1}{2}},$$

and since $a \equiv 1 \pmod 4$, by definition, this equals

$$\left(\frac{a}{2}\right)^b\left(\frac{d_1}{|a|}\right)(-1)^{\frac{1-a}{2}\cdot\frac{d_1-1}{2}} = \left(\frac{a}{2}\right)^b\left(\frac{d_1}{|a|}\right).$$

Also,

$$\left(\frac{a}{2}\right)^b = \left(\frac{2}{|a|}\right)^b.$$

Hence,

$$\left(\frac{a}{n}\right) = \left(\frac{2}{|a|}\right)^b\left(\frac{d_1}{|a|}\right) = \left(\frac{2^b d_1}{|a|}\right) = \left(\frac{n}{|a|}\right).$$

If $a > 0$, then the argument is similar.

Section 4.3

4.27 Let $n = 2^a \prod_{j=1}^k p_j^{a_j}$ be the canonical prime factorization of n, and let $b \in \mathbb{Z}$ with $\gcd(b, n) = 1$. Then since $\lambda(n)$ is divisible by all integers $\lambda(2^a), \phi(p_1^{a_1}), \ldots, \phi(p_k^{a_k})$, we have

$$b^{\lambda(n)} \equiv 1 \pmod{p_j^{a_j}} \text{ for all } j = 1, 2, \ldots, k \text{ and } b^{\lambda(n)} \equiv 1 \pmod{2^a}.$$

Therefore, by Theorem 2.5 on page 86, the Generalized Chinese Remainder Theorem,

$$b^{\lambda(n)} \equiv 1 \pmod{n},$$

which establishes $\lambda(n)$ as a universal exponent. Now we show that this is the *least* such. Let g_j be a primitve root modulo $p_j^{a_j}$ for $j = 1, 2, \ldots, k$. Then by the Chinese Remainder Theorem 2.4 on page 85, there is a unique solution c modulo n of the system of congruences,

$$x \equiv 5 \pmod{2^a} \text{ and } x \equiv g_j \pmod{p_j^{a_j}} \text{ for each } j = 1, 2, \ldots, k.$$

We need to show that $\text{ord}_n(c) = \lambda(n)$. To this end, let $d \in \mathbb{N}$ such that

$$c^d \equiv 1 \pmod{n}$$

so $\text{ord}_n(c) \mid d$. However, c satisfies the aforementioned system of congruences, so $\text{ord}_{p_j^{a_j}}(c) = \lambda(p_j^{a_j})$ for each $j = 1, 2, \ldots, k$ and $\text{ord}_{2^a}(c) = \lambda(2^a)$. Thus, by proposition 3.1 on page 140, $\lambda(p^t) \mid d$ for each prime power p^t dividing n. Thus, by the Generalized Chinese Remainder Theorem, again, $\lambda(n) \mid d$. Since $b^{\lambda(n)} \mid d$ whenever, $a^d \equiv 1 \pmod{n}$, then $\text{ord}_n(c) = \lambda(n)$, so $\lambda(n)$ is the least universal exponent.

4.29 $n = 223 \cdot 1181$.

4.31 $n = 139 \cdot 239$.

4.33 $n = 43 \cdot 827$.

4.35 $n = 23 \cdot 673$.

4.37 $n = 137 \cdot 1171$.

4.39 $n = 137 \cdot 439$.

Section 5.1

5.1 Since $\lfloor \alpha \rfloor = (n+1)/2$, then

$$\alpha_1 = \frac{1}{\alpha_0 - q_0} = \frac{2}{\sqrt{n^2+4} - n} = \frac{\sqrt{n^2+4} + n}{2} = n + \frac{\sqrt{n^2+4} - n}{2} = q_1 + \frac{1}{\alpha_2},$$

and

$$\alpha_2 = \frac{1}{\alpha_1 - q_1} = \frac{2}{\sqrt{n^2+4} - n} = \frac{\sqrt{n^2+4} + n}{2} = \alpha_1,$$

so we repeat the process, and get $\alpha = \langle (n+1)/2; n, n, n, \ldots \rangle$.

5.3 Let $\alpha_0 = \sqrt{n^2+2}$. Then $q_0 = \lfloor \alpha_0 \rfloor = n$, so

$$\alpha_1 = \frac{1}{\alpha_0 - q_0} = \frac{1}{\sqrt{n^2+2} - n} = \frac{\sqrt{n^2+2} + n}{2} = n + \frac{\sqrt{n^2+2} - n}{2} = q_1 + \frac{1}{\alpha_2},$$

$$\alpha_2 = \frac{1}{\alpha_1 - q_1} = \frac{2}{\sqrt{n^2+2} - n} = \sqrt{n^2+2} + n = 2n + (\sqrt{n^2+2} - n) = q_2 + \frac{1}{\alpha_3},$$

$$\alpha_3 = \frac{1}{\alpha_2 - q_2} = \frac{2}{\sqrt{n^2+2} - n} = \frac{\sqrt{n^2+2} + n}{2} = \alpha_1,$$

and the process repeats, so

$$\sqrt{n^2+2} = \langle n; n, 2n, n, 2n, \ldots \rangle.$$

5.5 (a) $\langle 4; 7, 7, \ldots \rangle$ (b) $\langle 5, 10, 10, \ldots \rangle$ (c) $\langle 10; 10, 20, 10, 20, \ldots \rangle$ (d) $\langle 9; 1, 18, 1, 18, \ldots \rangle$.

5.7 By Exercise 2.59 on page 112, $\lfloor -\alpha \rfloor = -\lfloor \alpha \rfloor - 1$. Thus,

$$-\alpha = \lfloor -\alpha \rfloor - \alpha + \lfloor \alpha \rfloor + 1 = \lfloor -\alpha \rfloor + \frac{1}{1 + \frac{\alpha - \lfloor \alpha \rfloor}{1 - \alpha + \lfloor \alpha \rfloor}}$$

$$= \langle \lfloor -\alpha \rfloor; 1, \frac{1}{\alpha - \lfloor \alpha \rfloor} - 1 \rangle = \langle \lfloor -\alpha \rfloor; 1, \alpha_1 - 1 \rangle,$$

and so we are done if $q_1 \neq 1$. If $q_1 = 1$, then $\alpha_2 = 1/(\alpha_1 - q_1) = 1/(\alpha_1 - 1)$, so $-\alpha = \langle -q_0 - 1; \alpha_2 + 1 \rangle$, and the result follows.

5.9 Since $\alpha - A_j/B_j$ and $\alpha - A_{j+1}/B_{j+1}$ have opposite sign, then

$$\left| \alpha - \frac{A_j}{B_j} \right| + \left| \alpha - \frac{A_{j+1}}{B_{j+1}} \right| = \frac{1}{B_j B_{j+1}}.$$

Therefore, if $|\alpha - A_k/B_k| \geq 1/(2B_k^2)$ for both $k = j$ and $k = j + 1$, then

$$\frac{1}{2B_j^2} + \frac{1}{2B_{j+1}^2} \leq \frac{1}{B_j B_{j+1}},$$

so by multiplying through by $2B_j B_{j+1}$, we get,

$$\frac{B_{j+1}}{B_j} + \frac{B_j}{B_{j+1}} \leq 2.$$

However, if we let $b = B_{j+1}/B_j$, then the latter inequality becomes, $b + 1/b \leq 2$, or by rewriting, $b^2 - 2b + 1 \leq 0$, or $(b-1)^2 \leq 0$, and this is possible only if $b = 1$. By the definition of the sequence, B_j, this can happen only if $j = 0$, contradicting that $j \in \mathbb{N}$.

5.11 Since there are infinitely many convergents, which are all rational, then the result is a direct consequence of Exercise 5.10.

5.13 If $\alpha = (1 + \sqrt{5})/2$ and $\beta = (1 - \sqrt{5})/2$, then $\alpha\beta = -1$. Also, if n is even, then

$$\beta^{n+1}(\beta^{n+1} - \alpha^{n+1}) = \beta^{2n+2} - (\beta\alpha)^{n+1} = \beta^{2n+2} + 1 > 1,$$

which, by Theorem 1.3, implies that

$$-\beta^{n+1} > \frac{1}{\alpha^{n+1} - \beta^{n+1}} = \frac{1}{\sqrt{5}F_{n+1}}, \tag{S6}$$

observing that $\beta^{n+1} < 0$. Thus, if n is even,

$$\alpha - \frac{F_{n+2}}{F_{n+1}} = \frac{\beta^{n+1}(\beta - \alpha)}{\alpha^{n+1} - \beta^{n+1}} = \frac{\beta^{n+1}(-F_1)}{F_{n+1}} = \frac{-\beta^{n+1}}{F_{n+1}} > \frac{1}{\sqrt{5}F_{n+1}^2},$$

where the inequality comes from (S6).

To make the final conclusion, we note that, if n is odd, then

$$\frac{F_{n+2}}{F_{n+1}} - \alpha = \frac{\beta^{n+1}(\alpha - \beta)}{\alpha^{n+1} - \beta^{n+1}} > 0,$$

since $\beta^{n+1} > 1$, for n odd, and $\alpha^j - \beta^j = \sqrt{5}F_j$ for any $\in \mathbb{N}$ by Theorem 1.3. Therefore, by Exercise 5.12,

$$\left|\frac{F_{n+2}}{F_{n+1}} - \alpha\right| < \frac{1}{\sqrt{5}F_{n+1}^2}.$$

Conversely, suppose that

$$\left|\frac{F_{n+2}}{F_{n+1}} - \alpha\right| < \frac{1}{\sqrt{5}F_{n+1}^2}.$$

If n is even, then since we know from the above that

$$\frac{F_{n+2}}{F_{n+1}} - \alpha < 0.$$

Thus, by this exercise,

$$\left|\frac{F_{n+2}}{F_{n+1}} - \alpha\right| = \alpha - \frac{F_{n+2}}{F_{n+1}} > 1/(\sqrt{5}F_{n+1}^2),$$

a contradiction. Hence, n is odd and the final conclusion is resolved.

Section 5.2

5.15 We have,

$$(\alpha_1\alpha_2)' = \left(\left[\frac{P_1 + \sqrt{D}}{Q_1}\right]\left[\frac{P_2 + \sqrt{D}}{Q_2}\right]\right)' = \left(\frac{P_1P_2 + D + (P_1 + P_2)\sqrt{D}}{Q_1Q_2}\right)' =$$

$$\left(\frac{P_1P_2 + D - (P_1 + P_2)\sqrt{D}}{Q_1Q_2}\right) = \left(\left[\frac{P_1 - \sqrt{D}}{Q_1}\right]\left[\frac{P_2 - \sqrt{D}}{Q_2}\right]\right) = \alpha_1'\alpha_2'.$$

Also,

$$(\alpha_1/\alpha_2)' = \left(\left[\frac{P_1 + \sqrt{D}}{Q_1}\right] \Big/ \left[\frac{P_2 + \sqrt{D}}{Q_2}\right]\right)' = \left(\frac{Q_1Q_2 + Q_2\sqrt{D}}{Q_1P_2 + Q_1\sqrt{D}}\right)' =$$

$$\left(\frac{(Q_1Q_2 + Q_2\sqrt{D})(Q_1P_2 - Q_1\sqrt{D})}{(Q_1P_2 + Q_1\sqrt{D})(Q_1P_2 - Q_1\sqrt{D})}\right)' =$$

$$\left(\frac{(Q_1Q_2P_2 - Q_2D) - (Q_2P_2 - Q_1Q_2)\sqrt{D}}{Q_1(P_2^2 - D)}\right),$$

whereas,

$$\alpha_1'\alpha_2' = \left(\left[\frac{P_1 - \sqrt{D}}{Q_1}\right] \Big/ \left[\frac{P_2 - \sqrt{D}}{Q_2}\right]\right) = \left(\frac{Q_1Q_2 - Q_2\sqrt{D}}{Q_1P_2 - Q_1\sqrt{D}}\right) =$$

$$\left(\frac{(Q_1Q_2 - Q_2\sqrt{D})(Q_1P_2 + Q_1\sqrt{D})}{(Q_1P_2 - Q_1\sqrt{D})(Q_1P_2 + Q_1\sqrt{D})}\right) =$$

$$\left(\frac{(Q_1Q_2P_2 - Q_2D) - (Q_2P_2 - Q_1Q_2)\sqrt{D}}{Q_1(P_2^2 - D)}\right) = (\alpha_1\alpha_2)',$$

as required.

5.17 Set $\beta = \lfloor\sqrt{ab}/b\rfloor + \sqrt{ab}/b$. Thus, by Theorem 5.9 on page 222, β is a quadratic irrational. Also, since $-1 < \beta' < 0$ and $\beta > 1$, β is reduced, so by Theorem 5.12 on page 228, there exists an $\ell \in \mathbb{N}$ such that

$$\beta = \langle \overline{2q_0; q_1, q_2, \ldots, q_{\ell-1}} \rangle,$$

where $\lfloor\sqrt{ab}/b\rfloor = q_0$. Hence, by Corollary 5.6,

$$\frac{1}{\sqrt{ab}/b - \lfloor\sqrt{ab}/b\rfloor} = \frac{-1}{\beta'} = \langle \overline{q_{\ell-1}; q_{\ell-2}, \ldots, q_1 2q_0} \rangle. \tag{S7}$$

However, we also have

$$\sqrt{ab}/b - \lfloor\sqrt{ab}/b\rfloor = \langle \overline{0; q_1, q_2, \ldots, q_{\ell-1}, 2q_0} \rangle,$$

so by inverting the latter we get,

$$\frac{1}{\sqrt{ab}/b - \lfloor\sqrt{ab}/b\rfloor} = \langle \overline{q_1; q_2, \ldots, q_{\ell-1}, 2q_0} \rangle. \tag{S8}$$

By equating (S7)–(S8), we get,

$$q_{\ell-1} = q_1, \quad q_{\ell-2} = q_2, \ldots, q_1 = q_{\ell-1}$$

as required.

5.19 (a) $\langle 4; \overline{8} \rangle$ (b) $\langle 5; \overline{2, 1, 1, 2, 10} \rangle$

5.21 (a) $\sqrt{29/3}$ (b) $\sqrt{31/2}$

5.23 By Corollary 5.5 on page 227, $\alpha_{\ell k+1} = \alpha_1$ for any nonnegative integer k. Thus,

$$\frac{P_{\ell k+1} + \sqrt{D}}{Q_{\ell k+1}} = \frac{P_1 + \sqrt{D}}{Q_1},$$

or by rewriting,

$$\sqrt{D}(Q_{\ell k+1} - Q_1) = P_{\ell k+1}Q_1 - P_1 Q_{\ell k+1}.$$

By the irrationality of $\sqrt{D}$, this means that

$$Q_{\ell k+1} = Q_1 \text{ and } P_{\ell k+1} = P_1.$$

Thus,

$$Q_1 = D - P_1^2 = D - P_{\ell k+1}^2 = Q_{\ell k}Q_{\ell k+1} = Q_{\ell k}Q_1,$$

so $Q_{\ell k} = 1$. We have shown that if $\ell \mid j$, then $Q_j = 1$.

Conversely, if $Q_j = 1$, then $\alpha_j = P_j + \sqrt{D}$. Therefore,

$$\lfloor \alpha_j \rfloor = P_j + \lfloor\sqrt{D}\rfloor = P_j + q_0.$$

However, by definition,

$$\alpha_j = \lfloor \alpha_j \rfloor + \frac{1}{\alpha_{j+1}} = P_j + q_0 + \frac{1}{\alpha_{j+1}}.$$

Hence,

$$q_0 + \frac{1}{\alpha_1} = \alpha_0 = \sqrt{D} = \alpha_j - P_j = q_0 + \frac{1}{\alpha_{j+1}}.$$

Thus, $\alpha_1 = \alpha_{j+1}$, which means that $q_1, q_2, \ldots, q_j$ repeats in the continued fraction expansion of $\sqrt{D}$. In other words, $j \equiv 0 \pmod{\ell}$.

Section 5.3

5.25 By Exercise 5.24, $P_{\ell/2} = P_{\ell/2+1}$. By Theorem 5.10 on page 223,

$$P_{\ell/2+1} = P_{\ell/2} = q_{\ell/2}Q_{\ell/2} - P_{\ell/2},$$

so $Q_{\ell/2} \mid 2P_{\ell/2}$. However by Theorem 5.10, again,

$$D = P_{\ell/2}^2 + Q_{\ell/2}Q_{\ell/2-1},$$

so $Q_{\ell/2} \mid 2D$.

5.27 Reduce $x^2 - Dy^2 = -1$ modulo 4 to get $x^2 \equiv -1 \,(\mathrm{mod}\ p)$ since $p \mid D$, which is impossible by Example 4.5 on page 182.

5.29 If ℓ is even and $Q_{\ell/2} = 2$, then by Theorem 5.14,

$$A_{\ell/2-1}^2 - B_{\ell/2-1}^2 p = (-1)^{\ell/2}Q_{\ell/2} = \pm 2,$$

which gives a solution to

$$x^2 - py^2 = \pm 2. \tag{S9}$$

Conversely, suppose that (S9) has a solution $r + s\sqrt{p}$, say. If ℓ is odd, then we may let $x_1 + y_1\sqrt{p}$ be the fundamental solution of $x^2 - py^2 = -1$. Since

$$N\left(\frac{(r + s\sqrt{p})^2}{2}\right) = \frac{(r^2 - s^2 p)^2}{4} = 1,$$

then by Theorem 5.16 on page 236,

$$\frac{(r + s\sqrt{p})^2}{2} = (x_1 + y_1\sqrt{p})^{2j} = x_{2j} + y_{2j}\sqrt{p},$$

for some $j \in \mathbb{N}$. Therefore, by equating coefficients of $\sqrt{p}$, we get that $rs = 2x_{2j}y_{2j}$, so rs is even. However, $r^2 - s^2 p = \pm 2$, so, if r is even for instance, then s is even so $4 \mid 2$, which is impossible. Hence, ℓ is even, so by Exercise 5.26 $Q_{\ell/2} = 2$.

5.31 If $x_1 \equiv 1 \,(\mathrm{mod}\ p)$, then by Exercise 5.30, $\ell \equiv 0 \,(\mathrm{mod}\ 4)$ and $Q_{\ell/2} = 2$. Therefore, by Theorem 5.14 on page 234,

$$A_{\ell/2-1}^2 - pB_{\ell/2-1}^2 = 2.$$

Thus, $A_{\ell/2-1}^2 \equiv 2 \,(\mathrm{mod}\ p)$, so by Corollary 4.2, $p \equiv \pm 1 \,(\mathrm{mod}\ 8)$. However, if $p \equiv 1 \,(\mathrm{mod}\ 8)$, then by Exercise 5.28, ℓ is odd. Thus, $p \equiv 7 \,(\mathrm{mod}\ 8)$.

Conversely, if $p \equiv 7 \,(\mathrm{mod}\ 8)$, then by Exercise 5.28, ℓ is even, and by Exercise 5.26, $Q_{\ell/2} = 2$. If $\ell/2$ is odd, then by Theorem 5.14,

$$A_{\ell/2-1}^2 - pB_{\ell/2-1}^2 = -2,$$

so $A_{\ell/2-1}^2 \equiv -2 \,(\mathrm{mod}\ p)$. By Exercise 4.3 on page 187, we have that

$$p \equiv 1, 3 \pmod{8},$$

a contradiction. Hence, $\ell \equiv 0 \,(\mathrm{mod}\ 4)$, and we may invoke Exercise 5.30 to conclude that $x_1 \equiv 1 \,(\mathrm{mod}\ p)$.

5.33 $0/1 = \langle 0 \rangle$, $1/j = \langle 0; j \rangle$ for $j = 10, 9, 8, 7, 6, 5, 4, 3, 2$; $2/9 = \langle 0; 4, 2 \rangle$; $2/7 = \langle 0; 3, 2 \rangle$; $3/10 = \langle 0; 3, 3 \rangle$; $3/8 = \langle 0; 2, 1, 2 \rangle$; $2/5 = \langle 0; 2, 2 \rangle$; $3/7 = \langle 0; 2, 3 \rangle$; $4/9 = \langle 0; 2, 4 \rangle$; $(3 + j)/(5 + k) = \langle 0; 1, 1, 2 + j \rangle$ for $(j, k) = (0, 0), (1, 2), (2, 4)$; $5/8 = \langle 0; 1, 1, 1, 2 \rangle$; $7/10 = \langle 0; 1, 2, 3 \rangle$; $5/7 = \langle 0; 1, 2, 2 \rangle$; $7/9 = \langle 0; 1, 3, 2 \rangle$; $(j - 1)/j = \langle 0; 1, j - 1 \rangle$, for $j = 3, 4, 5, 6, 7, 8, 9, 10$; and $1/1 = \langle 1 \rangle$.

5.35 This is an immediate consequence of Exercise 5.32.

Section 5.4

5.37 $1517 = 37 \cdot 41$.

5.39 We have that

$$(xu \pm yv)^2 - (yu \pm xv)^2 = x^2 u^2 + y^2 v^2 \pm 2xuyv - y^2 u^2 - x^2 v^2 \mp 2xuyv =$$

$$x^2 u^2 + y^2 v^2 - y^2 u^2 - x^2 v^2 = u^2(x^2 - y^2) - v^2(x^2 - y^2) = (u^2 - v^2)(x^2 - y^2) = ab.$$

Section 6.1

6.1 Clearly, if n is the sum of two integer squares, then it is the sum of two squares of rational numbers since all integers are rational. Conversely, if n is a sum of two squares of rational numbers: $n = (r/s)^2 + (v/w)^2$, then

$$n(sw)^2 = (rw)^2 + (vs)^2.$$

If n is not the sum of two integer squares, then by Theorem 6.3 on page 247, either $4 \mid n$ or there is a prime of the form $p \equiv 3 \pmod 4$ to an odd exponent dividing n. However, this would then be true for $(rw)^2 + (vs)^2$, a sum of two integer squares, which is a contradiction.

6.3 If p is an odd prime, then

$$p = \left(\frac{p + 1}{2} \right)^2 - \left(\frac{p - 1}{2} \right)^2.$$

By Exercise 5.40, since $d = 1$, then the representation is unique.

6.5 If $p = q^2 + r^2$ where q and r are primes, then clearly $p > 2$. Therefore, q and r have different parity, so one of q or r is even, namely one of them must be equal to 2 since both of them are prime.

6.7 This is a direct consequence of Lemma 6.2 on page 246.

6.9 $(11 + \sqrt{221})/10 = \langle \overline{2; 1, 1, 2} \rangle$ so $221 = 10^2 + 11^2$.

6.11 $\mathcal{F}_n = (2^{2^{n-1}})^2 + 1^2$.

Section 6.2

6.13 Let $z = -x$ in the equation. Then it becomes, $2x^2 + y^2 = y^3$ or by rewriting,

$$2x^2 = y^3 - y^2 = (y - 1)y^2,$$

so by setting $y - 1 = 2u^2$, we get that $x = u(1 + 2u^2)$ yielding infinitely many solutions for $u \in \mathbb{Z}$.

6.15 If $a^2 + (a+1)^2 = b^4 + (b+1)^4$, then we have

$$2a^2 + 2a + 1 = 2b^4 + 4b^3 + 6b^2 + 4b + 1,$$

so by subtracting 1 from both sides and dividing by 2,

$$a^2 + a = b^4 + 2b^3 + 3b^2 + 2b = (b^2 + b)^2 + 2(b^2 + b).$$

Therefore,

$$a^2 + a + 1 = (b^2 + b + 1)^2.$$

Also, since

$$a^2 < a^2 + a + 1 < (a+1)^2,$$

then $a^2 + a + 1$ cannot be a square. Hence, there are no solutions to $a^2 + (a+1)^2 = b^4 + (b+1)^4$.

6.17 We have that

$$x^3 + y^3 + z^3 - 3xyz = (x + y + z)(x^2 + y^2 + z^2 - xy - zx - yz).$$

Adding 8 to both sides and setting $z = 2$, assuming that $x^3 + y^3 - 6xy = 0$, we get,

$$8 = x^3 + y^3 + 8 - 3 \cdot 2 \cdot xy = (x + y + 2)(x^2 + y^2 + 4 - xy - 2x - 2y),$$

where the second factor on the right is always nonnegative. Therefore, one of $x + y + 2 = 1$, $x + y + 2 = 2$, $x + y + 2 = 4$, or $x + y + 2 = 8$ must hold. A check shows that the first two are impossible since $x, y \in \mathbb{N}$, and the third forces $x = y = 1$ which does not solve our equation. Hence, the only possible solution is given by the fourth, and only one variation solves our equation, namely $x = 3$ and $y = 3$.

6.19 Assume that $x^3 + 8x^2 - 6x + 8 = y^3$. Therefore,

$$y^3 - (x+1)^3 = 5x^2 - 9x + 7 > 0,$$

and

$$(x+3)^3 - y^3 = x^2 + 33x + 19 > 0.$$

It follows that $x^3 + 8x^2 - 6x + 8$ is a cube lying between $(x+1)^3$ and $(x+3)^3$, namely

$$x^3 + 8x^2 - 6x + 8 = (x+2)^3.$$

Hence, $2x(x - 9) = 0$ yielding the two solutions $(0, 2)$ and $(9, 11)$.

6.21 Let $n \in \mathbb{N}$. Since $4n + 2$ is not of the form given in condition (6.4) on page 252, then

$$4n + 2 = u^2 + v^2 + w^2$$

for nonnegative integers u, v, w. One of u, v, w must be odd since $4n + 2$ is not divisible by 4. Moreover, there must be an even number of the u, v, w that are odd since $4n + 2$ is even. Without loss of generality, let u, v be odd and w be even. Thus, $w = 2z$ and $u + v = 2x$ while $u - v = 2y$ from which we get $u = x + y$ and $v = x - y$. Hence,

$$4n + 2 = (x + y)^2 + (x - y)^2 + 4z^2,$$

from which we get that $2n + 1 = x^2 + y^2 + 2z^2$, where $x, y, z \in \mathbb{Z}$.

Section 6.3

6.23 Since $7 = 2^2 + 1^2 + 1^2 + 1^2$ and $5 = 2^2 + 1^2 + 0^2 + 0^2$, then using Lemma 6.3, $35 = 5^2 + 0^2 + 1^2 + 3^2$.

6.25 By Theorem 6.7 on page 257, $n - 1 = a^2 + b^2 + c^2 + d^2$ for some integers a, b, c, d. Therefore,

$$8n = (2a-1)^2 + (2a+1)^2 + (2b-1)^2 + (2b+1)^2 + (2c-1)^2 + (2c+1)^2 + (2d-1)^2 + (2d+1)^2.$$

6.27 By Example 4.8, there are infinitely many primes of the form $p = 8k + 7$, and by Theorem 6.5, there are integers a, b, c such that $p - 1 = 8k + 6 = a^2 + b^2 + c^2$, from which the result follows.

Section 6.4

6.29 Let $n = 6m$. Then $6m = (m + 1)^3 + (m - 1)^3 + (-m)^3 + (-m)^3$.

6.31 This follows from the identity

$$(3n^3 + 1)^3 = 2 + (3n^3 - 1)^3 + (3n^2)^3 + (3n^2)^3,$$

for any $n \in \mathbb{N}$.

6.33 Select any natural number of the form $n = u^3 - v^3$, of which there are infinitely many. Then the result follows from the identity

$$u^3 - v^3 + v^3 + m^3 = u^3 + m^3,$$

for any $m \in \mathbb{N}$.

Section 7.1

7.1 $(x, y) = (13, 19)$ is the fundamental solution of

$$x^2 - 19y^2 = -2$$

and $(x, y) = (170, 39)$ is the fundamental solution of

$$x^2 - 19y^2 = 1.$$

Thus,

$$(170 + 39\sqrt{19})(13 + 3\sqrt{19}) = 4433 + 1017\sqrt{19}$$

and

$$(170 + 39\sqrt{19})^2(13 + 3\sqrt{19}) = 1507207 + 345777\sqrt{19}$$

are two more solutions of $x^2 - 19y^2 = -2$.

7.3 $(x, y) = (5604, 569)$ is the smallest solution of

$$x^2 - 29y^2 = -1$$

using $\sqrt{97} = \langle 9; \overline{1, 5, 1, 1, 1, 1, 1, 1, 5, 1, 18} \rangle$ with period length $\ell = 11$, where $A_{\ell-1} = A_{10} = 5604$ and $B_{\ell-1} = B_{10} = 569$.

7.5 $3391 = 63^2 - 2 \cdot 17^2$.

7.7 $-9539 = 83^2 - 3 \cdot 74^2$.

Section 7.2

7.9 No, because $-ac = -33$ is not a quadratic residue modulo $b = 5$.

7.11 Yes, since all conditions in Theorem 7.3 are satisfied, trivially for $b = 1$, $c = -1$, and $a = 5$ since $-bc = 1$. Indeed, $(x, y, z) = (1, 2, 3)$ is such a solution.

Section 7.3

7.13 If x is even, then y is even so $x = 2a$ and $y = 2b$. Thus, $b^2 + 4 = 2a^3$, so $b = 2c$ and $a = 2d$. Therefore, $c^2 + 1 = 4d^3$, which is impossible since -1 is not a quadratic residue modulo 4. Hence, x and y are odd. Thus, $x^3 \equiv 1 \pmod 8$. An easy check shows that $x \equiv 1 \pmod 8$ is also forced. Hence, $x - 2 \equiv -1 \pmod 8$ and $(x - 2)|(x^3 - 8) = y^2 + 8$. Since $x - 2$ cannot be a product of primes p solely of the form $p \equiv 1, 3 \pmod 8$, then there is a prime p dividing $y^2 + 8$ such that either $p \equiv 5 \pmod 8$ or $p \equiv 7 \pmod 8$. In other words, the Legendre symbol

$$1 = \left(\frac{y^2}{p}\right) = \left(\frac{-8}{p}\right) = \left(\frac{-2}{p}\right),$$

contradicting Exercise 4.3 on page 187.

7.15 if x is even, then $y^2 \equiv -1 \pmod 8$ which is impossible. If $x \equiv -1 \pmod 4$, then $y^2 \equiv -1 \pmod 4$, again impossible. Therefore, $x \equiv 1 \pmod 4$. Thus, $x^2 - 3x + 9 \equiv 3 \pmod 4$. However,

$$y^2 + 4 = (x + 3)(x^2 - 3x + 9),$$

so $x^2 - 3x + 9$ divides $y^2 + 4$ meaning that there is a prime $p \equiv 3 \pmod 4$ dividing $y^2 + 4$ yielding

$$1 = \left(\frac{y^2}{p}\right) = \left(\frac{-2^2}{p}\right) = \left(\frac{-1}{p}\right).$$

Thus, -1 is a quadratic residue modulo p, which is impossible.

7.17 This is an application of Theorem 7.5 with $a = 7$, $b = 4$, and $k = 3$.

7.19 $(3a^2 - 1)^2 + (a^3 - 3a)^2 = 3a^4 + 3a^2 + 1 + a^6 = (a^2 + 1)^3$.

7.21 For any $m \in \mathbb{N}$ we have

$$(1 + m^n)^n + (m(1 + m^n))^n = (1 + m^n)^{n+1}.$$

Section 7.4

7.23 If $x^2 + y^2 = z^2$ and $xy = 2w^2$, then both

$$(x + y)^2 = z^2 + 4w^2 \text{ and } (x - y)^2 = z^2 - 4w^2.$$

Multiplying these two equations together, we get that

$$(x^2 - y^2)^2 = z^4 - 16w^4 = z^4 - (2w)^4,$$

which contradicts Exercise 7.22.

7.25 We have,

$$\left(\frac{x^4 - y^4}{2}\right)^2 = \frac{x^8 + y^8 - 2x^4y^4}{4} = \frac{(x^4 + y^4)^2 - 4x^4y^4}{4} =$$

$$\left(\frac{x^4 + y^4}{2}\right)^2 - (xy)^4 = z^4 - (xy)^4, \qquad (\text{S10})$$

which contradicts Exercise 7.22, if all summands are natural numbers. Thus, $x = y = z = 1$, thereby reducing (S10) to the equation $0^2 = 1^4 - 1^4$, and we have our result.

Bibliography

[1] M. Agrawal, N. Kayal, and N. Saxena, *Primes is in P*, Ann. of Math. **160** (2004), 781–793. (*Cited on pages 320–322.*)

[2] M. Aigner and G.M. Ziegler, **Proofs from the Book**, Springer, Berlin, Heidelberg, New York, Tokyo (2002). (*Cited on page 37.*)

[3] W.R. Alford, A. Granville, and C. Pomerance, *There are infinitely many Carmichael numbers*, Ann. Mth. **140** (1994), 703–722. (*Cited on page 95.*)

[4] N.C. Ankeny, *Sums of three squares*, Proc. Amer. Math. Soc. **8** (1957), 316–319. (*Cited on page 252.*)

[5] M. Bellare and P. Rogaway, *Optimal asymmetric encryption* in **Advances in Cryptology**, EUROCRPT '94, Springer-Verlag, Berlin, LNCS **950** (1994), 92–111. (*Cited on page 172.*)

[6] J.P.M. Binet, *Mémoire sur l'integration des équations linéaires aux différences finies, d'un ordre quelconque, à coefficients variables*, Comptes Rendus Acad. des Sciences, Paris, **17**, (1843), 559–567. (*Cited on page 5.*)

[7] D. Boneh, A. Joux, and P.Q. Nguyen, *Why textbook ElGamal and RSA encryption are insecure*, in **Advances in Cryptology**, ASIACRYPT 2000 (Kyoto), Springer-Verlag, Berlin, LNCS **1976** (2000), 30–43. (*Cited on page 172.*)

[8] D. Boneh and G. Durfee, *Cryptanalysis of RSA with private key d less than $N^{0.292}$*, IEEE Transactions on Information Theory **46** (2000), 1339–1349. (*Cited on page 174.*)

[9] J.W. Bruce, *A really trivial proof of the Lucas-Lehmer test*, American Math. Monthly, **100** (1993), 370–371. (*Cited on page 60.*)

[10] J. Chen, *On the representation of a large even integer as the sum of a prime and the product of at most two primes*, Sci. Sinica **16** (1973), 157–176. (*Cited on page 66.*)

[11] A. Claesson and T.K. Petersen, *Conway's Napkin Problem*, The American Math. Monthly, **114** (2007), 217–231. (*Cited on page 59.*)

[12] D. Coppersmith, *Small solutions to polynomial equations, and low exponent RSA vulnerabilities*, J. Cryptol. **10** (1997), 233–260. (*Cited on page 174.*)

[13] A. de Moivre, *Miscellanea Analytica de Seriebus et Quadrataris*, J. Tonson and J. Watts, London, 1730. (*Cited on page 5.*)

[14] L.E. Dickson, **History of the Theory of Numbers**, Vol. 1, Chelsea, New York, (1992). (*Cited on pages 64, 111.*)

[15] P. Finsler, *Über die Primzahlen zwischen n und 2n*, Festschrift zum 60. Geburtstag von Prof. Dr. Andreas Speiser, (1945), 118–122. (*Cited on page 70.*)

[16] C.F. Gauss, **Disquisitiones Arithmeticae** (English edition), Springer-Verlag, Berlin, Heidelberg, New York, Tokyo (1985). (*Cited on pages 33, 35, 90, 146, 189, 205.*)

[17] A. Granville, *Some conjectures related to Fermat's last theorem* in **Number Theory** (R.A. Mollin, ed.), Walter de Gruyter, Berlin, New York (1990). (*Cited on page 319.*)

[18] R.K. Guy, **Unsolved Problems in Number Theory**, Vol. **1**, Second Edition, Springer-Verlag, Berlin (1994). (*Cited on pages 95, 101, 147.*)

[19] G.H. Hardy, **A Mathematician's Apology**, Cambridge University Press (1940). (*Cited on page 60.*)

[20] G.H. Hardy and E.M. Wright, **An Introduction to the Theory of Numbers**, Oxford University Press, Fifth Edition (1980). (*Cited on page 59.*)

[21] J. Hastad, *Solving simultaneous modular equations of low degree*, Siam J. Comput. **17** (1988), 336–341. (*Cited on page 174.*)

[22] D.E. Knuth, **The Art of Computer Programming**, Volume **2: Seminumerical Algorithms**, Third Edition, Addison-Wesley, Reading, Paris (1998). (*Cited on pages 163, 209.*)

[23] E. Landau, *Handbuch der Lehre von der Verteilung der Primzahlen*, Teubner, Leipzig (1909), reprinted (both volumes) by Chelsea Publishing Co., New York (1953). (*Cited on page 65.*)

[24] R.S. Lehman, *Factoring large integers*, Math. Comp. **28** (1974), 637–646. (*Cited on page 203.*)

[25] D.H. Lehmer, **Selected Papers of D.H. Lehmer**, Volumes I–III, D. McCarthy (ed.), The Charles Babbage Research Centre, St. Pierre, Canada (1981). (*Cited on pages 64, 203.*)

[26] W.J. Levesque, **Fundamentals of Number Theory**, Addison-Wesley, Reading, Menlo Park, London, Amsterdam, Don Mills, Sydney (1977).

[27] A.A. Martínez, *Euler's "mistake"? The radical product rule in historical perspective*, American Math. Monthly **114** (2007), 273–285. (*Cited on page 292.*)

[28] P.V.A. Mohan, **Residue Number Systems: Algorithms and Architectures**, Springer, Berlin, Heidelberg, New York, Tokyo, (2002). (*Cited on page 89.*)

[29] R.A. Mollin, **Quadratics**, CRC Press, Boca Raton, London, Tokyo (1995). (*Cited on pages 13, 272.*)

[30] R.A. Mollin, **Algebraic Number Theory**, Chapman and Hall/CRC Press, Boca Raton, London, Tokyo (1999). (*Cited on page 151.*)

[31] R.A. Mollin, **RSA and Public-Key Cryptography**, Chapman and Hall/CRC, Boca Raton, London, New York (2003). (*Cited on page 201.*)

[32] R.A. Mollin, **Codes — The Guide to Secrecy from Ancient to Modern Times**, CRC, Taylor and Francis Group, Boca Raton, London, New York (2005). (*Cited on pages 119, 136, 173.*)

[33] R.A. Mollin, *Lagrange, central norms, and quadratic Diophantine equations*, Internat. J. Math. and Math. Sci. **7** (2005), 1039–1047. (*Cited on page 238.*)

[34] R.A. Mollin, **An Introduction to Cryptography**, Second Edition, CRC Press, Taylor and Francis Group, Boca Raton, London, New York (2007). (*Cited on pages 11, 127, 169, 201, 205.*)

[35] H.L. Montgomery and S. Wagon, *A heuristic for the prime number theorem*, The Math. Intelligencer, **28** (2006), 6–9. (*Cited on page 65.*)

[36] M.A. Morrison and J. Brillhart, *A method of factoring and the factorization of F_7*, Math. Comp. **29** (1975). 183–205. (*Cited on page 240.*)

[37] P. van Oorschot, *A comparison of practical public-key cryptosystems based on integer factorization and discrete logarithms*, in **Contemporary Cryptography: The Science of Information Integrity**, G. Simmons, ed., IEEE Press, Piscatoway, N.J. (1992), 289–322. (*Cited on page 167.*)

[38] PKCS1, *Public key cryptography standard no. 1, version 2.0*, RSA Labs. (*Cited on page 172.*)

[39] S. Pohlig and M. Hellman, *An improved algorithm for computing logarithms over $GF(p)$ and its cryptographic significance*, IEEE Transactions on Information Theory, Volume 24 (1978), 106–110. (*Cited on page 166.*)

[40] J.M. Pollard, *An algorithm for testing primality of any integer*, Bull. London Math. Soc. **3** (1971), 337–340. (*Cited on page 205.*)

[41] B. Riemann, *Uber die Anzahl der Primzahlen unter einer gegeben Grösse*, Monatsberichte der Berliner Akademie, (1859). (*Cited on page 72.*)

[42] R.L. Rivest, A. Shamir, and L. Adleman, *A method for obtaining digital signatures and public-key cryptosystems*, Communications of the A.C.M. **21** (1978), 120–126. (*Cited on pages 167–168.*)

[43] M.I. Rosen, *A proof of the Lucas-Lehmer test*, American Math. Monthly **95** (1988), 855–856. (*Cited on page 181.*)

[44] L. Rosenhead, *Henry Cabourn Pocklington*, Obituary Notices of the Royal Society (1952), 555–565. (*Cited on page 125.*)

[45] J.B. Rosser and L. Schoenfeld, *Approximate formulas for functions of prime numbers*, Illinois J. Math. **6** (1962), 64–89. (*Cited on page 71.*)

[46] A. Selberg, *An elementary proof of the prime number theorem*, Annals of Math. **50** (1949), 305–313. (*Cited on page 65.*)

[47] P.L. Tchebychev, *Mémoire sur les nombres premiers*, J. de Math. Pures ppl. **17** (1852), 366–390. (*Cited on page 69.*)

[48] G. Tenenbaum and M. Mendès France, **The prime numbers and their distribution**, Amer. Math. Soc., Providence (2000). (*Cited on page 65.*)

[49] S.S. Wagstaff and J.W. Smith, *Methods of factoring large integers*, in **Number Theory**, New York, 1984–1985, LNM, **1240**, Springer-Verlag, Berlin (1987), 281–303. (*Cited on page 242.*)

[50] P.G. Walsh, **The Pell Equation and Powerful Numbers**, Master's Thesis, University of Calgary, Canada (1988). (*Cited on page 319.*)

[51] M. Weiner, *Cryptanalysis of short RSA secret exponents*, IEEE Transactions on Information Theory **36** (1990), 555–558. (*Cited on page 174.*)

[52] H.C. Williams, *Primality testing on a computer*, Ars Combin. **5** (1978), 127–185. (*Cited on page 121.*)

[53] H.C. Williams, **Lucas and Primality Testing**, Wiley (1997). (*Cited on page 63.*)

[54] D. Zagier, *Newman's proof of the prime number theorem*, Amer. Math. Monthly **104** (1997), 705–708. (*Cited on page 65.*)

Index

The Author

Richard Anthony Mollin received his Bachelor's and Master's degrees from the University of Western Ontario in 1971 and 1972, respectively. His Ph.D. was obtained from Queen's University in 1975 in Kingston, Ontario, where he was born. Since then he has held various positions including at Montreal's Concordia University, the University of Victoria, the University of Toronto, York University, McMaster University in Hamilton, the University of Lethbridge, and Queen's University in Kingston, where he was one of the first NSERC University Research Fellows. He is currently a full professor in the Mathematics Department of the University of Calgary, where he has been employed since 1982. He has over 180 publications, including 10 books, in algebra, number theory, and computational mathematics. He has been awarded 5 separate Killam awards over the past quarter century, including one in 2005, to complete his eighth book *Codes—The Guide to Secrecy from Ancient to Modern Times*, [32]. He is a member of the Mathematical Association of America, past member of both the Canadian and American Mathematical Societies, and a member of various editorial boards. Moreover, he has been invited to lecture at numerous universities, conferences, and society meetings, as well as holding numerous research grants from universities and governmental agencies. Furthermore, he is the founder of the Canadian Number Theory Association, and held its first conference in Banff in 1988, immediately preceding his NATO Advanced Study Institute.

Printed in the United States
by Baker & Taylor Publisher Services